LANDSLIDES

Analysis and Control

Special Report 176

Robert L. Schuster
Raymond J. Krizek
editors

Transportation Research Board
Commission on Sociotechnical Systems
National Research Council

NATIONAL ACADEMY OF SCIENCES   Washington, D.C.   1978

Transportation Research Board Special Report 176
Edited for TRB by Mildred Clark

Additional copies of Figure 2.1 (in pocket in back of book) are available for $5.00 each.

modes
1 highway transportation
3 rail transportation

subject area
63 mechanics (earth mass)

Transportation Research Board publications are available by ordering directly from the board. They may also be obtained on a regular basis through organizational or individual supporting membership in the board; members or library subscribers are eligible for substantial discounts. For further information, write to the Transportation Research Board, National Academy of Sciences, 2101 Constitution Avenue, N.W., Washington, D.C. 20418.

Library of Congress Cataloging in Publication Data
National Research Council. Transportation Research Board.
    Landslides, analysis and control.

    (Special report - Transportation Research Board, National Research Council; 176)
    Includes bibliographies and index.
    1. Slopes (Soil mechanics) 2. Landslides. I. Schuster, Robert L. II. Krizek, Raymond J. III. Title. IV. Series: National Research Council. Transportation Research Board. Special report - Transportation Research Board, National Research Council; 176.
TA710.N32   1978   624'.151   78-27034
ISBN 0-309-02804-3

First Printing, 1979
Second Printing, November 1979
Third Printing, October 1981
Fourth Printing, Octorer 1985
Fifth Printing, August 1988

Sponsorship of the Papers in This Special Report

DIVISION A–REGULAR TECHNICAL ACTIVITIES
Kurt W. Bauer, Southeastern Wisconsin Regional Planning Commission, chairman

GROUP 2–DESIGN AND CONSTRUCTION OF TRANSPORTATION FACILITIES
Eldon J. Yoder, Purdue University, chairman

Task Force on Review of Special Report 29–Landslides
Robert L. Schuster, U.S. Geological Survey, Denver, chairman
David S. Gedney, Federal Highway Administration
Raymond J. Krizek, Northwestern University
David L. Royster, Tennessee Department of Transportation
Dwight A. Sangrey, Cornell University
William G. Weber, Jr., Washington Department of Transportation, Yakima
Tien H. Wu, Ohio State University

John W. Guinnee, Transportation Research Board staff

Contents

The Editors and Authors

David S. Gedney. Acting director, Northeast Corridor Project, Federal Railroad Administration. Formerly chief, Construction and Maintenance Division, Federal Highway Administration; regional engineer, Region 15, Federal Highway Administration; and chief, Soil and Rock Mechanics Branch, Construction and Maintenance Division, Federal Highway Administration. His field of specialization is geotechnical engineering as applied to transportation problems.

Raymond J. Krizek. Professor of civil engineering, Northwestern University, Evanston, Illinois. Formerly faculty member at the University of Maryland and lecturer at the Catholic University of America. He is a specialist in soil mechanics and foundation engineering and has a variety of teaching, research, and professional engineering experience.

Ta Liang. Professor of civil and environmental engineering and director, Remote Sensing Program, Cornell University, Ithaca, New York. He has long been engaged in teaching and research in the fields of aerial photograph interpretation, remote sensing, and geotechnical and geological engineering and has been a consultant to industrial, national, and international organizations on engineering and economic development projects in many parts of the world.

P. Erik Mikkelsen. Senior associate engineer, Shannon and Wilson, Inc., Seattle. During his 15 years of geotechnical engineering experience, he has become an expert in instrumentation used in landslides, embankments, deep excavations, and deep foundations.

N. R. Morgenstern. Professor of civil engineering, University of Alberta, Edmonton. He has been a consultant in geotechnical engineering and applied earth sciences since 1961. His area of professional specialization is geotechnical engineering; he has made noteworthy contributions on the sub-jects of landslides, rock mechanics, soil-structure interaction, and soil engineering problems in cold regions.

F. Lionel Peckover. Geotechnical consultant, Vaudreuil, Quebec. Formerly in charge of geotechnical engineering work with Canadian National Railways (1959-1976), Canadian St. Lawrence Seaway Authority (1953-1959), and Division of Building Research, National Research Council of Canada (1947-1953). His field of specialization is geotechnical engineering with emphasis on railway application, particularly the treatment of unstable rock slopes.

Douglas R. Piteau. President, Piteau and Associates, Vancouver, British Columbia. He has extensive experience in engineering geology and rock mechanics related to railways and highways, mine development and operations, and site evaluation of dams and tunnels.

Harold T. Rib. Chief, Aerial Surveys Branch, Federal Highway Administration. Formerly chief, Exploratory Techniques Group, and highway research engineer, Federal Highway Administration. His field of specialization is application of remote-sensing techniques to transportation engineering and terrain analysis.

David L. Royster. Chief, Soils and Geological Engineering Division, Tennessee Department of Transportation, Nashville. He has held various positions with Tennessee Department of Transportation since 1958. His field of specialization is engineering geology as applied to highway design and construction.

Dwight A. Sangrey. Professor of civil and environmental engineering, Cornell University, Ithaca, New York. In addition to teaching and research at Cornell, he has been active in practice and as a consultant on projects involving instability of slopes. Major areas of professional activity involve dynamic loading of soils, soil sensitivity, and marine geotechnical engineering.

Robert L. Schuster. Chief, Engineering Geology Branch, U.S. Geological Survey, Denver. Formerly professor and chairman, Department of Civil Engineering, University of Idaho, and associate professor and professor, Department of Civil Engineering, University of Colorado. Major areas of professional specialization are engineering geology and geotechnical engineering.

George F. Sowers. Senior geotechnical consultant and senior vice president, Law Engineering Testing Company, Marietta, Georgia, and regents professor of civil engineering, Georgia Institute of Technology, Atlanta. He is a geotechnical engineer specializing in the interrelation of geology, engineering design, and construction and has been a worldwide consultant on earthfill dams, foundations, and earth and rock construction.

David J. Varnes. Geologist, Engineering Geology Branch, U.S. Geological Survey, Denver. He has had 37 years of experience as an engineering geologist with USGS, including being chief of the Engineering Geology Branch. Fields of specialization are slope stability in rock and soil and engineering geologic mapping.

William G. Weber, Jr. Washington Department of Transportation, Yakima. Formerly research engineer, Bureau of Materials, Pennsylvania Department of Transportation, and senior materials and research engineer, Materials and Research Department, California Division of Highways. He has expertise in geotechnical design of structures involving soft soils, groundwater flow, and stability of slopes.

Stanley D. Wilson. Consulting engineer. Seattle. Formerly executive vice president of Shannon and Wilson, Inc. His significant contributions in the development of field operational equipment, his advocacy of its use in dams and other civil engineering projects, and his analyses of the results have led to better understanding of the mechanisms of landslides and of the performance of embankments and dams.

Tien H. Wu. Professor of civil engineering, Ohio State University, Columbus. Formerly professor, Department of Civil Engineering, Michigan State University, and visiting professor, Norwegian Geotechnical Institute and National University of Mexico. He has specialized in the relation of the engineering properties of soils to their mechanical behavior and has been a consultant on a variety of problems in soil mechanics and geotechnical engineering.

Chapter 1

Introduction

Robert L. Schuster

This book is a successor to Highway Research Board Special Report 29, *Landslides and Engineering Practice* (*1.8*). Special Report 29, which was written by the Highway Research Board Committee on Landslide Investigations and published in 1958, achieved an excellent reputation, both in North America and abroad, as a text on landslides. Because of its popularity, the original printing was sold out within a few years after publication. Since then, there has been a continuing interest in reissuing the original text or publishing a worthwhile successor.

In 1972 the Highway Research Board organized the Task Force for Review of Special Report 29—Landslides. The membership of this task force was selected from several committees within the HRB Soils and Geology Group: its charge was

To review the out-of-print Special Report 29—Landslides— and to recommend what action should be taken in response to the high interest in revising this publication.

This task force was further instructed to act as the coordinating unit to implement its recommendations.

After considerable study of the original report, the task force concluded that, because of the large amount of new technical information that had become available since 1958 on landslides and related engineering, the best course of action would be to completely rewrite the book rather than to reprint it or to revise it in part. The task force decided that the general format of the original volume would be retained but that the contents would be expanded to include concepts and methods not available in 1958. To achieve that objective, the task force secured the aid of authors who have broad geotechnical expertise. They were drawn from the fields of civil engineering and geology and have specializations in soil mechanics, engineering geology, and ·interpretation of aerial photographs.

SCOPE OF THIS VOLUME

The scope of this volume is the same as that for Special Report 29: to bring together in coherent form and from a wide range of experience such information as may be useful to those who must recognize, avoid, control, design for, or correct landslide movement.

This new version, however, introduces geologic concepts and engineering principles and techniques that have been developed since publication of Special Report 29 so that both the analysis and the control of soil and rock slopes are addressed. For example, included are new methods of stability analysis and the use of computer techniques in implementing these methods. In addition, rock-slope engineering and the selection of shear-strength parameters for slope-stability analyses are two topics that were poorly understood in 1958 and therefore were given scant attention in Special Report 29. Since that time, these two subjects have received a significant amount of study and have become fairly well understood; thus, they are presented as separate chapters in the present volume.

The book is divided into two general parts. The first part deals principally with the definition and assessment of the landslide problem. It includes chapters on slope-movement types and processes, recognition and identification of landslides, field investigations, instrumentation, and evaluation of strength properties. The second part of the book deals with solutions to the landslide problem. Chapters are included on methods of slope-stability analysis, design techniques, and remedial measures that can be applied to both soil and rock-slope problems.

Although considerable effort has been made to eliminate presentation of the same material in different chapters, some repetition has been necessary to provide continuity of thought and to allow adequate explanation of specific topics. We consider this repetition to be more acceptable than constant referral in the text to other chapters.

DEFINITIONS AND RESTRICTIONS

In Special Report 29 the term landslide is defined as the downward and outward movement of slope-forming materials—natural rock, soils, artificial fills, or combinations of these materials. Today the term deserves further refinement because, as shown in Chapter 2, slope movements can now be divided into five groups: falls, topples, slides, spreads, and flows. As used in this text, a landslide constitutes the group of slope movements wherein shear failure occurs along a specific surface or combination of surfaces.

Although this volume deals primarily with slope failures belonging to the group designated as slides, some attention is given to the other four types of slope movements. The use of the term landslide in the title is somewhat inaccurate in that theoretically it does not cover the five basic failure modes described above; however, the decision was made to use this term because it is popular and easily recognized and because the book is mainly devoted to landslides.

In keeping with the practice followed in Special Report 29, surficial creep was excluded from consideration; however, creep of a more deep-seated nature is considered in discussions dealing with slope movements. Also excluded are subsidence not occurring on slopes and most types of movement primarily due to freezing and thawing of water. In addition, snow and ice avalanches and mass wasting due to slope-failure phenomena in tropic and arctic climates are not considered. Although a few examples are drawn from other parts of the world, most of the descriptions of slope movements and engineering techniques involve slopes in North America.

Of the five groups of potential slope movements considered, only slides are currently susceptible to quantitative stability analysis by use of the conventional sliding-wedge or circular-arc techniques. These methods of slope-stability analysis are not applicable to falls, topples, spreads, or flows. However, enough is now known about the kinematics and nature of development of such failures that qualitative or statistical approaches or both can be used to make reasonable assessments in problem areas or potential problem areas. Research dealing with such problems is currently being undertaken to enable at least crude quantitative stability analyses to be performed on slopes subject to spreads and flows and possibly even to falls and topples.

Although slope-stability problems related to transportation facilities are stressed, most of the examples apply equally well to all cases of slope failure, such as those relating to coastlines, mining, housing developments, and farmlands. As noted by Eckel in the Introduction in Special Report 29 (*1.7*, p. 2):

The factors of geology, topography, and climate that interact to cause landslides are the same regardless of the use to which man puts a given piece of land. The methods for examination of landslides are equally applicable to problems in all kinds of natural or human environment. And the known methods for prevention or correction of landslides are, within economic limits, independent of the use to which the land is put. It is hoped, therefore, that despite the narrow range of much of its exemplary material,

this volume will be found useful to any engineer whose practice leads him to deal with landslides.

ECONOMICS OF SLOPE MOVEMENTS

Although individual slope failures generally are not so spectacular or so costly as certain other natural catastrophes such as earthquakes, major floods, and tornadoes, they are more widespread and the total financial loss due to slope failures probably is greater than that for any other

Figure 1.1. Damage to embankment on I-75 in Campbell County, Tennessee, from a landslide that occurred April 1972.

Figure 1.2. Homes and street damaged in October 1978 Laguna Beach, California, landslide.

Figure 1.3. Damage to railway facilities from 1972 Shigeto landslide in Japan.

single geologic hazard to mankind. In addition, much of the damage occurring in conjunction with earthquakes and floods is due to landslides instigated by shaking or water.

Reliable estimates of the overall costs of landslides are difficult to obtain for geographic entities as large as the United States or Canada. In 1958, Smith (*1.32*) stated that "the average yearly cost of landslides in the United States runs to hundreds of millions of dollars," an estimate that was probably realistic at that time. However, in the 20 years since Smith assembled his data, a combination of inflation, increased construction in landslide-prone areas, and use of larger cuts and fills in construction has resulted in considerably increased annual costs of landslides. For example, environmental and political considerations and right-of-way costs control the selection of highway routing today to a much greater degree than was the case 20 years ago; thus, highway planners often cannot avoid construction in landslide-prone areas. Landslide costs include both direct and indirect losses from landslides affecting highways, railroads, industrial installations, mines, homes, and other public and private properties. Direct costs are those losses incurred in actual damages to installations or property; examples of such damages are shown in Figures 1.1, 1.2, and 1.3. Examples of indirect costs are (a) loss of tax revenues on properties devalued as a result of landslides, (b) reduced real estate values in areas threatened by landslides, (c) loss of productivity of agricultural or forest lands affected by landslides, and (d) loss of industrial productivity due to interruption of transportation systems by landslides. Indirect costs of landslides are difficult to evaluate, but they may be larger than the direct costs.

In 1976, Krohn and Slosson (*1.16*) estimated the annual landslide damage to buildings and their sites in the United States to be $400 million (1971 dollars). This figure does not include other damages, such as those to transportation facilities and mines, or indirect costs. In the same year, Jones (*1.13*) estimated the direct landslide damage losses to buildings and their sites to be about $500 million annually. Based on the above estimates plus indirect costs and estimated damages to facilities not classed as buildings, a reasonable estimate of present-day direct and indirect costs of slope failures in the United States exceeds $1 billion/year.

Somewhat more accurate cost estimates can be made for individual landslides or for landslides occurring in relatively small geographic areas. For instance, the Portuguese Bend landslide in Palos Verdes Hills, California, has been estimated to have cost more than $10 million in damage to roads, houses, and other structures between 1956 and 1959 (*1.23*). Jones, Embody, and Peterson (*1.14*) noted that the filling of the reservoir behind Grand Coulee Dam in the state of Washington cost taxpayers and private property owners at least $20 million to avoid and correct the damage due to landslides that occurred between 1934 and 1952.

Within the United States, greater effort at detailing the costs of slope movements has been expended in California than in any other state. In a classic study of slope-movement costs in the San Francisco Bay area, Taylor and Brabb (*1.35*)

3

documented information on these costs for nine Bay-area counties during the winter of 1968-1969. The data were derived largely from interviews with planners and assessors in the county government and engineers and geologists in city, county, and state governments. Costs of slope movements totaled at least $25 million, of which about $9 million was direct loss or damage to private property (due mainly to drop in market value); $10 million was direct loss or damage to public property (chiefly for repair or relocation of roads and utilities); and about $6 million consisted of miscellaneous costs that could not be easily classified in either the public or the private sector. This is a tremendous expense for the relatively small area involved. In addition, Taylor and Brabb noted that their data are incomplete in that they were not able to obtain costs on many of the slope movements. They felt, therefore, that the total cost of the 1968-1969 slope movements for the San Francisco Bay area may possibly have been several times greater than the estimated $25 million.

A survey conducted by the Federal Highway Administration indicates that approximately $50 million is spent annually to repair major landslides on the federally financed portion of the national highway system (1.3, 1.4). This system includes federal and state highways but does not include most county and city roads or streets, private roads and streets, or roads built by other governmental agencies such as the U.S. Forest Service. Distribution of the direct costs of major landslides for 1973 by Federal Highway Administration regions within the United States is shown in Figure 1.4 (1.3). The cost for an individual region is based on both the landslide risk and the amount of highway construction in the area. In addition, the given costs represent a single year; the average annual cost for a particular region could vary significantly from the given cost.

Total annual costs of landslides to highways in the United States are difficult to determine precisely because of the difficulty in defining the following factors: (a) costs of smaller slides that are routinely handled by maintenance forces; (b) costs of slides on non-federal-aid routes; and (c) indirect costs that are related to landslide damage, such as traffic disruption and delays, inconvenience to motorists, engineering costs for investigation, and analysis and design of mitigation measures. If these factors are included, Chassie and Goughnour (1.3) of the

Federal Highway Administration believe that $100 million is a conservative estimate of the total annual cost of landslide damage to highways and roads in the United States.

For planning purposes, other studies have attempted to project costs of slope movements. In a study predicting the cost of geologic hazards in California from 1970 to 2000, the California Division of Mines and Geology (1.1) estimated that the costs of slope movements throughout the state during that period would be nearly $10 billion, or an average of more than $300 million a year. This estimate is based on the assumption that loss-reduction practices in use in California in 1970 for slope failures will remain unchanged. Figure 1.5 (1.1) shows a comparison of the estimated losses due to slope movements and losses due to other geologic hazards and urbanization. Of the so-called "catastrophic" geologic hazards included in the study, losses due to slope movements exceed those due to floods and, in turn, are exceeded by those due to earthquakes. California, however, is particularly prone to earthquake activity, and in most other parts of the United States and Canada losses due to slope movements probably would be greater than those due to earthquakes.

Various studies have shown that most damaging landslides are human related; thus, the degree of hazard can be reduced beforehand by introduction of measures such as improved grading ordinances, land-use controls, and drainage or runoff controls (1.37). For example, Nilsen and Turner (1.25) showed that in Contra Costa County, California, approximately 80 percent of the landslides have been caused by human activity. Briggs, Pomeroy, and Davies (1.2) noted that more than 90 percent of the landslides in Allegheny County, Pennsylvania, have been related to human activities. The study by the California Division of Mines and Geology (1.1) indicated that the $9.9 billion estimated losses due to slope movements can be reduced 90 percent or more by a combination of measures involving adequate geologic investigations, good engineering practice, and effective enforcement of legal restraints on land use and disturbance.

Chassie and Goughnour (1.4) further substantiated the concept that improved geologic and geotechnical studies can significantly reduce the landslide hazard. They noted that

Figure 1.4. Costs of landslide repairs to federal-aid highways in United States for 1973 (1.3).

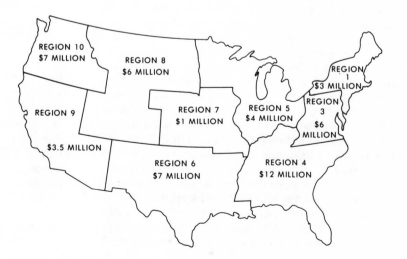

Figure 1.5. Predicted economic losses from geologic
hazards and urbanization in California from 1970 to
2000 (1.1).

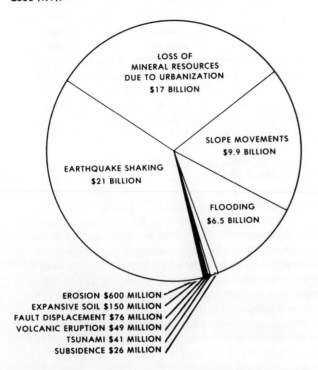

a debris avalanche, which started as an ice avalanche from
a glacier high on the north peak of Mt. Huascaran but soon
became a mixture of ice, water, rock, and soil, roared
through valley villages, killing some 4000 to 5000 people
(1.5, 1.22). An even greater number of people were killed
in a repeat of this tragedy 8 years later, when an earthquake
of magnitude 7.75 occurred off the coast of Peru and
triggered another disastrous debris avalanche on the slopes
of Huascarán (1.5, 1.28). This debris avalanche descended
at average speeds of roughly 320 km/h (200 mph) into the
same valley but over a much larger area and killed more
than 18 000 people. The village of Ranrahirca, which had
been rebuilt after being destroyed by the 1962 debris
avalanche in which 2700 of its people were killed, was partially destroyed by the 1970 avalanche. The 1962 avalanche
was prevented from flowing into the town of Yungay by a
protective ridge, but the 1970 avalanche overtopped this
ridge and buried the town along with an estimated 15 000
of its 17 000 inhabitants. Only the tops of a few palm
trees in the central plaza and parts of the walls of the main
cathedral were left protruding above the mud to mark the
site of this formerly prosperous and picturesque city (1.28).

In 1974, another massive landslide in the Andes Mountains of Peru killed approximately 450 people (1.18). This
landslide, which occurred in the valley of the Mantaro
River, had a volume of 1.6 Gm³ (2.1 billion yd³), making
it one of the largest in recorded history. It temporarily
dammed the Mantaro River, forming a lake with a depth
of about 170 m (560 ft) and a length of about 31 km (19
miles). In overtopping this landslide dam, the river caused
extensive damage downstream, destroying approximately
20 km (12 miles) of road, three bridges, and many farms.

On October 9, 1963, the most disastrous landslide in
European history—the Vaiont Reservoir slide—occurred
in northeastern Italy. A mass of rock and soil having a
volume of about 250 Mm³ (330 million yd³) slid into the
reservoir, sending a wave 260 m (850 ft) up the opposite
slope and at least 100 m (330 ft) over the crest of the dam
into the valley below, where it destroyed five villages and
took 2000 to 3000 lives (1.15, 1.17).

Japan has also suffered continuing large loss of life and
property from landslides and other slope movements. Although some slope failures in Japan have been triggered
by earthquakes, most are a direct result of heavy rains
during the typhoon season. Data from a Japan Ministry
of Construction publication (1.12) and a written communication in 1974 are given in Table 1.1 and show the number of deaths and damaged houses caused by slope failures
in Japan for the 4-year period from 1969 through 1972.

improved geotechnical techniques in New York State reduced landslide repair costs by as much as 90 percent in the
7 years prior to 1976. Slosson (1.31) showed that landslide
losses sustained by the city of Los Angeles as a result of the
1968-1969 winter storm were 97 percent lower for those
sites developed under modern grading codes by using modern geotechnical methods than for sites developed before
1952, when no grading codes existed and engineering geology and geotechnical engineering studies were not required.
For the state of California, Leighton (1.19) estimated that
reductions of 95 to 99 percent in landslide losses can be
obtained by means of preventive measures that incorporate
thorough preconstruction investigation, analysis, and design and that are followed by careful construction procedures.

In addition to the economic losses due to slope movements, a significant loss of human life is directly attribut-
able to landslides and other types of slope failures. Fatalities due to catastrophic slope failures have been recorded since people began to congregate in areas subject
to such failures. One such catastrophe (probably a debris
flow) was noted by Spanish conquistadors in Bolivia in
the sixteenth century (1.30). According to the priest Padre
Calancha, who observed the event from a distance, Hanco-
Hanco, a community of about 2000 inhabitants, disap-
peared "in a few minutes and was swallowed by the earth
without more evidence of its former existence than a cloud
of dust which arose where the village had been situated."

In the twentieth century many individual slope failures
have resulted in large numbers of fatalities. Probably the
best known of these catastrophic failures are the debris
avalanches of 1962 and 1970 on the slopes of Mt. Huascaran in the Andes Mountains of Peru. In January 1962,

Table 1.1. Deaths and damage due to recent
slope-failure disasters in Japan.

Year	Houses Damaged	Deaths	
		Number	Percent[a]
1969	521	82	50
1970	38	27	26
1971	5205	171	54
1972	1564	239	44

[a]Of deaths due to slope failure in relation to deaths
due to all other natural disasters.

5

Of particular interest is the high ratio of deaths due to slope failures to deaths from all other natural disasters, including earthquakes.

North American slope failures have not commonly resulted in major losses of life, because most catastrophic slope failures have occurred in nonpopulated areas. However, there have been several notable exceptions in this century. The first was in Canada in 1903, when a great landslide killed approximately 70 people in the coal mining town of Frank, Alberta (*1.21*). More recently, the Hebgen Lake earthquake struck southwestern Montana in 1959 and triggered the Madison Canyon landslide. That catastrophic landslide, shown in Figure 1.6 (*1.36*), had a volume of 28 Mm³ (37 million yd³) and buried 26 people who were camped along the banks of the Madison River (*1.10, 1.39*).

Probably the worst natural disaster in central Virginia's recorded history was the 1969 flooding and associated debris flows resulting from hurricane Camille (*1.38*). Although no exact number of deaths due to slope movements can be ascertained, estimates are that a substantial percentage of the 150 people who died in Virginia as a result of hurricane Camille were victims of debris flows resulting from the hurricane.

Another recent catastrophic slope failure in North America was the debris flow that occurred in 1971 in Champlain clay in the Canadian town of Saint-Jean-Vianney, Quebec (*1.34*). That flow carried 40 homes to destruction and 31 persons to their deaths.

The most recent major catastrophe involving slope failure in the United States was the Buffalo Creek dam failure at Saunders, West Virginia, in 1972 (*1.6*). Heavy rains led to the failure of three coal-refuse impoundments. The resulting debris flow consisting of released water, coal wastes, and sludge traveled 24 km (15 miles) downstream, killing 125 people and leaving 4000 homeless.

The landslides and other types of slope movements

Figure 1.6. Madison Canyon landslide of August 21, 1959, in southwestern Montana.

mentioned above can all be classified as major disasters. In addition to catastrophes of this magnitude, however, slope failures of lesser importance occur continually throughout the world. Because no systematic records of these day-to-day slope failures have been maintained in the United States and Canada (as contrasted to Japan, Table 1.1), ascertaining the number of deaths per year owing to slope failures is not possible. However, Krohn and Slosson (*1.16*) estimated that the total loss of life in the United States from all forms of landslide activity exceeds approximately 25 lives per year, a greater total than the average number of deaths due to earthquakes.

LEGAL ASPECTS OF SLOPE MOVEMENTS

In Special Report 29, Smith (*1.32*, p. 13) stated:

Few legal precedents have been established to guide the courts in determining responsibility for landslides or in assessing the damages caused by them. This dearth of specific laws and legal decisions is perhaps due to two main factors—many, if not most, cases that involve private companies are settled out of court; most cases against state or federal agencies are settled out of court or the public agency exercises its sovereign right of refusal to consent to be sued.

In the United States during the 20 years since this statement was made, the number of legal cases resulting from property damage due to landslides has been ever increasing; the cases involve private companies and landholders as well as public agencies. For that reason it is important that those who undertake activities that involve the use of slopes have an understanding of the legal implications of that use. This section deals briefly with the legal aspects of landslides and provides references for those wishing to explore the subject further. The discussion is based on perusal of current literature and on substantial information provided in a written communication in 1975 from C. L. Love, attorney for the Legal Division of the California Department of Transportation. Because of the constantly changing status of litigation involving natural hazards, some of these concepts will likely change within the next few years.

Landslides and Transportation Routes

Since most litigation involving landslides on transportation routes relates to construction and maintenance of public highways or roads, we will assume in this discussion that a public entity is the defendant. As noted by Love, the law relating to public agencies is based on the concept of sovereign immunity; thus, the consequent liability of public agencies for landslides is generally more limited than the liability of private individuals under similar circumstances.

Love further states that, when liability for a landslide is discussed, it must, of course, be assumed that a landslide has caused injury to some legally protected interest of a party, thus enabling an action against the public entity. The legally protected interest of the injured party may be his or her personal property, real estate, or physical well-being. It also must be assumed that the public entity is in some way responsible for the landslide. Such responsibility, or liability, can be based on construction or maintenance operations that create or activate a landslide on public property or on mere public ownership of property that either contains or is in the immediate vicinity of an active or potentially active landslide.

Love notes that there have been numerous cases in which private property has been damaged or personal injury has resulted from landslides or rock falls or both on public highways in the United States. In these instances the liability of the public entity having jurisdiction over the highway has varied from state to state. Some states, for all intents and purposes, bar suits against public entities because of sovereign immunity; however, many states have established statutory provisions under which recovery can be realized. Such statutes generally delineate specific duties and responsibilities of public agencies, specific circumstances of the slope failure, procedural requirements for bringing action against the public entity, and specific defenses available to the public entity.

Although the protection of sovereign immunity commonly has been invoked successfully in cases in which a reasonable degree of prudence has been exercised by those who have designed and constructed the works, Professor G. F. Sowers of the Department of Civil Engineering, Georgia Institute of Technology, stated in a written communication in 1976:

While it is presently true that states and the federal government as owners invoke the protection of "sovereign immunity," there are many indications that this sheltered position will not always be the case. Public sympathy has generated pressure on state legislatures that has caused them to admit liability. In cases where the doctrine of sovereign immunity has held, the injured party and, too often, some overly zealous members of the legal profession have sought other sources of relief. These involve the designers of the works, the builders, and even private maintenance forces. While presently employees of the governmental bodies appear to be held harmless from legal action, there are indications that it is possible to bring personal suits for negligence against such employees. While such lawsuits may eventually be lost, there are enough lawyers who will take the statistical chance that some cases will not be thrown out of court that we should expect to see personal suits against political administrators, public employees, and everyone who has anything to do with construction, whether they are responsible or not.

Lewis and others (*1.20*) divided the legal rights of private citizens against public agencies with regard to landslides into two categories:

1. A property owner's rights in response to invasion of the property by sliding material or interference with the lateral support of the property by construction or maintenance of a public way and
2. A highway traveler's rights in tort against a public entity for injuries sustained from a landslide that resulted in part from the negligent construction or maintenance of a public way.

The extent of such rights varies among states, but a general discussion is given below.

Liability in Invasion of Property or Loss of Support

When a landslide results in damage to property, either by invasion of the property or loss of its lateral support, the liability of a public entity is not necessarily based on statutes. Under the fifth amendment to the U.S. Constitution, just compensation must be paid when public works or other governmental activities result in the taking of private property; that concept can be extended to the damaging of property as a result of an action of a public agency. The owner of the property brings an action known as an inverse condemnation suit to recover damages. According to Love, many state constitutions contain provisions similar to those of the fifth amendment, but, even in the absence of such a limitation in a state constitution, the courts have held that a state cannot take (or damage) private property for public use without just compensation.

The case of *Albers v. County of Los Angeles* [62 Cal. 2d 250, 42 Cal. Rptr. 89 (1965)], which established, or broadened, the concept of inverse condemnation in California, provides an outstanding example of the manner in which the state courts have interpreted constitutional provisions for the payment of just compensation in the event that private property is either taken or damaged. In that litigation, which was concerned with the Portuguese Bend landslide on the Palos Verdes Peninsula in southern California, the plaintiffs alleged that the county of Los Angeles had constructed Crenshaw Boulevard through an ancient landslide area and that in the course of carrying out the construction program had placed some 134 000 m³ (175 000 yd³) of earth at a critical spot in the landslide area, causing reactivation of movement and consequent damage to the plaintiffs' properties (*1.26, 1.29*). The constitution of the state of California guarantees that property of a private property owner will not be damaged or taken for public use unless just compensation for it is given to the property owner. This constitutional protection is the basis on which the property owners recovered $5 360 000 from the county of Los Angeles (*1.24*).

It can be concluded that, if public-works activities result in the creation of a new landslide or the reactivation of an old landslide that causes damage to private property, the public entity is liable to the full extent of such damage if the particular state has such a constitutional provision. According to Love, even in jurisdictions that do not have a provision relating directly to damage of private property, the courts have tended to find that the damage that resulted to the private property constitutes a taking for which just compensation must be paid.

Liability for Injuries Sustained From a Landslide

Love states that, although the courts have made it clear that a public entity is not an insurer of the safety of persons using its highways, in certain circumstances travelers are protected by law from landslides. In general, the public entity will not be held liable for injuries if it can be shown that the acts or omissions that created the dangerous condition were reasonable or that the action taken to protect against such injuries or the failure to take such action was reasonable. The reasonableness of action or inaction is determined by considering the time and opportunity that the public employees had to take action and by weighing the probability and gravity of potential injury to persons foreseeably exposed to the risk of injury against the practicality and cost of protecting against such injury.

In California, most actions involving personal injury to highway travelers as a result of landslides are based on a statute that imposes liability for the dangerous condition of public property. The injured person must prove that the public property was in dangerous condition at the time of the injury, that the injury resulted from that dangerous condition, and that the dangerous condition created a foreseeable risk of the kind of injury incurred. Love points out that, in addition, the dangerous condition must be the result of negligence, a wrongful act, or failure of an employee of the public entity to act within the scope of his or her employment, and the public entity must have had notice of the dangerous condition in sufficient time prior to the injury to have taken measures to protect against it. Thus, liability depends on whether circumstances and conditions were such that the danger was reasonably foreseeable in the exercise of ordinary care and, if so, on whether reasonable measures were taken by the public entity to prevent injury (39 Am. Jur. 2d *Highways* § 532, p. 939).

A public entity, since it is not an insurer of the safety of travelers on its highways, need only maintain highways in reasonably safe condition for ordinary travel under ordinary conditions or under such conditions as should reasonably be expected (*1.20*). In the case of *Boskovich v. King County* [188 Wash. 63, 61 P. 2d 1299 (1936)], the court held that a motorist was not entitled to recover from the highway department for injuries sustained when a landslide broke loose from a steep hillside bordering a highway and struck his automobile because there was no proof that negligence in construction or maintenance of the highway was the cause of the landslide.

Landslides and Property Development

This section discusses liability related to damages from landslides caused by the development of private property. Detailed information on litigation related to landslides in property developments is given by Sutter and Hecht (*1.33*).

The current trend in public policy is toward protection of the consumer, a reversal of the days when caveat emptor (let the buyer beware) reflected public policy (*1.27*). This trend has extended to home purchasers since they are protected by law against losses due to improper workmanship or poor planning, including certain losses due to landslides.

The trend toward increased protection for the homeowner has resulted in a drastic increase in the number of legal cases involving landslides on private property. After consulting with an attorney, the owner of a home that has suffered damage from a landslide typically files legal action against the developer, the civil engineer who laid out the development, the geotechnical engineer, the

geologist, the grading contractor, the city, the builder, the lending agency, the insurance company that insured the home, the former owner, and the real estate agent handling the sale if the property was not purchased directly from the developer (*1.11*). A typical complaint may seek recovery on theories of strict liability (i.e., liability without fault), negligence, breach of warranty, negligent misrepresentation, fraud, and, if a public entity is involved, inverse condemnation (*1.27*). In most cases, the developer is the prime target because he is subject to strict liability for "defects" in the construction of the house or grading of the lot; to establish strict liability against any of the other parties such as the soils engineer or the geologist is much more difficult.

Liability of Engineers and Geologists for Landslide Damages

There has been a certain amount of variability in legal interpretations of liability of geotechnical engineers and geologists in regard to landslide losses; such liability is discussed below, but the conclusions reached are general in nature and not necessarily valid in any specific court.

Liability of geotechnical engineers and engineering geologists for landslide damages to home sites is based most often on the theory of negligence and occasionally on negligent misrepresentation. Although allegations seeking to recover damages from geotechnical engineers and geologists on the basis of strict liability, breach of warranty, or intentional misrepresentation are often included in a complaint, they are not usually applicable under normal circumstances. Patton (*1.27*) discussed each of these theories of liability as it applies to engineering geologists as follows, and it is felt that Patton's line of reasoning can be extended to include geotechnical engineers involved in development of private property.

1. Negligence

The most common theory of liability alleged against engineering geologists is negligence. Negligence is the omission to do something which an ordinarily prudent person would have done under similar circumstances or the doing of something which an ordinarily prudent person would not have done under those circumstances. An engineering geologist is required to exercise that degree of care and skill ordinarily exercised in like cases by reputable members of his profession practicing in the same or similar locality at the same time under similar conditions. He has the duty to exercise ordinary care in the course of performing his duties for the protection of any person who foreseeably and with reasonable certainty may be injured by his failure to do so.

Although failing to comply with a state statute or with county or municipal ordinances normally is considered to be negligence per se, the mere compliance with the letter of the law in such cases does not necessarily relieve one of liability since it generally is recognized that statutes and ordinances set forth only minimum requirements and circumstances may require more than the minimum. An engineering geologist cannot rely upon the approval of a project by an inspector for a governmental agency to relieve him of liability.

2. Negligent misrepresentation

Negligent misrepresentation is a species of fraud along with intentional misrepresentation and concealment. Negligent misrepresentation is simply the assertion, as a fact, of that which is not true by one who has no reasonable ground for believing it to be true. Although misrepresentations of opinions generally are not actionable, they become actionable where the person making the alleged misrepresentation holds himself to be specially qualified to render the opinion. A statement of opinion by an engineering geologist that no unsupported bedding occurs in a particular slope, could be actionable as negligent misrepresentation if he has no basis for that opinion.

3. Intentional misrepresentation and concealment

Intentional misrepresentation (the assertion, as a fact, of that which is not true by one who does not believe it to be true) and concealment (the suppression of a fact or condition by one who is bound to disclose it) are species of fraud which are seldom if ever applicable to engineering geologists. Such conduct on the part of an engineering geologist is not only legally actionable but raises serious doubts about the professional integrity of the geologist involved.

4. Breach of warranty

Breach of warranty is generally not available as a viable theory of recovery against an engineering geologist.

5. Strict liability

Although the theory of strict liability is still developing and its limits are not as yet clearly defined, it would appear now that an engineering geologist would not be liable on the theory of strict liability lacking some participation as the developer of mass produced property. Other cases indicate that the theory is available only against the developers of mass-produced property and would not be available against the developer of a single lot or building site. There is no clear indication as to at what point between development of a single lot and development of a tract the theory becomes applicable. It does seem clear at this time that if an engineering geologist offers only professional services in connection with the development of even a large tract he will not subject himself to strict liability.

In regard to negligence, Sowers noted:

Unfortunately, the legal profession has been expanding the definition of negligence to any act committed by the public official, engineer or contractor. Some courts have applied the most extravagant standards of professional knowledge to average run-of-the-mill design and construction. In other words, some courts would presume that every engineer must possess the wisdom and expertise of a Terzaghi.

In voicing an opinion somewhat different from Patton's in regard to strict liability, Fife, another California attorney active in litigation involving geologists and geotechnical en-gineers, stated in 1973 that professional liability of the technical professional was approaching a major crossroad in its development (*1.9*). Fife felt that the scope of professional liability in the technical disciplines was at the point where it would proceed either toward strict liability under pressure from skilled plaintiff's counsel or toward a "reasonableness standard" by which adherence to the average standards of the profession involved would constitute a complete defense. However, unless professional groups become more actively involved in the process of shaping the future scope of their professional liability, eventual application of strict liability to technical professionals seems inevitable.

In their book, *Landslide and Subsidence Liability* (*1.33*), Sutter and Hecht present considerable information on strict liability in California. Since November 1973, the cutoff date for the cases included in this reference, certain California court decisions have changed the liability of geologists and geotechnical engineers from strict liability to liability for negligence only. The publisher, California Contunuing Education at the Bar, plans periodic supplements to cover changes that have occurred since Sutter and Hecht's book was published.

ACKNOWLEDGMENTS

Sincere appreciation is expressed to all those who contributed information, both published and unpublished, without which this volume could not have been written. To list all contributors would be impossible; to list the most important would be unfair to the others. Thus, the most reasonable and equitable approach is simply to acknowledge the help and cooperation of all who contributed information in the form of data, photographs, ideas, and advice.

Appreciation is also due the entire staff of the Transportation Research Board. Our special thanks go to John W. Guinnee, engineer of soils, geology, and foundations, for his constant encouragement and advice, and to Mildred Clark, senior editor, who aided immensely in the technical aspects of the editoral process.

REFERENCES

1.1 Alfors, J. T., Burnett, J. L., and Gay, T. E., Jr. Urban Geology: Master Plan for California. California Division of Mines and Geology, Bulletin 198, 1973, 112 pp.

1.2 Briggs, R. P., Pomeroy, J. S., and Davies, W. E. Landsliding in Allegheny County, Pennsylvania. U.S. Geological Survey, Circular 728, 1975, 18 pp.

1.3 Chassie, R. G., and Goughnour, R. D. National Highway Landslide Experience. Highway Focus, Vol. 8, No. 1, Jan. 1976, pp. 1-9.

1.4 Chassie, R. G., and Goughnour, R. D. States Intensifying Efforts to Reduce Highway Landslides. Civil Engineering, Vol. 46, No. 4, April 1976, pp. 65-66.

1.5 Cluff, L. S. Peru Earthquake of May 31, 1970: Engineering Geology Observations. Seismological Society of America Bulletin, Vol. 61, No. 3, June 1971, pp. 511-521.

1.6 Davies, W. E. Buffalo Creek Dam Disaster: Why it Happened. Civil Engineering, Vol. 43, No. 7, July 1973, pp. 69-72.

1.7 Eckel, E. B. Introduction. In Landslides and Engineering Practice, Highway Research Board, Special Rept. 29, 1958, pp. 1-5.

1.8 Eckel, E. B., ed. Landslides and Engineering Practice. Highway Research Board, Special Rept. 29, 1958, 232 pp.

1.9 Fife, P. K. Professional Liability and the Public Interest. In Geology, Seismicity, and Environmental Impact, Association of Engineering Geologists, Special Publ., 1973, pp. 9-14.

1.10 Hadley, J. B. Landslides and Related Phenomena Accompanying the Hebgen Lake Earthquake of August 17, 1959. U.S. Geological Survey, Professional Paper 435, 1964, pp. 107-138.

1.11 Hays, W. V. Panel discussion. Proc., Workshop on Physical Hazards and Land Use: A Search for Reason. Department of Geology, San Jose State Univ., California, 1974, pp. 8-14.

1.12 Japan Ministry of Construction. Dangerous Slope Failure. Division of Erosion Control, Department of River Works, 1972, 14 pp.

1.13 Jones, D. E. Handout for Roundtable Discussions. National Workshop on Natural Hazards, June 30-July 2, 1976, Univ. of Colorado, Boulder, Institute of Behavioral Science, 1976.

1.14 Jones, F. O., Embody, D. R., and Peterson, W. L. Landslides Along the Columbia River Valley, Northeastern Washington. U.S. Geological Survey, Professional Paper 367, 1961, 98 pp.

1.15 Kiersch, G. A. Vaiont Reservoir Disaster. Civil Engineering, Vol. 34, No. 3, 1964, pp. 32-39.

1.16 Krohn, J. P., and Slosson, J. E. Landslide Potential in the United States. California Geology, Vol. 29, No. 10, Oct. 1976, pp. 224-231.

1.17 Lane, K. S. Stability of Reservoir Slopes. In Failure and Breakage of Rock (Fairhurst, C., ed.), Proc., 8th Symposium on Rock Mechanics, Univ. of Minnesota, 1966, American Institute of Mining, Metallurgical and Petroleum Engineers, New York, 1967, pp. 321-336.

1.18 Lee, K. L., and Duncan, J. M. Landslide of April 25, 1974, on the Mantaro River, Peru. National Academy of Sciences, Washington, D.C., 1975, 72 pp.

1.19 Leighton, F. B. Urban Landslides: Targets for Land-Use Planning in California. In Urban Geomorphology, Geological Society of America, Special Paper 174, 1976, pp. 37-60.

1.20 Lewis, H., McDaniel, A. H., Peters, R. B., and Jacobs, D. M. Damages Due to Drainage, Runoff, Blasting, and Slides. National Cooperative Highway Research Program, Rept. 134, 1972, 23 pp.

1.21 McConnell, R. G., and Brock, R. W. The Great Landslide at Frank, Alberta. In Annual Report of the Canada Department of the Interior for the Year 1902-03, Sessional Paper 25, 1904, pp. 1-17.

1.22 McDowell, B. Avalanche! In Great Adventures With National Geographic, National Geographic Society, Washington, D.C., 1963, pp. 263-269.

1.23 Merriam, R. Portuguese Bend Landslide, Palos Verdes Hills, California. Journal of Geology, Vol. 68, No. 2, March 1960, pp. 140-153.

1.24 Morton, D. M., and Streitz, R. Landslides: Part Two. California Division of Mines and Geology, Mineral Information Service, Vol. 20, No. 11, 1967, pp. 135-140.

1.25 Nilsen, T. H., and Turner, B. L. Influence of Rainfall and Ancient Landslide Deposits on Recent Landslides (1950-71) in Urban Areas of Contra Costa County, California. U.S. Geological Survey, Bulletin 1388, 1975, 18 pp.

1.26 Nordin, J. G. The Portuguese Bend Landslide, Palos Verdes Hills, Los Angeles County, California. In Landslides and Subsidence, Resources Agency of California, 1965, pp. 56-62.

1.27 Patton, J. H., Jr. The Engineering Geologist and Professional Liability. In Geology, Seismicity, and Environmental Impact, Association of Engineering Geologists, 1973, pp. 5-8.

1.28 Plafker, G., Ericksen, G. E., and Fernandez Concha, J. Geological Aspects of the May 31, 1970, Peru Earthquake. Seismological Society of America Bulletin, Vol. 61, No. 3, June 1971, pp. 543-578.

1.29 Pollock, J. P. Discussion. In Landslides and Subsidence, Resources Agency of California, 1965, pp. 74-77.

1.30 Sanjines, A. G. Sintesis Historica de la Vida de la Ciudad, 1548-1948. Primer Premio de la Alcaldia, 2, La Paz, Bolivia, 1948, 86 pp.

1.31 Slosson, J. E. The Role of Engineering Geology in Urban Planning. In The Governors' Conference on Environmental Geology, Colorado Geological Survey, Special Publ. 1, 1969, pp. 8-15.

1.32 Smith, R. Economic and Legal Aspects. In Landslides and Engineering Practice, Highway Research Board, Special Rept. 29, 1958, pp. 6-19.

1.33 Sutter, J. H., and Hecht, M. L. Landslide and Subsidence Liability. California Continuing Education at the Bar, Berkeley, California Practice Book 65, 1974, 240 pp.

1.34 Tavenas, F., Chagnon, J. Y., and La Rochelle, P. The Saint-Jean-Vianney Landslide: Observations and Eyewitnesses Accounts. Canadian Geotechnical Journal, Vol. 8, No. 3, 1971, pp. 463-478.

1.35 Taylor, F. A., and Brabb, E. E. Distribution and Cost by Counties of Structurally Damaging Landslides in the San Francisco Bay Region, California, Winter of 1968-69. U.S. Geological Survey, Miscellaneous Field Studies Map MF-327, 1972.

1.36 U.S. Geological Survey. The Hebgen Lake, Montana, Earthquake of August 17, 1959. U.S. Geological Survey, Professional Paper 435, 1964, 242 pp.

1.37 Wiggins, J. H., Slosson, J. E., and Krohn, J. P. Natural Hazards: An Expected Building Loss Assessment. J. H. Wiggins Co., Redondo Beach, California, Draft rept. to National Science Foundation, 1977, 134 pp.

1.38 Williams, G. P., and Guy, H. P. Erosional and Depositional Aspects of Hurricane Camille in Virginia, 1969. U.S. Geological Survey, Professional Paper 804, 1973, 80 pp.

1.39 Witkind, I. J. Events on the Night of August 17, 1959: The Human Story. U.S. Geological Survey, Professional Paper 435, 1964, pp. 1-4.

PHOTOGRAPH CREDITS

Figure 1.1   Courtesy of Tennessee Department of Transportation
Figure 1.2   Courtesy of The Register, Santa Ana, California
Figure 1.3   Courtesy of Japan Society of Landslide
Figure 1.6   J. R. Stacy, U.S. Geological Survey

Chapter 2

Slope Movement Types and Processes

(LANDSLIDES)

David J. Varnes

This chapter reviews a fairly complete range of slope-movement processes and identifies and classifies them according to features that are also to some degree relevant to their recognition, avoidance, control, or correction. Although the classification of landslides presented in Special Report 29 (*2.182*) has been well received by the profession, some deficiencies have become apparent since that report was published in 1958; in particular, more than two dozen partial or complete classifications have appeared in various languages, and many new data on slope processes have been published.

One obvious change is the use of the term slope movements, rather than landslides, in the title of this chapter and in the classification chart. The term landslide is widely used and, no doubt, will continue to be used as an all-inclusive term for almost all varieties of slope movements, including some that involve little or no true sliding. Nevertheless, improvements in technical communication require a deliberate and sustained effort to increase the precision associated with the meaning of words, and therefore the term slide will not be used to refer to movements that do not include sliding. However, there seems to be no single simple term that embraces the range of processes discussed here. Geomorphologists will see that this discussion comprises what they refer to as mass wasting or mass movements, except for subsidence or other forms of ground sinking.

The classification described in Special Report 29 is here extended to include extremely slow distributed movements of both rock and soil; those movements are designated in many classifications as creep. The classification also includes the increasingly recognized overturning or toppling failures and spreading movements. More attention is paid to features associated with movements due to freezing and thawing, although avalanches composed mostly of snow and ice are, as before, excluded.

Slope movements may be classified in many ways, each

having some usefulness in emphasizing features pertinent to recognition, avoidance, control, correction, or other purpose for the classification. Among the attributes that have been used as criteria for identification and classification are type of movement, kind of material, rate of movement, geometry of the area of failure and the resulting deposit, age, causes, degree of disruption of the displaced mass, relation or lack of relation of slide geometry to geologic structure, degree of development, geographic location of type examples, and state of activity.

The chief criteria used in the classification presented here are, as in 1958, type of movement primarily and type of material secondarily. Types of movement (defined below) are divided into five main groups: falls, topples, slides, spreads, and flows. A sixth group, complex slope movements, includes combinations of two or more of the other five types. Materials are divided into two classes: rock and engineering soil; soil is further divided into debris and earth. Some of the various combinations of movements and materials are shown by diagrams in Figure 2.1 (in pocket in back of book); an abbreviated version is shown in Figure 2.2. Of course, the type of both movement and

Figure 2.2. Abbreviated classification of slope movements. (Figure 2.1 in pocket in back of book gives complete classification with drawings and explanatory text.)

TYPE OF MOVEMENT			TYPE OF MATERIAL		
			BEDROCK	ENGINEERING SOILS	
				Predominantly coarse	Predominantly fine
FALLS			Rock fall	Debris fall	Earth fall
TOPPLES			Rock topple	Debris topple	Earth topple
SLIDES	ROTATIONAL	FEW UNITS	Rock slump	Debris slump	Earth slump
	TRANSLATIONAL		Rock block slide	Debris block slide	Earth block slide
		MANY UNITS	Rock slide	Debris slide	Earth slide
LATERAL SPREADS			Rock spread	Debris spread	Earth spread
FLOWS			Rock flow (deep creep)	Debris flow	Earth flow (soil creep)
COMPLEX			Combination of two or more principal types of movement		

materials may vary from place to place or from time to time, and nearly continuous gradation may exist in both; therefore, a rigid classification is neither practical nor desirable. Our debts to the earlier work of Sharpe (*2.146*) remain and are augmented by borrowings from many other sources, including, particularly, Skempton and Hutchinson (*2,154*), Nemčok, Pašek, and Rybář (*2.116*), de Freitas and Watters (*2.37*), Záruba and Mencl (*2.193*), and Zischinsky (*2.194*). Discussions with D. H. Radbruch-Hall of the U.S. Geological Survey have led to significant beneficial changes in both content and format of the presentation.

The classification presented here is concerned less with affixing short one- or two-word names to somewhat complicated slope processes and their deposits than with developing and attempting to make more precise a useful vocabulary of terms by which these processes and deposits may be described. For example, the word creep is particularly troublesome because it has been used long and widely, but with differing meanings, in both the material sciences, such as metallurgy, and in the earth sciences, such as geomorphology. As the terminology of physics and materials science becomes more and more applied to the behavior of soil and rock, it becomes necessary to ensure that the word creep conveys in each instance the concept intended by the author. Similarly, the word flow has been used in somewhat different senses by various authors to describe the behavior of earth materials. To clarify the meaning of the terms used here, verbal definitions and discussions are employed in conjunction with illustrations of both idealized and actual examples to build up descriptors of movement, material, morphology, and other attributes that may be required to characterize types of slope movements satisfactorily.

TERMS RELATING TO MOVEMENT

Kinds of Movement

Since all movement between bodies is only relative, a description of slope movements must necessarily give some attention to identifying the bodies that are in relative motion. For example, the word slide specifies relative motion between stable ground and moving ground in which the vectors of relative motion are parallel to the surface of separation or rupture; furthermore, the bodies remain in contact. The word flow, however, refers not to the motions of the moving mass relative to stable ground, but rather to the distribution and continuity of relative movements of particles within the moving mass itself.

Falls

In falls, a mass of any size is detached from a steep slope or cliff, along a surface on which little or no shear displacement takes place, and descends mostly through the air by free fall, leaping, bounding, or rolling. Movements are very rapid to extremely rapid (see rate of movement scale, Figure 2.1u) and may or may not be preceded by minor movements leading to progressive separation of the mass from its source.

Rock fall is a fall of newly detached mass from an area of bedrock. An example is shown in Figure 2.3. Debris fall is a fall of debris, which is composed of detrital fragments prior to failure. Rapp (*2.131*, p. 104) suggested that falls of newly detached material be called primary and those involving earlier transported loose debris, such as that from shelves, be called secondary. Among those termed debris falls here, Rapp (*2.131*, p. 97) also distinguished pebble falls (size less than 20 mm), cobble falls (more than 20 mm, but less than 200 mm), and boulder falls (more than 200 mm). Included within falls would be the raveling of a thin colluvial layer, as illustrated by Deere and Patton (*2.36*), and of fractured, steeply dipping weathered rock, as illustrated by Sowers (*2.162*).

The falls of loess along bluffs of the lower Mississippi River valley, described in a section on debris falls by Sharpe (*2.146*, p. 75), would be called earth falls (or loess falls) in the present classification.

Topples

Topples have been recognized relatively recently as a distinct type of movement. This kind of movement consists of the forward rotation of a unit or units about some pivot point, below or low in the unit, under the action of gravity and forces exerted by adjacent units or by fluids in cracks. It is tilting without collapse. The most detailed descriptions have been given by de Freitas and Watters (*2.37*), and some of their drawings are reproduced in Figure 2.1d1 and d2. From their studies in the British Isles, they concluded that toppling failures are not unusual, can develop in a variety of rock types, and can range in volume from 100 m³ to more than 1 Gm³ (130 to 1.3 billion yd³). Toppling may or may not culminate in either falling or sliding, depending on the geometry of the failing mass and the orientation and extent of the discontinuities. Toppling failure has been pictured by Hoek (*2.61*), Aisenstein (*2.1*, p. 375), and Bukovansky, Rodríquez, and Cedrún (*2.16*) and studied in detail in laboratory experiments with blocks by Hofmann (*2.63*). Forward rotation was noted in the Kimbley copper pit by Hamel (*2.56*), analyzed in a high rock cut by Piteau and others (*2.125*), and described among the prefailure movements at Vaiont by Hofmann (*2.62*).

Slides

In true slides, the movement consists of shear strain and displacement along one or several surfaces that are visible or may reasonably be inferred, or within a relatively narrow zone. The movement may be progressive; that is, shear failure may not initially occur simultaneously over what eventually becomes a defined surface of rupture, but rather it may propagate from an area of local failure. The displaced mass may slide beyond the original surface of rupture onto what had been the original ground surface, which then becomes a surface of separation.

Slides were subdivided in the classification published in 1958 (*2.182*) into (a) those in which the material in motion is not greatly deformed and consists of one or a few units and (b) those in which the material is greatly deformed or consists of many semi-independent units. These subtypes were further classed into rotational slides and planar slides. In the present classification, emphasis is put on the distinction between rotational and translational slides, for that

Figure 2.4. Slope failure in uniform material (2.182).

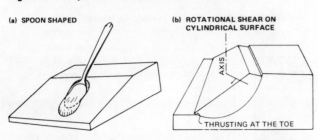

(a) SPOON SHAPED

(b) ROTATIONAL SHEAR ON CYLINDRICAL SURFACE

AXIS

THRUSTING AT THE TOE

difference is of at least equal significance in the analysis of stability and the design of control methods. An indication of degree of disruption is still available by use of the terms block or intact for slides consisting of one or a few moving units and the terms broken or disrupted for those consisting of many units; these terms avoid a possible source of confusion, pointed out by D. H. Radbruch-Hall, in the use of the term debris slide, which is now meant to indicate only a slide originating in debris material, which may either proceed as a relatively unbroken block or lead to disruption into many units, each consisting of debris.

Rotational Slides

The commonest examples of rotational slides are little-deformed slumps, which are slides along a surface of rupture that is curved concavely upward. Slumps, and slumps combined with other types of movement, make up a high proportion of landslide problems facing the engineer. The movement in slumps takes place only along internal slip surfaces. The exposed cracks are concentric in plan and concave toward the direction of movement. In many slumps the underlying surface of rupture, together with the exposed scarps, is spoon-shaped (Figure 2.4). If the slide extends for a considerable distance along the slope perpendicular to the direction of movement, much of the rupture surface may approach the shape of a sector of a cylinder whose axis is parallel to the slope (Figure 2.4). In slumps, the movement is more or less rotational about an axis that is parallel to the slope. In the head area, the movement may be almost wholly downward and have little apparent rotation; however, the top surface of each

unit commonly tilts backward toward the slope (Figures 2.1g, 2.1i, 2.4, 2.5, 2.6, and 2.7), but some blocks may tilt forward.

Figure 2.5 shows some of the commoner varieties of slump failure in various kinds of materials. Figure 2.7 shows the backward tilting of strata exposed in a longitudinal section through a small slump in lake beds. Although the rupture surface of slumps is generally concave upward, it is seldom a spherical segment of uniform curvature. Often the shape of the surface is greatly influenced by faults, joints, bedding, or other preexisting discontinuities of the material. The influence of such discontinuities must be considered carefully when the engineer makes a slope-stability analysis that assumes a certain configuration for the surface of rupture. Figures 2.7 and 2.8 show how the surface of rupture may follow bedding planes for a considerable part of its length. Upward thrusting and slickensides along the lateral margin of the toe of a slump are shown in Figure 2.9.

The classic purely rotational slump on a surface of smooth curvature is relatively uncommon among the many types of gravitational movement to which geologic materials are subject. Since rotational slides occur most frequently in fairly homogeneous materials, their incidence among constructed embankments and fills, and hence their interest to engineers, has perhaps been high relative to other types of failure, and their methods of analysis have in the past been more actively studied. Geologic materials are seldom uniform, however, and natural slides tend to be complex or at least significantly controlled in their mode of movement by internal inhomogeneities and discontinuities. Moreover, deeper and deeper artificial cuts for damsites, highways, and other engineering works have increasingly produced failures not amenable to analysis by the methods appropriate to circular arc slides and have made necessary the development of new methods of analytical design for prevention or cure of failures in both bedrock and engineering soils.

The scarp at the head of a slump may be almost vertical. If the main mass of the slide moves down very far, the steep scarp is left unsupported and the stage is set for a new failure (similar to the original slump) at the crown of the slide. Occasionally, the scarps along the lateral margins of the upper part of the slide may also be so high and steep that slump blocks break off along the sides and move downward and inward toward the middle of the main slide. Figure 2.10 (2.183) shows a plan view of slump units along the upper margins of a slide; the longest dimensions of these units are parallel with, rather than perpendicular to, the direction of movement of the main slide. Any water that finds its way into the head of a slump may be ponded by the backward tilt of the unit blocks or by other irregularities in topography so that the slide is kept wet constantly. By the successive creation of steep scarps and trapping of water, slumps often become self-perpetuating areas of instability and may continue to move and enlarge intermittently until a stable slope of very low gradient is attained.

Translational Slides

In translational sliding the mass progresses out or down and out along a more or less planar or gently undulatory surface and has little of the rotary movement or backward tilting characteristic of slump. The moving mass commonly slides out on the original ground surface. The distinction between rotational and translational slides is useful in planning control measures. The rotary movement of a slump, if the surface of rupture dips into the hill at the foot of the slide, tends to restore equilibrium in the unstable mass; the driving moment during movement decreases and the slide may stop moving. A translational slide, however, may progress indefinitely if the surface on which it rests is sufficiently inclined and as long as the shear resistance along this surface remains lower than the more or less constant driving force. A translational slide in which the moving mass consists of a single unit that is not greatly deformed or a few closely related units may be called a block slide. If the moving mass consists of many semi-independent units, it is termed a broken or disrupted slide.

The movement of translational slides is commonly controlled structurally by surfaces of weakness, such as faults, joints, bedding planes, and variations in shear strength between layers of bedded deposits, or by the contact between firm bedrock and overlying detritus (Figure 2.11). Several examples of block slides are shown in Figures 2.1j2, 2.1l, 2.12, 2.13 (*2.136*), 2.14 (*2.107*), and 2.15. In many translational slides, the slide mass is greatly deformed or breaks up into many more or less independent units. As deformation and disintegration continue, and especially as water content or velocity or both increase, the broken or disrupted slide mass may change into a flow; however, all gradations exist. Broken translational slides of rock are shown in Figure 2.1j3 and of debris in Figures 2.1k and 2.16 (*2.83*).

Lateral Spreads

In spreads, the dominant mode of movement is lateral extension accommodated by shear or tensile fractures. Two types may be distinguished.

1. Distributed movements result in overall extension

Figure 2.5. Varieties of slump (*2.182*).

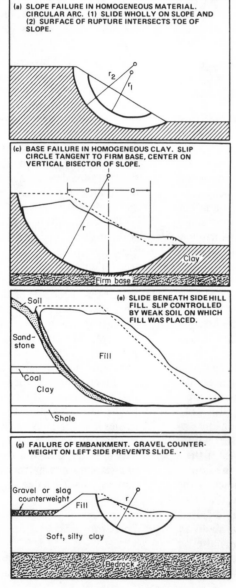

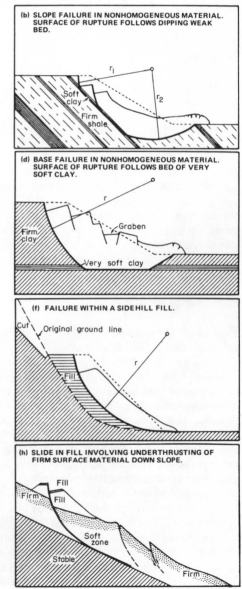

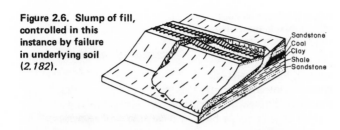

Figure 2.6. Slump of fill, controlled in this instance by failure in underlying soil (2.182).

Sandstone
Coal
Clay
Shale
Sandstone

Figure 2.7. Slump in thinly bedded lake deposits of silt and clay in Columbia River valley (note backward tilting of beds above surface of rupture) (2.182).

Figure 2.8. Slump in bedded deposits similar to those shown in Figure 2.7 (note that surface of rupture follows horizontal bedding plane for part of its length) (2.182).

Figure 2.9. Slickensides in foot area of shallow slide in Pennington shale residuum (highly weathered clay shale) along I-40 in Roane County, Tennessee.

but without a recognized or well-defined controlling basal shear surface or zone of plastic flow. These appear to occur predominantly in bedrock, especially on the crests of ridges (Figure 2.1m1). The mechanics of movement are not well known.

2. Movements may involve fracturing and extension of coherent material, either bedrock or soil, owing to liquefaction or plastic flow of subjacent material. The coherent upper units may subside, translate, rotate, or disintegrate, or they may liquefy and flow. The mechanism of failure can involve elements not only of rotation and translation but also of flow; hence, lateral spreading failures of this type may be properly regarded as complex. They form, however, such a distinctive and dominant species in certain geologic situations that specific recognition seems worthwhile.

Examples of the second type of spread in bedrock are shown in Figure 2.1m2 and 2.1m3. In both examples, taken from actual landslides in the USSR and Libya respectively, a thick layer of coherent rock overlies soft shale and

Figure 2.10. Ames slide near Telluride, Colorado (2.182, 2.183). This slump-earth flow landslide occurred in glacial till overlying Mancos shale. Repeated slumping took place along upper margins after main body of material had moved down. Long axes of slump blocks B and B' are parallel with rather than perpendicular to direction of movement of main part of slide. Blocks B and B' moved toward left, rather than toward observer.

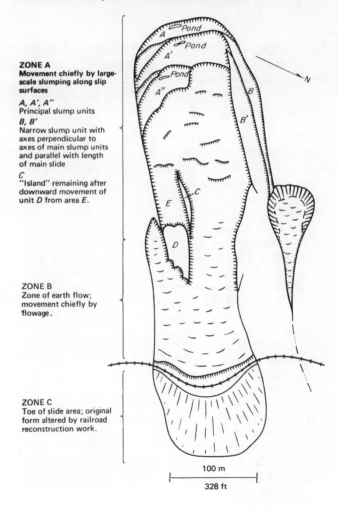

ZONE A
Movement chiefly by large-scale slumping along slip surfaces

A, A', A''
Principal slump units

B, B'
Narrow slump unit with axes perpendicular to axes of main slump units and parallel with length of main slide

C
"Island" remaining after downward movement of unit D from area E.

ZONE B
Zone of earth flow; movement chiefly by flowage.

ZONE C
Toe of slide area; original form altered by railroad reconstruction work.

100 m
328 ft

15

Figure 2.11. Thin layer of residual debris that slid on inclined strata of metasiltstone along I-40 in Cocke County, Tennessee.

Figure 2.12. Block slide at quarry (2.182).

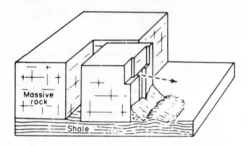

Figure 2.13. Development of landslides in horizontal sequence of claystone and coal caused by relaxation of horizontal stresses resulting from reduction in thickness of overlying strata (2.136).

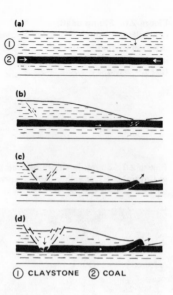

① CLAYSTONE ② COAL

Figure 2.14. Section view of translational slide at Point Fermin, near Los Angeles, California (see also Figure 2.15) (2.107, 2.182). Maximum average rate of movement was 3 cm/week (1.2 in).

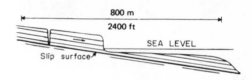

claystone. The underlying layer became plastic and flowed to some extent, allowing the overlying firmer rock to break into strips and blocks that then became separated. The cracks between the blocks were filled with either soft material squeezed up from below or detritus from above. The lateral extent of these slides is remarkable, involving bands several to many kilometers wide around the edges of plateaus and escarpments. The rate of movement of most lateral spreads in bedrock is apparently extremely slow.

Laterally spreading slope movements also form in fine-grained earth material on shallow slopes, particularly in sensitive silt and clay that loses most or all of its shear strength on disturbance or remolding. The failure is usually progressive; that is, it starts in a local area and spreads. Often the initial failure is a slump along a stream bank or shore, and the progressive failure extends retrogressively back from the initial failure farther and farther into the bank. The principal movement is translation rather than rotation. If the underlying mobile zone is thick, the blocks at the head may sink downward as grabens, not necessarily with backward rotation, and there may be upward and outward extrusion and flow at the toe. Movement generally begins suddenly, without appreciable warning, and proceeds with rapid to very rapid velocity.

These types appear to be members of a gradational series of landslides in surficial materials ranging from block slides at one extreme, in which the zone of flow beneath the sliding mass may be absent or very thin, to earth flows or completely liquefied mud flows at the other extreme, in which the zone of flow includes the entire mass. The form that is

taken depends on local factors. Most of the larger landslides in glacial sediments of northern North America and Scandinavia lie somewhere within this series.

Lateral spreads in surficial deposits have been destructive of both life and property and have, therefore, been the subject of intensive study. Examples may be cited from Sweden (Caldenius and Lundstrom, 2.18), Canada (Mitchell and Markell, 2.108), Alaska (Seed and Wilson, 2.144), and California (Youd, 2.191). Most of the spreading failures in the western United States generally involve less than total liquefaction and seem to have been mobilized only by seismic shock. For example, there were damaging failures in San Fernando Valley, California, during the 1969 earthquake because of liquefaction of underlying sand and silt and spreading of the surficial, firmer material. The spreading failure of Bootlegger Cove clay beneath the Turnagain Heights residential district at Anchorage, Alaska, during the 1964 great earthquake resulted in some loss of life and extensive damage. In some areas within the city of San Francisco, the principal damage due to the 1906 earthquake resulted from spreading failures that not only did direct damage to structures but also severed principal water-supply lines and thereby hindered firefighting.

All investigators would agree that spreading failures in glacial and marine sediments of Pleistocene age present some common and characteristic features: Movement often occurs for no apparent external reason, failure is generally sudden, gentle slopes are often unstable, dominant movement is translatory, materials are sensitive, and pore-water pressure is important in causing instability. All de-

Figure 2.16. Debris slide of disintegrating soil slip variety (*2.83, 2.182*).

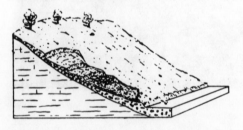

grees of disturbance of the masses have been observed; some failures consist almost entirely of one large slab or "flake," but others liquefy almost entirely to small chunks or mud.

Flows

Many examples of slope movement cannot be classed as falls, topples, slides, or spreads. In unconsolidated materials, these generally take the form of fairly obvious flows, either fast or slow, wet or dry. In bedrock, the movements most difficult to categorize include those that are extremely slow and distributed among many closely spaced, noninterconnected fractures or those movements within the rock mass that result in folding, bending, or bulging. In many instances, the distribution of velocities resembles that of viscous fluids; hence, the movements may be described as a form of flow of intact rock.

Much of what is here described under flowlike distributed movements has been classified as creep, both of rock and soil. But creep has come to mean different things to different persons, and it seems best to avoid the term or to use it in a well-defined manner. As used here, creep is considered to have a meaning similar to that used in mechanics of materials; that is, creep is simply deformation that continues under constant stress. Some of the creep deforma-

tion may be recoverable over a period of time upon release of the stress, but generally most of it is not. The movement commonly is imperceptible (which is usually one of the essential attributes of creep as defined in geomorphology), but increasingly sophisticated methods of measurement make this requirement difficult to apply. Furthermore, the usual partition of creep into three stages—primary (decelerating), secondary (steady or nearly so), and tertiary (accelerating to failure)—certainly includes perceptible deformation in the final stages. Laboratory studies show that both soil and rock, as well as metals, can exhibit all three stages of creep. Observations in the field, such as those reported by Müller (*2.112*) at Vaiont, embrace within the term creep perceptible movements that immediately preceded catastrophic failure.

There is disagreement also as to whether creep in rock and soil should be restricted to those movements that are distributed through a mass rather than along a defined fracture. Authorities are about equally divided on this point but, in keeping with the use of the term in engineering mechanics, the acceptance of this restriction is not favored. Creep movements can occur in many kinds of topples, slides, spreads, and flows, and the term creep need not be restricted to slow, spatially continuous deformation. Therefore, spatially continuous deformations are classified as various types of flow in rock, debris, and earth.

Flows in Bedrock

Flow movements in bedrock include deformations that are distributed among many large or small fractures, or even microfractures, without concentration of displacement along a through-going fracture. The movements are generally extremely slow and are apparently more or less steady in time, although few data are available. Flow movements may result in folding, bending, bulging, or other manifestations of plastic behavior, as shown in Figure 2.1p1, 2.1p2, 2.1p3, and 2.1p4. The distribution of velocities may roughly simulate that of viscous fluids.

These kinds of movements have come under close study only within the last decade or so and are being recognized more and more frequently in areas of high relief in many parts of the world. They are quite varied in character, and several kinds have been described as creep by Nemčok, Pašek, and Rybář (*2.116*) in a general classification of landslides and other mass movements, as gravitational slope deformation by Nemcok (*2.114, 2.115*), by the term Sackung (approximate translation: sagging) by Zischinsky (*2.194, 2.195*), as depth creep of slopes by Ter-Stepanian (*2.172*), and as gravitational faulting by Beck (*2.5*). In the United States, ridge-top depressions due to large-scale creep have been described by Tabor (*2.166*). A review of gravitational creep (mass rock creep) together with descriptions of examples from the United States and other countries has been prepared by Radbruch-Hall (*2.130*). The significance of these relatively slow but pervasive movements to human works on and within rock slopes is only beginning to be appreciated.

Flows in Debris and Earth

Distributed movements within debris and earth are often

more accurately recognized as flows than those in rocks because the relative displacements within the mass are commonly larger and more closely distributed and the general appearance is more obviously that of a body that has behaved like a fluid. Moreover, the fluidizing effect of water itself is, as a rule, a part of the process. Slip surfaces within the moving mass are usually not visible or are short lived, and the boundary between moving mass and material in place may be a sharp surface of differential movement or a zone of distributed shear.

There is complete gradation from debris slides to debris flows, depending on water content, mobility, and character of the movement, and from debris slide to debris avalanche as movement becomes much more rapid because of lower cohesion or higher water content and generally steeper slopes. Debris slides and, less commonly, debris avalanches may have slump blocks at their heads. In debris slides, the moving mass breaks up into smaller and smaller parts as it advances toward the foot, and the movement is usually slow. In debris avalanches, progressive failure is more rapid, and the whole mass, either because it is quite wet or because it is on a steep slope, liquefies, at least in part, flows, and tumbles downward, commonly along a stream channel, and may advance well beyond the foot of the slope. Debris avalanches are generally long and narrow and often leave a serrate or V-shaped scar tapering uphill at the head, as shown in Figures 2.1q3 and 2.17, in contrast to the horseshoe-shaped scarp of a slump.

Debris flows, called mud flows in some other classifications, are here distinguished from the latter on the basis of particle size. That is, the term debris denotes material that contains a relatively high percentage of coarse fragments, whereas the term mud flow is reserved for an earth flow consisting of material that is wet enough to flow rapidly and that contains at least 50 percent sand-, silt-, and clay-sized particles. Debris flows commonly result from unusually heavy precipitation or from thaw of snow or frozen soil. The kind of flow shown in Figure 2.1q1 often occurs during torrential runoff following cloudbursts. It is favored by the presence of soil on steep mountain slopes from which the vegetative cover has been removed by fire or other means, but the absence of vegetation is not a prerequisite. Once in motion, a small stream of water heavily laden with soil has transporting power that is disproportionate to its size, and, as more material is added to the stream by caving of its banks, its size and power increase. These flows commonly follow preexisting drainageways, and they are often of high density, perhaps 60 to 70 percent solids by weight, so that boulders as big as automobiles may be rolled along. If such a flow starts on an unbroken hillside it will quickly cut a V-shaped channel. Some of the coarser material will be heaped at the side to form a natural levee, while the more fluid part moves down the channel (Figure 2.17). Flows may extend many kilometers, until they drop their loads in a valley of lower gradient or at the base of a mountain front. Some debris flows and mud flows have been reported to proceed by a series of pulses in their lower parts; these pulses presumably are caused by periodic mobilization of material in the source area or by periodic damming and release of debris in the lower channel.

The term avalanche, if unmodified, should refer only to

Figure 2.17. Debris avalanche or very rapid debris flow at Franconia Notch, New Hampshire, June 24, 1948, after several days of heavy rainfall (*2.182*). Only soil mantle 2 to 5 m (7 to 16 ft) thick, which lay over bedrock on a slope of about 1:1, was involved. Scar is about 450 m (1500 ft) long; natural levees can be seen along sides of flow. US-3 is in foreground.

slope movements of snow or ice. Rapp (*2.132*) and Temple and Rapp (*2.169*), with considerable logic, recommend that, because the term debris avalanche is poorly defined, it should be abandoned, and that the term avalanche should be used only in connection with mass movements of snow, either pure or mixed with other debris. The term debris avalanche, however, is fairly well entrenched and in common usage (Knapp, *2.86*); hence, its appearance in the classification as a variety of very rapid to extremely rapid debris flow seems justified.

Recent studies have contributed much to a better understanding of the rates and duration of rainfall that lead to the triggering of debris flows, the physical properties of the material in place, the effect of slope angle, the effect of pore-water pressure, the mobilization of material and mechanism of movement, and the properties of the resulting deposit. The reader is referred especially to the works of Campbell (*2.20*), Daido (*2.34*), Fisher (*2.46*), Hutchinson (*2.70*), Hutchinson and Bhandari (*2.72*), Johnson and Rahn (*2.76*), Jones (*2.78*), Prior, Stephens, and Douglas (*2.129*), Rapp (*2.131*), K. M. Scott (*2.141*), R. C. Scott (*2.142*), and Williams and Guy (*2.188*). Flowing movements of surficial debris, including creep of the mantle of weathered rock and soil, are shown in Figure 2.1q2, 2.1q4, and 2.1q5. Soil flow, or solifluction, which in areas of perennially or permanently frozen ground is better termed gelifluction, takes many forms and involves a variety of mechanisms that can be treated adequately only in works devoted to this special field, which is of great significance to engineering works at high latitudes and altitudes. The reader is referred to summaries by Dylik (2.40), Washburn (2.187), and Corte (2.27); the proceedings of the International Conference on Permafrost (2.111); and recent work by McRoberts and Morgenstern (2.104, 2.105) and Embleton and King (2.42).

Subaerial flows in fine-grained materials such as sand, silt, or clay are classified here as earth flows. They take a variety of forms and range in water content from above saturation to essentially dry and in velocity from extremely rapid to extremely slow. Some examples are shown in Figure 2.1r1 through 2.1r5. At the wet end of the scale are mud flows, which are soupy end members of the family of predominantly fine-grained earth flows, and subaqueous flows or flows originating in saturated sand or silt along shores.

In a recent paper reviewing Soviet work on mud flows, Kurdin (2.91) recommended a classification of mud flows based on (a) the nature of the water and solid-material supply; (b) the structural-rheological model, that is, whether the transporting medium is largely water in the free state or is a single viscoplastic mass of water and fine particles; (c) the composition of the mud flow mass, that is, whether it consists of mud made up of water and particles less than 1 mm (0.04 in) in size or of mud plus gravel, rubble, boulders, and rock fragments; and (d) the force of the mud flow as defined by volume, rate of discharge, and observed erosive and destructive power. In the Soviet literature mud flows include not only what are here classified as debris flows but also heavily laden flows of water-transported sediment.

According to Andresen and Bjerrum (2.3), subaqueous flows are generally of two types: (a) retrogressive flow slide or (b) spontaneous liquefaction, as shown in Figure 2.18. The retrogressive flows, as shown in Figure 2.1r1, occur mostly along banks of noncohesive clean sand or silt. They are especially common along tidal estuaries in the coastal provinces of Holland, where banks of sand are subject to scour and to repeated fluctuations in pore-water pressure because of the rise and fall of the tide (Koppejan,

Figure 2.18. Retrogressive flow slide and spontaneous liquefaction (2.3).

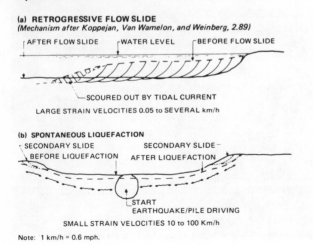

(a) **RETROGRESSIVE FLOW SLIDE**
(Mechanism after Koppejan, Van Wamelon, and Weinberg, 2.89)

SCOURED OUT BY TIDAL CURRENT
LARGE STRAIN VELOCITIES 0.05 to SEVERAL km/h

(b) **SPONTANEOUS LIQUEFACTION**

SMALL STRAIN VELOCITIES 10 to 100 Km/h

Note: 1 km/h = 0.6 mph.

Figure 2.19. Earth flow near Greensboro, Florida (2.80, 2.182). Material is flat-lying, partly indurated clayey sand of the Hawthorn formation (Miocene). Length of slide is 275 m (900 ft) from scarp to edge of trees in foreground. Vertical distance is about 15 m (45 ft) from top to base of scarp and about 20 m (60 ft) from top of scarp to toe. Slide occurred in April 1948 after year of unusually heavy rainfall, including 40 cm (16 in) during 30 d preceding slide.

Van Wamelon, and Weinberg, 2.89). When the structure of the loose sand breaks down along a section of the bank, the sand flows rapidly along the bottom, and, by repeated small failures, the slide eats into the bank and enlarges the cavity. Sometimes the scarp produced is an arc, concave toward the water, and sometimes it enlarges greatly, retaining a narrow neck or nozzle through which the sand flows. An extensive discussion and classification of subaqueous mass-transport processes and the resulting deposits have been presented by Carter (2.21).

Rapid earth flows also occur in fine-grained silt, clay, and clayey sand, as shown in Figures 2.1r2 and 2.19 (2.80). These flows form a complete gradation with slides involving failure by lateral spreading, but they involve not only liquefaction of the subjacent material but also retrogressive failure and liquefaction of the entire slide mass. They usually take place in sensitive materials, that is, in those materials whose shear strength on remolding at constant water content is decreased to a small fraction of its original value. Rapid earth flows have caused loss of life and immense destruction of property in Scandinavia, the St. Lawrence River valley in Canada, and Alaska during the 1964 earthquake. The properties of the material involved, which is usually a marine or estuarine clay of late Pleistocene age, have been thoroughly studied by many investigators during the last 15 years. Summary papers have been written by Bjerrum and others (2.12) on flows in Norway and by Mitchell and Markell (2.108), and Eden and Mitchell (2.41) on flows in Canada. Shoreline flows produced by the Alaskan earthquake at Valdez and Seward have been described by Coulter and Migliaccio (2.28) and Lemke (2.98). The large failure on the Reed Terrace near Kettle Falls, Washington, shown in Figures 2.20 and 2.21 (2.79), resembles in some respects the earth flow at Riviere Blanche, Quebec, shown in Figure 2.1r2 (2.146).

The somewhat drier and slower earth flows in plastic earth are common in many parts of the world wherever there is a combination of clay or weathered clay-bearing rocks, moderate slopes, and adequate moisture; Figure 2.22 shows a typical example. A common elongation of the flow, channelization in depression in the slope, and spreading of the toe are illustrated in Figure 2.1r3 and also shown in an actual debris flow in Figure 2.23.

The word flow naturally brings water to mind, and some content of water is necessary for most types of flow movement. But small dry flows of granular material are common, and a surprising number of large and catastrophic flow movements have occurred in quite dry materials. Therefore, the classification of flows indicates the complete range of water content—from liquid at the top to dry at the bottom. Tongues of rocky debris on steep slopes moving extremely slowly and often fed by talus cones at the head are called block streams (Figure 2.1q5). Because of rainwash, a higher proportion of coarse rocks may be in the surface layers than in the interior. Dry flows of sand are common along shores or embankments underlain by dry granular material. In form, they may be channelized, as shown in Figures 2.1r4 and 2.24, or sheetlike, as shown in Figure 2.25 (2.79). Small flows of dry silt, powered by impact

on falling from a cliff, have been recognized, but so far as is known none has been studied in detail (Figure 2.26).

Flows of loess mobilized by earthquake shock have been more destructive of life than any other type of slope failure. Those that followed the 1920 earthquake in Kansu Province, China (Close and McCormick, 2.23), shown in Figure 2.1r5, took about 100 000 lives. Apparently the normal, fairly coherent internal structure of the porous silt was destroyed by earthquake shock, so that, for all practical purposes, the loess became a fluid suspension of silt in air and flowed down into the valleys, filling them and overwhelming villages. The flows were essentially dry, according to the report. Extensive flows of loess accompanied the Chait earthquake of July 10, 1949, in Tadzhikistan, south-central Asia, and buried or destroyed 33 villages as the flows covered the bottoms of valleys to depths of several tens of meters for many kilometers (Gubin, 2.54).

Complex

More often than not, slope movements involve a combina-

Figure 2.20. Reed Terrace area, right bank of Lake Roosevelt reservoir on Columbia River, near Kettle Falls, Washington, May 15, 1951 (2.182). Landslide of April 10, 1952, involving about 11 Mm³ (15 million yd³) took place by progressive slumping, liquefaction, and flowing out of glaciofluvial sediments through narrow orifice into bottom of reservoir.

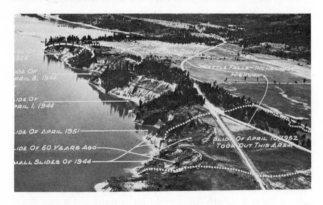

Figure 2.22. Earth flow developing from slump near Berkeley, California (2.182).

Figure 2.21. Reed Terrace area, Lake Roosevelt, Washington, August 1, 1952, after landslide of April 10, 1952 (2.79, 2.182).

Figure 2.23. Old debris flow in altered volcanic rocks west of Pahsimeroi River in south central Idaho.

Figure 2.24. Dry sand flow in Columbia River valley (2.182). Material is sand over lake-bed silt; dry sand from upper terrace flowed like liquid through notch in more compact sand and silt below.

Figure 2.25. Shallow, dry, sand flow along shore of Lake Roosevelt, Washington (2.79). Wave erosion or saturation of sediment by lake water caused thin skin of material to lose support and ravel off terrace scarp.

Figure 2.26. Dry flow of silt (2.182). Material is lake-bed silt of Pleistocene age from high bluff on right bank of Columbia River, 4 km (2.5 miles) downstream from Belvedere, Washington. Flow was not observed while in motion, but is believed to result from blocks of silt falling down slope, disintegrating, forming a single high-density solid-in-air suspension, and flowing out from base of cliff.

tion of one or more of the principal types of movement described above, either within various parts of the moving mass or at different stages in development of the movements. These are termed complex slope movements, and a few examples of the many possible types are illustrated in Figure 2.1s1 through 2.1s5.

Of particular interest regarding hazards of landslides to life and property are large, extremely rapid rock fall-debris flows, referred to as rock-fragment flow (variety rock-fall avalanche) in the 1958 classification (2.182). Rock slide- and rock fall-debris flows are most common in rugged mountainous regions. The disaster at Elm, Switzerland (Heim, 2.58, pp. 84, 109-112), which took 115 lives, started with small rock slides at each side of a quarry on the mountainside. A few minutes later the entire mass of rock above the quarry crashed down and shot across the valley. The movement of the rock fragments, which had to that moment been that of a rock slide and rock fall, appears to have taken on the character of a flow. The mass rushed up the other side of the small valley, turned and streamed into the main valley, and flowed for nearly 1.5 km (1 mile) at high velocity before stopping (Figure 2.1s1). About 10 Mm³ (13 million yd³) of rock descended an average of 470 m (1540 ft) vertically in a total elapsed time of about 55 s. The kinetic energy involved was enormous. A similar and even larger rock-fall avalanche occurred at Frank, Alberta, in 1903 and also caused great loss of life and property (McConnell and Brock, 2.103; Cruden and Krahn, 2.33).

These rock fall-debris flows are minor, however, compared with the cataclysmic flow that occurred at the time of the May 31, 1970, earthquake in Peru, which buried the city of Yungay and part of Ranrahirca, causing a loss of more than 18 000 lives. According to Plafker, Ericksen, and Fernandez Concha (2.126), the movement started high on Huascarán Mountain at an altitude of 5500 to 6400 m and involved 50 Mm³ to 100 Mm³ (65 million to 130 million yd³) of rock, ice, snow, and soil that traveled 14.5 km (9 miles) from the source to Yungay at a velocity between 280 and 335 km/h (175 to 210 mph). They reported strong evidence that the extremely high velocity and low friction of the flow were due, at least in part, to lubrication by a cushion of air entrapped beneath the debris. Pautre, Sabarly, and Schneider (2.122) suggested that the mass may have ridden on a cushion of steam. A sketch of the area affected is shown in Figure 2.27, taken from a paper by Cluff (2.24) on engineering geology observations. Crandell and Fahnestock (2.29) cited evidence for an air cushion beneath one or more rock fall-debris flows that occurred in December 1963 at Little Tahoma Peak and Emmons Glacier on the east flank of Mt. Rainier volcano, Washington.

Such flows probably cannot be produced by a few thousand or a few hundred thousand cubic meters of material. Many millions of megagrams are required; and, when that much material is set in motion, perhaps even slowly, predictions of behavior based on past experience with small failures become questionable. The mechanics of large, extremely rapid debris flows, many of which appear to have been nearly dry when formed, have come under much recent study. The large prehistoric Blackhawk landslide (Figure 2.28) shows so little gross rearrangement within the sheet of debris of which it is composed that Shreve (*2.148*) believed the broken material was not fluidized but slid on an ephemeral layer of compressed air. He reported, similarly, that the large landslide that was triggered by the Alaska earthquake of 1964 and fell onto the Sherman Glacier showed little large-scale mixing and did not flow like a viscous fluid but instead slid like a flexible sheet (Shreve, *2.149*). On the other hand, Johnson and Ragle (*2.77*) reported,

Many rock-snow slides that followed from the Alaska earthquake of March 27, 1964, illustrated a variety of flow mechanics. The form of some slides suggests a complete turbulence during flow, while the form of others gives evidence for steady-state flow or for controlled shearing.

From a detailed analysis of the kinematics of natural rock fall-debris flows and from model studies, Hsü (*2.66*) disputed Shreve's hypothesis that some slid as relatively undeformed sheets on compressed air and concluded, rather, that they flowed.

Obviously, there is much yet to be learned about these processes, particularly as similar features indicating mass movements of huge size have been recognized in Mariner 9 photographs of the surface of Mars (Sharp, *2.145*), where it is yet uncertain that significant amounts of either liquid

or gas were available for fluidization.

Getting back to Earth, we note self-explanatory examples of complex movements in Figure 2-1: slump-topple in Figure 2.1s2, rock slide-rock fall in Figure 2.1s3, and the common combination of a slump that breaks down into an earth flow in its lower part in Figure 2.1s5.

The illustration of cambering and valley bulging in Figure 2.1s4 is adapted from the classical paper by Hollingworth and Taylor (*2.65*) on the Northampton Sand Ironstone in England, their earlier paper on the Kettering district (*2.64*), and a sketch supplied by J. N. Hutchinson. The complex movements were described by Hutchinson (*2.68*) as follows:

Cambering and Valley Bulging. These related features were first clearly recognized in 1944 by Hollingworth, Taylor, and Kellaway (see reference in Terzaghi, 1950) in the Northampton Ironstone field of central England, where they are believed to have a Late Pleistocene origin. The ironstone occurs in the near-horizontal and relatively thin Northampton Sands, which are the uppermost solid rocks in the neighborhood. These are underlain, conformably, by a great thickness of the Lias, into which shallow valleys, typically 1200 to 1500 meters wide and 45 meters deep, have been eroded. Excavations for dam trenches in the valley bottoms have shown the Lias there to be thrust strongly upward and contorted, while opencast workings in the Northampton Sands

Figure 2.27. Area affected by May 31, 1970, Huascarán debris avalanche, which originated at point A (*2.24*). Yungay was protected from January 10, 1962, debris avalanche by 180 to 240-m (600 to 800-ft) high ridge (point B), but a portion of May 31, 1970, debris avalanche diverted from south side of canyon wall, topped "protective" ridge, and descended on Yungay below. Only safe place in Yungay was Cemetery Hill (point C), where some 93 people managed to run to before debris avalanche devastated surrounding area. Moving at average speed of 320 km (200 mph), debris arrived at point D, 14.5 km (9 miles) distant and 3660 m (12 000 ft) lower, within 3 to 4 min after starting from north peak of Huascarán. Debris flowed upstream along course of Rio Santa (point E) approximately 2.5 km (1.5 miles). Debris continued to follow course of Rio Santa downstream to Pacific Ocean, approximately 160 km (100 miles), devastating villages and crops occupying floodplain.

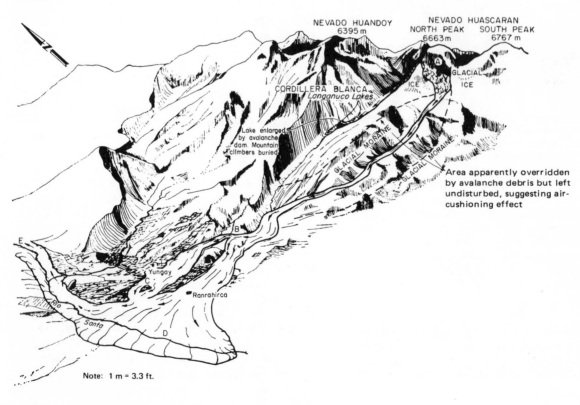

occupying the interfluves reveal a general valleyward increase of dip of "camber" of this stratum, often passing into dip and fault structure, suggesting corresponding downward movements along the valley margins. In adjusting to these movements, the rigid cap-rock has been dislocated by successive, regularly spaced fissures which parallel the valley and are known as "gulls." Similar features have been recognized in other parts of England and in Bohemia. The mechanisms by which cambering and valley bulging have been formed remain to be established.

Hutchinson (*2.70*) also pointed out that Sharpe's definition of flow (*2.146*), which requires zero relative displacement at the boundary of the flow (flow adheres to the stable material), does not fit the observed distribution of velocities at Beltinge, where a mud flow developed in a temperate climate on a 30-m-high (98-ft) coastal cliff of stiff, fissured London clay subject to moderate marine erosion. Here the mud flow was bounded both on the sides and on the bottom by discrete surfaces along which shear displacements occurred. For these kinds of movements Hutchinson and Bhandari (*2.72*) proposed the term mudslides. These can be regarded as complex movements in which the internal distribution of velocities within the moving mass may or may not resemble that of viscous fluids, but the movement relative to stable ground is finite discontinuous shear. It would seem that the material of the sliding earth flow is behaving as a plastic body in plug flow, as suggested by Hutchinson (*2.69*, pp. 231-232) and as analyzed in detail by Johnson (*2.75*).

Sequence or Repetition of Movement

The term retrogressive has been used almost consistently for slides or flow failures that begin at a local area, usually along a slope, and enlarge or retreat opposite to the direction of movement of the material by spreading of the failure surface, successive rotational slumps, falls, or liquefaction of the material. Kojan, Foggin, and Rice (*2.87*, pp. 127-128) used the term for failure spreading downslope.

On the other hand, the term progressive has been used to indicate extension of the failure (a) downslope (Blong, *2.13*; Kjellman, *2.84*; Ter-Stepanian, *2.170*, *2.171*; Thomson and Hayley, *2.177*), (b) upslope (but not specifically upslope only) (Seed, *2.143*; Tavenas, Chagnon, and La Rochelle, *2.168*), and (c) either upslope or downslope, or unspecified (Terzaghi and Peck, *2.176*; Bishop, *2.7*; Romani, Lovell, and Harr, *2.135*; Lo, *2.100*; Frölich, *2.51*; Ter-Stepanian and Goldstein, *2.173*, and many others).

I suggest that the term progressive be used for failure that is either advancing or retreating or both simultaneously, that the term retrogressive be used only for retreating failures, and that failures that enlarge in the direction of movement be referred to simply as advancing failures.

The terms complex, composite, compound, multiple, and successive have been used in different ways by various authors. I suggest the following definitions.

1. Complex refers to slope movements that exhibit more than one of the major modes of movement. This is the sense of the meaning suggested by Blong (*2.14*). The term is synonymous with composite, as used by Prior, Stephens, and

Figure 2.28. Blackhawk landslide (*2.147*). Upslope view, southward over lobe of dark marble breccia spread beyond mouth of Blackhawk Canyon on north flank of San Bernardino Mountains in southern California. Maximum width of lobe is 3.2 km (2 miles); height of scarp at near edge is about 15 m (50 ft).

Figure 2.29. Main types of rotational slide (*2.68*).

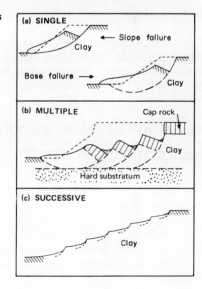

Archer (*2.128*) to describe sliding mud flows.

2. Compound refers to movements in which "the failure surface is formed of a combination of curved and planar elements and the slide movements have a part-rotational, part-translational character" (Skempton and Hutchinson, *2.154*).

3. Multiple refers to manifold development of the same mode of movement. As applied to retrogressive rotational sliding, the term refers to the production of "two or more slipped blocks, each with a curved slip surface tangential to a common, generally deep-seated slip sole [Figure 2.29]. Clearly, as the number of units increases, the overall character of the slip becomes more translational, though in failing, each block itself rotates backwards" (Hutchinson, *2.68*). Leighton (*2.97*) distinguished two types of multiple slide blocks: superposed, in which each slide block rides out on the one below, and juxtaposed, in which adjacent

23

moving units have a common basal surface of rupture, as shown in Figure 2.30.

4. Successive refers to any type of multiple movements that develop successively in time. According to Skempton and Hutchinson (*2.154*), "Successive rotational slips consist of an assembly of individual shallow rotational slips. The rather sparse data available suggest that successive slips generally spread up a slope from its foot." Hutchinson (*2.67*) states,

Below a slope inclination of about 13° [in London Clay], rotational slips of type R are replaced by successive rotational slips (type S). These probably develop by retrogression from a type R slip in the lower slope. Each component slip is usually of considerable lateral extent, forming a step across the slope. Irregular successive slips, which form a mosaic rather than a stepped pattern in plan are also found.

Figure 2.31 (*2.67*) shows the main types of landslides in London clay.

Landslides that develop one on top of another are called multistoried by Ter-Stepanian and Goldstein (*2.173*). Figure 2.32 shows their illustration of a three-storied landslide in Sochi on the coast of the Black Sea.

Rate of Movement

The rate-of-movement scale used in this chapter is shown at the bottom of the classification chart in Figure 2.1u. Metric equivalents to the rate scale shown in the 1958 classification have been derived by Yemel'ianova (*2.190*), and these should now be regarded as the primary definitions.

TERMS RELATING TO MATERIAL

Principal Divisions

The following four terms have been adopted as descriptions of material involved in slope movements.

1. Bedrock designates hard or firm rock that was intact and in its natural place before the initiation of movement.
2. Engineering soil includes any loose, unconsolidated, or poorly cemented aggregate of solid particles, generally of natural mineral, rock, or inorganic composition and either transported or residual, together with any interstitial gas or liquid. Engineering soil is divided into debris and earth.

a. Debris refers to an engineering soil, generally surficial, that contains a significant proportion of coarse material. According to Shroder (*2.150*), debris is used to specify material in which 20 to 80 percent of the fragments are greater than 2 mm (0.08 in) in size and the remainder of the fragments less than 2 mm.

b. Earth (again according to Shroder) connotes material in which about 80 percent or more of the fragments are smaller than 2 mm; it includes a range of materials from nonplastic sand to highly plastic clay.

This division of material that is completely gradational is admittedly crude; however, it is intended mainly to enable a name to be applied to material involved in a slope

Figure 2.30. Two types of multiple slide blocks (*2.97*).

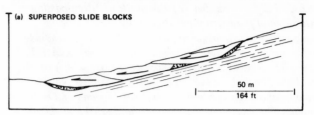

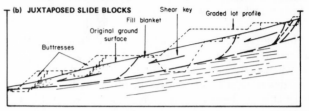

Figure 2.31. Main types of landslides in London clay (*2.67*).

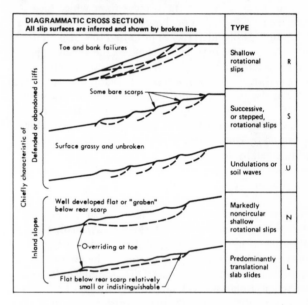

movement on the basis of a limited amount of information.

Water Content

By modifying the suggestions of Radbruch-Hall (*2.130*), we may define terms relating to water content simply as (a) dry, contains no visible moisture; (b) moist, contains some water but no free water and may behave as a plastic solid but not as a liquid; (c) wet, contains enough water to behave in part as a liquid, has water flowing from it, or supports significant bodies of standing water; and (d) very wet, contains enough water to flow as a liquid under low gradients.

Texture, Structure, and Special Properties

As amounts of information increase, more definite designation can be made about slope movements. For example, a bedrock slump may be redesignated as a slump in sandstone

Figure 2.32. Three-storied landslide in Sochi on coast of Black Sea, USSR (2.173). Boundaries of three stories of sliding are shown in section and plan by three types of lines.

4 BLOCKS OF ARGILLITE AND SANDSTONE
5 CRUSHED ARGILLITE
6 SLOW EARTH FLOW IN COLLUVIUM

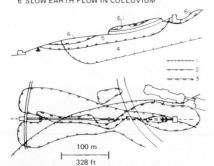

100 m
328 ft

Figure 2.33. Shallow translational slide that developed on shaly slope in Puente Hills of southern California (2.147). Slide has low D/L ratio (note wrinkles in surface).

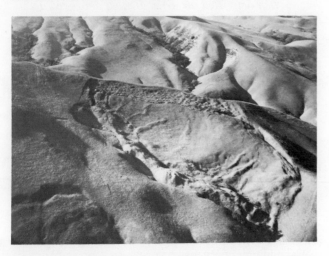

over stiff-fissured clay shale, or an earth slide may be given more precise definition as a block slide in moist sensitive clay.

TERMS RELATING TO SIZE OR GEOMETRY

A rather large body of descriptive terms has been built up relating to the size, shape, and morphology of slope movements and their deposits. Some of these have already been mentioned, such as the relation of rotational slides to curved surfaces of rupture and translational slides to planar surfaces of rupture. The close association between the morphology of a slope movement and its dominant genetic process, which is evident in a qualitative way from the foregoing text and illustrations, has been tested quantitatively through the use of refined measures of the parts and geometric attributes of landslides by Crozier (2.32) and by Blong (2.14, 2.15). These authors, together with Snopko (2.157), Klengel and Pašek (2.85), Shroder (2.150), and Laverdière (2.94), have made available a terminology that

is adequate to describe almost any feature of a slump earth flow. In addition, Skempton and Hutchinson (2.154) used the ratio of D/L, where D is the maximum thickness of the slide and L is the maximum length of the slide upslope. From Skempton's figures showing original use of this ratio (2.152), it seems probable that the intended length is that of a chord of the rupture surface (L_c), rather than the total length (L), as shown in Figure 2.1t. Skempton and Hutchinson gave a range of D/L_c values of 0.15 to 0.33 for rotational slides in clay and shale, and they stated that slab slides, which commonly occur in a mantle of weathered or colluvial material on clayey slopes, rarely if ever have D/L_c ratios greater than 0.1. Figure 2.33 illustrates such a shallow slab slide. In a statistical study of the forms of landslides along the Columbia River valley, Jones, Embody, and Peterson (2.79) made extensive use of the horizontal component (HC) or distance from the foot of the landslide to the crown, measured in a longitudinal section of the landslide, and the vertical component (VC) or difference in altitude between the foot and crown, measured in the same section.

TERMS RELATING TO GEOLOGIC, GEOMORPHIC, GEOGRAPHIC, OR CLIMATIC SETTING

The classification of landslides proposed by Savarensky (2.140) and followed to some degree in eastern Europe makes the primary division of types on the basis of the relation of slope movements to the geologic structure of the materials involved. Accordingly, asequent slides are those in which the surface of rupture forms in homogeneous material; consequent slides are those in which the position and geometry of the surface of rupture are controlled by preexisting discontinuities such as bedding, jointing, or contact between weathered and fresh rock; and insequent slides are those in which the surface of rupture cuts across bedding or other surfaces of inhomogeneity. The Japanese have used a classification of landslides separated into (a) tertiary type, involving incompetent tertiary sedimentary strata (Takada, 2.167); (b) hot-spring-volcanic type, which is in highly altered rocks; and (c) fracture-zone type, which occurs in fault zones and highly broken metamorphic rocks. Sharpe (2.146, pp. 57-61) distinguished three types of mud flows: semiarid, alpine, and volcanic, to which Hutchinson (2.68) has added a fourth variety, temperate.

Types of landslides are sometimes identified by the geographic location at which the type is particularly well developed. For example, Sokolov (2.159) refers to block slides of the Angara type (similar to that shown in Figure 2.1m2), of the Tyub-Karagan type (similar to that shown in Figure 2.1m3), and of the Ilim and Crimean types, all named after localities. Reiche (2.133) applied the term Toreva-block (from the village of Toreva on the Hopi Indian Reservation in Arizona) to "a landslide consisting essentially of a single large mass of unjostled material which, during descent, has undergone backward rotation toward the parent cliff about a horizontal axis which roughly parallels it" (Figure 2.1g). Shreve (2.149), in summarizing data on landslides that slid on a cushion of compressed air, referred to these landslides as being of the Blackhawk type, from the rock fall-debris slide-debris flow at Blackhawk Mountain in southern California (2.148). Although the use of locality terms may occasionally be a convenience, it is not recommended as a general practice, for the terms themselves are not informative to a reader who lacks knowledge of the locality.

TERMS RELATING TO AGE OR STATE OF ACTIVITY

Active slopes are those that are either currently moving or that are suspended, the latter term implying that they are not moving at the present time but have moved within the last cycle of seasons. Active slides are commonly fresh; that is, their morphological features, such as scarps and ridges, are easily recognizable as being due to gravitational movement, and they have not been significantly modified by surficial processes of weathering and erosion. However, in arid regions, slides may retain a fresh appearance for many years.

Inactive slopes are those for which there is no evidence that movement has taken place within the last cycle of seasons. They may be dormant, in which the causes of failure remain and movement may be renewed, or they may be stabilized, in which factors essential to movement have been removed naturally or by human activity. Slopes that have long-inactive movement are generally modified by erosion and weathering or may be covered with vegetation so that the evidence of the last movement is obscure. They are often referred to as fossil (Záruba and Mencl, *2.193*; Klengel and Pašek, *2.85*; Nossin, *2.118*) or ancient (Popov, *2.127*) landslides in that they commonly have developed under different geomorphological and climatic conditions thousands or more years ago and cannot repeat themselves at present.

FORMING NAMES

The names applied to slope movements can be made progressively more informative, as more data are obtained, by building up a designation from several descriptor words, each of which has a defined meaning. For example, a slow, moist, translational debris slab slide means material moving along a planar surface of a little-disturbed mass of fragmented material having a D/L_c ratio of 0.1 or less, containing some water but none free, and moving at a rate between 1.5 m/month and 1.5 m/year (5 ft/month or year). Once all these particulars are established in the description, the movement could be referred to thereafter simply as a debris slide.

CAUSES OF SLIDING SLOPE MOVEMENTS

The processes involved in slides, as well as in other slope movements, comprise a continuous series of events from cause to effect. An engineer faced with a landslide is primarily interested in preventing the harmful effects of the slide. In many instances the principal cause of the slide cannot be removed, so it may be more economical to alleviate the effects continually or intermittently without attempting to remove the cause. Some slides occur in a unique environment and may last only a few seconds. The damage can be repaired, and the cause may be of only academic interest unless legal actions are to be taken. More often, however, landslides take place under the influence of geologic, topographic, or climatic factors that are common to large areas. The causes must then be understood if other similar slides are to be avoided or controlled.

Seldom, if ever, can a landslide be attributed to a single definite cause. As clearly shown by Zolotarev (*2.196*), the process leading to the development of the slide has its beginning with the formation of the rock itself, when its basic physical properties are determined, and includes all the subsequent events of crustal movements, erosion, and weathering. Finally, some action, perhaps trivial, sets a mass of material in motion downhill. The last action cannot be regarded as the only cause, even though it was necessary in the chain of events. As Sowers and Sowers (*2.161*, p. 506) point out,

In most cases a number of causes exist simultaneously and so attempting to decide which one finally produced failure is not only difficult but also incorrect. Often the final factor is nothing more than a trigger that set in motion an earth mass that was already on the verge of failure. Calling the final factor the cause is like calling the match that lit the fuse that detonated the dynamite that destroyed the building the cause of the disaster.

In this connection, however, the determination of all the geologic causes of a landslide should not be confused with determination of legal responsibility. The interrelations of landslide causes are lucidly and graphically presented by Terzaghi (*2.175*). His work, that of Sharpe (*2.146*), Ladd (*2.92*), and Bendel (*2.6*), and that of more recent researchers, such as Záruba and Mencl (*2.193*), Skempton and Hutchinson (*2.154*), Krinitzsky and Kolb (*2.90*), Rapp (*2.131*), and Legget (*2.96*) were used in the preparation of this section.

All slides involve the failure of earth materials under shear stress. The initiation of the process can therefore be reviewed according to (a) the factors that contribute to increased shear stress and (b) the factors that contribute to low or reduced shear strength. Although a single action, such as addition of water to a slope, may contribute to both an increase in stress and a decrease in strength, it is helpful to separate the various physical results of such an action. The principal factors contributing to the sliding of slope-forming materials are outlined in the following discussion. The operation of many factors is self-evident and needs no lengthy description; some factors are only discussed briefly, or reference is made to literature that gives examples or treats the subject in detail.

Factors That Contribute to Increased Shear Stress

Removal of Lateral Support

The removal of lateral support is the commonest of all factors leading to instability, and it includes the following actions:

1. Erosion by (a) streams and rivers, which produce most natural slopes that are subject to sliding (Hutchinson, *2.67*; Jones, Embody, and Peterson, *2.79*; Eyles, *2.43*; Fleming, Spencer, and Banks, *2.48*; California Division of

Highways, *2.19*), (b) glaciers, which have deeply cut and oversteepened many valleys in mountainous regions that have been the sites of large slides and debris flows (Plafker, Ericksen, and Fernandez Concha, *2.126*), (c) waves and longshore or tidal currents (Wood, *2.189*; Ward, *2.186*; Hutchinson, *2.71*; Koppejan, Van Wamelon, and Weinberg, *2.89*), and (d) subaerial weathering, wetting and drying, and frost action;

2. Previous rock fall, slide (Kenney and Drury, *2.81*), subsidence, or large-scale faulting that create new slopes; and

3. Work of human agencies in which (a) cuts, quarries, pits, and canals (Van Rensburg, *2.181*; Piteau, *2.124*; Patton, *2.121*; Cording, *2.26*) are established, (b) retaining walls and sheet piling are removed, and (c) lakes and reservoirs are created and their levels altered (Müller, *2.112*; Jones, Embody, and Peterson, *2.79*; Lane, *2.93*; Dupree and Taucher, *2.39*).

Surcharge

Surcharge also results from both natural and human agencies. The surcharge from natural agencies may be

1. Weight of rain, hail, snow, and water from springs;
2. Accumulation of talus overriding landslide materials;
3. Collapse of accumulated volcanic material, producing avalanches and debris flows (Francis and others, *2.50*);
4. Vegetation (Gray, *2.53*; Pain, *2.120*); and
5. Seepage pressures of percolating water.

The surcharge from human agencies may be

1. Construction of fill;
2. Stockpiles of ore or rock;
3. Waste piles (Bishop, *2.8*; Davies, *2.35*; Smalley, *2.156*);
4. Weight of buildings and other structures and trains; and
5. Weight of water from leaking pipelines, sewers, canals, and reservoirs.

Transitory Earth Stresses

Earthquakes have triggered a great many landslides, both small and extremely large and disastrous. Their action is complex, involving both an increase in shear stress (horizontal accelerations may greatly modify the state of stress within slope-forming materials) and, in some instances, a decrease in shear strength (Seed, *2.143*; Morton, *2.110*; Solonenko, *2.160*; Lawson, *2.95*; Hansen, *2.57*; Newmark, *2.117*; Simonett, *2.151*; Hadley, *2.55*; Gubin, *2.54*). Vibrations from blasting, machinery, traffic, thunder, and adjacent slope failures also produce transitory earth stresses.

Regional Tilting

A progressive increase in the slope angle through regional tilting is suspected as contributing to some landslides (Terzaghi, *2.175*). The slope must obviously be on the point of failure for such a small and slow-acting change to be effective.

Removal of Underlying Support

Examples of removal of underlying support include

1. Undercutting of banks by rivers (California Division of Highways, *2.19*) and by waves;
2. Subaerial weathering, wetting and drying, and frost action;
3. Subterranean erosion in which soluble material, such as carbonates, salt, or gypsum is removed and granular material beneath firmer material is worked out (Ward, *2.186*; Terzaghi, *2.174*);
4. Mining and similar actions by human agencies;
5. Loss of strength or failure in underlying material; and
6. Squeezing out of underlying plastic material (Záruba and Mencl, *2.193*, pp. 68-78).

Lateral Pressure

Lateral pressure may be caused by

1. Water in cracks and caverns,
2. Freezing of water in cracks,
3. Swelling as a result of hydration of clay or anhydrite, and
4. Mobilization of residual stress (Bjerrum, *2.9*; Krinitzsky and Kolb, *2.90*).

Volcanic Processes

Stress patterns in volcanic edifices and crater walls are modified by general dilation due to inflation or deflation of magma chambers, fluctuation in lava-lake levels, and increase in harmonic tremors (Tilling, Koyanagi, and Holcomb, *2.178*; Moore and Krivoy, *2.109*; Fiske and Jackson, *2.47*).

Factors That Contribute to Low or Reduced Shear Strength

The factors that contribute to low or reduced shear strength of rock or soil may be divided into two groups. The first group includes factors stemming from the initial state or inherent characteristics of the material. They are part of the geologic setting that may be favorable to landslides, exhibit little or no change during the useful life of a structure, and may exist for a long period of time without failure. The second group includes the changing or variable factors that tend to lower the shear strength of the material.

Initial State

Factors in the initial state of the material that cause low shear strength are composition, texture, and gross structure and slope geometry.

Composition

Materials are inherently weak or may become weak upon change in water content or other changes. Included especially are organic materials, sedimentary clays and shales, decomposed rocks, rocks of volcanic tuff that may weather to clayey material, and materials composed dominantly of soft platy minerals, such as mica, schist, talc, or serpentine.

Texture

The texture is a loose structure of individual particles in sensitive materials, such as clays, marl, loess, sands of low density, and porous organic matter (Aitchison, *2.2*; Bjerrum and Kenney, *2.11*; Cabrera and Smalley, *2.17*). Roundness of grain influences strength as compressibility and internal friction increase with angularity.

Gross Structure and Slope Geometry

Included in gross structure and slope geometry are

1. Discontinuities, such as faults, bedding planes, foliation in schist, cleavage, joints, slickensides, and brecciated zones (Skempton and Petley, *2.155*; Fookes and Wilson, *2.49*; Komarnitskii, *2.88*; St. John, Sowers, and Weaver, *2.138*; Van Rensburg, *2.181*; Jennings and Robertson, *2.74*; Bjerrum and Jørstad, *2.10*);

2. Massive beds over weak or plastic materials (Záruba and Mencl, *2.193*; Nemčok, *2.113*);

3. Strata inclined toward free face;

4. Alternation of permeable beds, such as sand or sandstone, and weak impermeable beds, such as clay or shale (Henkel, *2.59*); and

5. Slope orientation (Rice, Corbett, and Bailey, *2.134*; Shroder, *2.150*).

Changes Due to Weathering and Other Physicochemical Reactions

The following changes can occur because of weathering and other physicochemical reactions:

1. Softening of fissured clays (Skempton, *2.153*; Sangrey and Paul, *2.139*; Eden and Mitchell, *2.41*);

2. Physical disintegration of granular rocks, such as granite or sandstone, under action of frost or by thermal expansion (Rapp, *2.131*);

3. Hydration of clay minerals in which (a) water is absorbed by clay minerals and high water contents decrease cohesion of all clayey soils, (b) montmorillonitic clays swell and lose cohesion, and (c) loess markedly consolidates upon saturation because of destruction of the clay bond between silt particles;

4. Base exchange in clays, i.e., influence of exchangeable ions on physical properties of clays (Sangrey and Paul, *2.139*; Liebling and Kerr, *2.99*; Torrance, *2.179*);

5. Migration of water to weathering front under electrical potential (Veder, *2.184*);

6. Drying of clays that results in cracks and loss of cohesion and allows water to seep in;

7. Drying of shales that creates cracks on bedding and shear planes and reduces shale to chips, granules, or smaller particles; and

8. Removal of cement by solution.

Changes in Intergranular Forces Due to Water Content and Pressure in Pores and Fractures

Buoyancy in saturated state decreases effective intergranular pressure and friction. Intergranular pressure due to capillary tension in moist soil is destroyed upon saturation. Simple softening due to water and suffusion and slaking are discussed by Mamulea (*2.101*).

Changes can occur because of natural actions, such as rainfall and snowmelt, and because of a host of human activities, such as diversion of streams, blockage of drainage, irrigation and ponding, and clearing of vegetation and deforestation.

Crozier (*2.30, 2.31*), Shroder (*2.150*), and Spůrek (*2.163*) discuss the general effect of climate; Temple and Rapp (*2.169*) Williams and Guy (*2.188*), Jones (*2.78*), and So (*2.158*), catastrophic rainfall; Conway (*2.25*), Denness (*2.38*), and Piteau (*2.124*), effect of groundwater; Gray (*2.53*), Bailey (*2.4*), Cleveland (*2.22*), Rice, Corbett, and Bailey (*2.134*), and Swanston (*2.165*), deforestation; Peck (*2.123*) and Hirao and Okubo (*2.60*), correlation of rainfall and movement; and Shreve (*2.148*), Voight (*2.185*), Kent (*2.82*), and Goguel and Pachoud (*2.52*), gaseous entrainment or cushion.

Changes in Structure

Changes in structure may be caused by fissuring of shales and preconsolidated clays and fracturing and loosening of rock slopes due to release of vertical or lateral restraints in valley walls or cuts (Bjerrum, *2.9*; Aisenstein, *2.1*; Ferguson, *2.45*; Matheson and Thomson, *2.102*; Mencl, *2.106*). Disturbance or remolding can affect the shear strength of materials composed of fine particles, such as loess, dry or saturated loose sand, and sensitive clays (Gubin, *2.54*; Youd, *2.191*; Smalley, *2.156*; Mitchell and Markell, *2.108*).

Miscellaneous Causes

Other causes of low shear strength are (a) weakening due to progressive creep (Suklje, *2.164*; Ter-Stepanian, *2.172*; Trollope, *2.180*; Piteau, *2.124*) and actions of tree roots (Feld, *2.44*) and burrowing animals.

REFERENCES

2.1 Aisenstein, B. Some Additional Information About Deconsolidation Fissures of Rock on Steep Slopes. Proc., 2nd Congress, International Society of Rock Mechanics, Belgrade, Vol. 3, 1970, pp. 371-376.

2.2 Aitchison, G. D. General Report on Structurally Unstable Soils. Proc., 8th International Conference on Soil Mechanics and Foundation Engineering, Moscow, Vol. 3, 1973, pp. 161-190.

2.3 Andresen, A., and Bjerrum, L. Slides in Subaqueous Slopes in Loose Sand and Silt. In Marine Geotechnique (Richards, A. F., ed.), Univ. of Illinois Press, Urbana, 1967, pp. 221-239.

2.4 Bailey, R. G. Landslide Hazards Related to Land Use Planning in Teton National Forest, Northwest Wyoming. Intermountain Region, U.S. Forest Service, 1971, 131 pp.

2.5 Beck, A. C. Gravity Faulting as a Mechanism of Topographic Adjustment. New Zealand Journal of Geology and Geophysics, Vol. 11, No. 1, 1967, pp. 191-199.

2.6 Bendel, L. Ingenieurgeologie: Ein Handbuch fuer Studium und Praxis. Springer-Verlag, Vienna, Vol. 2, 1948, 832 pp.

2.7 Bishop, A. W. Progressive Failure, With Special Reference to the Mechanism Causing It. Proc., Geotechnical Conference on Shear Strength Properties of Natural

Soils and Rocks, Norwegian Geotechnical Institute, Oslo, Vol. 2, 1967, pp. 142-150.

2.8 Bishop, A. W. The Stability of Tips and Spoil Heaps. Quarterly Journal of Engineering Geology, Vol. 6, No. 4, 1973, pp. 335-376.

2.9 Bjerrum, L. Progressive Failure in Slopes of Over-Consolidated Plastic Clay and Clay Shales. Journal of Soil Mechanics and Foundations Division, American Society of Civil Engineers, Vol. 93, No. SM5, 1967, pp. 3-49.

2.10 Bjerrum, L., and Jørstad, F. A. Stability of Rock Slopes in Norway. Norwegian Geotechnical Institute, Publ. 79, 1968, pp. 1-11.

2.11 Bjerrum, L., and Kenney, T. C. Effect of Structure on the Shear Behavior of Normally Consolidated Quick Clays. Proc., Geotechnical Conference on Shear Strength Properties of Natural Soils and Rocks, Oslo, Norwegian Geotechnical Institute, Vol. 2, 1967, pp. 19-27.

2.12 Bjerrum, L., Løken, T., Heiberg, S., and Foster, R. A Field Study of Factors Responsible for Quick Clay Slides. Norwegian Geotechnical Institute, Publ. 85, 1971, pp. 17-26.

2.13 Blong, R. J. The Underthrust Slide: An Unusual Type of Mass Movement. Geografiska Annaler, Vol. 53A, No. 1, 1971, pp. 52-58.

2.14 Blong, R. J. A Numerical Classification of Selected Landslides of the Debris Slide-Avalanche-Flow Type. Engineering Geology, Vol. 7, No. 2, 1973, pp. 99-114.

2.15 Blong, R. J. Relationships Between Morphometric Attributes of Landslides. Zeitschrift fuer Geomorphologie, Supp. Band 18, 1973, pp. 66-77.

2.16 Bukovansky, M., Rodríquez, M. A., and Cedrún, G. Three Rock Slides in Stratified and Jointed Rocks. Proc., 3rd Congress, International Society of Rock Mechanics, Denver, Vol. 2, Part B, 1974, pp. 854-858.

2.17 Cabrera, J. G., and Smalley, I. J. Quickclays as Products of Glacial Action: A New Approach to Their Nature, Geology, Distribution, and Geotechnical Properties. Engineering Geology, Vol. 7, No. 2, 1973, pp. 115-133.

2.18 Caldenius, C., and Lundstrom, R. The Landslide at Surte on the River Göta älv. Sweden Geologiska undersökningen, Series Ca, No. 27, 1955, 64 pp.

2.19 California Division of Highways. Bank and Shore Protection in California Highway Practice. California Division of Highways, Sacramento, 1960, 423 pp.

2.20 Campbell, R. H. Soil Slips, Debris Flows, and Rainstorms in the Santa Monica Mountains and Vicinity, Southern California. U.S. Geological Survey, Professional Paper 851, 1975, 51 pp.

2.21 Carter, R. M. A Discussion and Classification of Subaqueous Mass-Transport With Particular Application to Grain-Flow, Slurry-Flow, and Fluxoturbidites. Earth-Science Reviews, Vol. 11, No. 2, 1975, pp. 145-177.

2.22 Cleveland, G. B. Fire + Rain = Mudflows: Big Sur 1972. California Geology, Vol. 26, No. 6, 1973, pp. 127-135.

2.23 Close, U., and McCormick, E. Where the Mountains Walked. National Geographic Magazine, Vol. 41, 1922, pp. 445-464.

2.24 Cluff, L. S. Peru Earthquake of May 31, 1970: Engineering Geology Observations. Seismological Society of America Bulletin, Vol. 61, No. 3, 1971, pp. 511-521.

2.25 Conway, B. W. The Black Ven Landslip, Charmouth, Dorset. Institute of Geological Sciences, United Kingdom, Rept. 74/3, 1974, 4 pp.

2.26 Cording, E. J., ed. Stability of Rock Slopes. Proc., 13th Symposium on Rock Mechanics, Univ. of Illinois at Urbana-Champaign, American Society of Civil Engineers, New York, 1972, 912 pp.

2.27 Corte, A. E. Geocryology and Engineering. In Reviews in Engineering Geology (Varnes, D. J. and Kiersch, G. A., eds.), Geological Society of America, Vol. 2, 1969, pp. 119-185.

2.28 Coulter, H. W., and Migliaccio, R. R. Effects of the Earthquake of March 27, 1964, at Valdez, Alaska. U.S. Geological Survey, Professional Paper 542-C, 1966, 36 pp.

2.29 Crandell, D. R., and Fahnestock, R. K. Rockfalls and Avalanches From Little Tahoma Peak on Mount Rainier Volcano, Washington. U.S. Geological Survey, Bulletin 1221-A, 1965, 30 pp.

2.30 Crozier, M. J. Earthflow Occurrence During High Intensity Rainfall in Eastern Otago (New Zealand). Engineering Geology, Vol. 3, No. 4, 1969, pp. 325-334.

2.31 Crozier, M. J. Some Problems in the Correlation of Landslide Movement and Climate. In International Geography, Proc., 22nd International Geographical Congress, Montreal, Univ. of Toronto Press, Ontario, Vol. 1, 1972, pp. 90-93.

2.32 Crozier, M. J. Techniques for the Morphometric Analysis of Landslips. Zeitschrift fuer Geomorphologie, Vol. 17, No. 1, 1973, pp. 78-101.

2.33 Cruden, D. M., and Krahn, J. A Reexamination of the Geology of the Frank Slide. Canadian Geotechnical Journal, Vol. 10, No. 4, 1973, pp. 581-591.

2.34 Daido, A. On the Occurrence of Mud-Debris Flow. Bulletin of the Disaster Prevention Research Institute, Kyoto Univ., Japan, Vol. 21, Part 2, No. 187, 1971, pp. 109-135.

2.35 Davies, W. E. Coal Waste Bank Stability. Mining Congress Journal, Vol. 54, No. 7, 1968, pp. 19-24.

2.36 Deere, D. U., and Patton, F. D. Slope Stability in Residual Soils. Proc., 4th Pan-American Conference on Soil Mechanics and Foundation Engineering, San Juan, American Society of Civil Engineers, New York, Vol. 1, 1971, pp. 87-170.

2.37 de Freitas, M. H., and Watters, R. J. Some Field Examples of Toppling Failure. Geotechnique, Vol. 23, No. 4, 1973, pp. 495-514.

2.38 Denness, B. The Reservoir Principle of Mass Movement. Institute of Geological Sciences, United Kingdom, Rept. 72/7, 1972, 13 pp.

2.39 Dupree, H. K., and Taucher, G. J. Bighorn Reservoir Landslides, South-Central Montana. In Rock Mechanics, The American Northwest (Voight, B., ed.), Expedition Guide for 3rd Congress, International Society of Rock Mechanics, Denver, Pennsylvania State Univ., 1974, pp. 59-63.

2.40 Dylik, J. Solifluxion, Congelifluxion, and Related Slope Processes. Geografiska Annaler, Vol. 49A, No. 2-4, 1967, pp. 167-177.

2.41 Eden, W. J., and Mitchell, R. J. The Mechanics of Landslides in Leda Clay. Canadian Geotechnical Journal, Vol. 7, No. 3, 1970, pp. 285-296.

2.42 Embleton, C., and King, C. A. M. Periglacial Geomorphology. Wiley, New York, 1975, 203 pp.

2.43 Eyles, R. J. Mass Movement in Tangoio Conservation Reserve, Northern Hawke's Bay. Earth Science Journal, Vol. 5, No. 2, 1971, pp. 79-91.

2.44 Feld, J. Discussion of Slope Stability in Residual Soils (Session 2). Proc., 4th Pan-American Conference on Soil Mechanics and Foundation Engineering, San Juan, American Society of Civil Engineers, New York, 1971, Vol. 3, p. 125.

2.45 Ferguson, H. F. Valley Stress Release in the Allegheny Plateau. Bulletin, Association of Engineering Geologists, Vol. 4, No. 1, 1967, pp. 63-71.

2.46 Fisher, R. V. Features of Coarse-Grained, High-Concentration Fluids and Their Deposits. Journal of Sedimentary Petrology, Vol. 41, No. 4, 1971, pp. 916-927.

2.47 Fiske, R. S., and Jackson, E. D. Orientation and Growth of Hawaiian Volcanic Rifts: The Effect of Regional Structure and Gravitational Stresses. Proc., Royal Society of London, Series A, Vol. 329, 1972, pp. 299-326.

2.48 Fleming, R. W., Spencer, G. S., and Banks, D. C. Empirical Study of Behavior of Clay Shale Slopes. U.S. Army Engineer Nuclear Cratering Group, Technical Rept. 15, Vols. 1 and 2, 1970, 397 pp.

2.49 Fookes, P. G., and Wilson, D. D. The Geometry of Discontinuities and Slope Failures in Siwalik Clay. Geotechnique, Vol. 16, No. 4, 1966, pp. 305-320.

2.50 Francis, P. W., Roobol, M. J., Walker, G. P. L., Cobbold, P. R., and Coward, M. The San Pedro and San Pablo Volcanoes of Northern Chile and Their Hot Avalanche Deposits. Geologische Rundschau, Vol. 63, No. 1, 1974, pp. 357-388.

2.51 Fröhlich, O. K. General Theory of Stability of Slopes. Geotechnique, Vol. 5, No. 1, 1955, pp. 37-47.

2.52 Goguel, J., and Pachoud, A. Géologie et Dynamique de l'écroulement du Mont Granier, dans le Massif de Chartreuse, en Novembre 1248. France Bureau de Recherches Géologiques et Minières Bulletin, Section 3—Hydrogéologie-Géologie de l'Ingénieur, No. 1, 1972, pp. 29-38.

2.53 Gray, D. H. Effects of Forest Clear-Cutting on the Stability of Natural Slopes. Bulletin, Association of Engineering Geologists, Vol. 7, No. 1-2, 1970, pp. 45-66.

2.54 Gubin, I. Y. Regularity of Seismic Manifestations of the Tadzhikistana Territory. USSR Academy of Sciences, Moscow, 1960, 463 pp. (in Russian).

2.55 Hadley, J. B. Landslides and Related Phenomena Accompanying the Hebgen Lake Earthquake of August 17, 1959. U.S. Geological Survey, Professional Paper 435, 1964, pp. 107-138.

2.56 Hamel, J. V. Kimbley Pit Slope Failure. Proc., 4th Pan-American Conference on Soil Mechanics and Foundation Engineering, San Juan, American Society of Civil Engineers, New York, Vol. 2, 1971, pp. 117-127.

2.57 Hansen, W. R. Effects of the Earthquake of March 27, 1964, at Anchorage, Alaska. U.S. Geological Survey, Professional Paper 542-A, 1965, 68 pp.

2.58 Heim, A. Bergsturz und Menschenleben. Fretz and Wasmuth Verlag, Zürich, 1932, 218 pp.

2.59 Henkel, D. J. Local Geology and the Stability of Natural Slopes. Journal of Soil Mechanics and Foundations Division, American Society of Civil Engineers, Vol. 93, No. SM4, 1967, pp. 437-446.

2.60 Hirao, K., and Okubo, S. Studies on the Occurrence and Expansion of Landslides Caused by Ebino-Yoshimatsu Earthquake. Report of Cooperative Research for Disaster Prevention, Japan Science and Technology Agency, No. 26, 1971, pp. 157-189 (in Japanese with English summary).

2.61 Hoek, E. Recent Rock Slope Stability Research of the Royal School of Mines, London. Proc., 2nd International Conference on Stability in Open Pit Mining, Vancouver, 1971, Society of Mining Engineers, American Institute of Mining, Metallurgical and Petroleum Engineers, New York, 1972, pp. 23-46.

2.62 Hofmann, H. The Deformation Process of a Regularly Jointed Discontinuum During Excavation of a Cut. Proc., 2nd Congress, International Society of Rock Mechanics, Belgrade, Vol. 3, 1970, pp. 267-273 (in German).

2.63 Hofmann, H. Modellversuche zur Hangtektonik. Geologische Rundschau, Vol. 62, No. 1, 1973, pp. 16-29.

2.64 Hollingworth, S. E., and Taylor, J. H. An Outline of the Geology of the Kettering District. Proc., Geologists' Association, Vol. 57, 1946, pp. 204-233.

2.65 Hollingworth, S. E., and Taylor, J. H. The Northampton Sand Ironstone: Stratigraphy, Structure, and Reserves. U.K. Geological Survey, Memoirs, 1951, 211 pp.

2.66 Hsu, K. J. Catastrophic Debris Streams (Sturzstroms) Generated by Rockfalls. Geological Society of America Bulletin, Vol. 86, No. 1, 1975, pp. 129-140.

2.67 Hutchinson, J. N. The Free Degradation of London Clay Cliffs. Proc., Geotechnical Conference on Shear Strength of Natural Soils and Rocks, Norwegian Geotechnical Institute, Olso, Vol. 1, 1967, pp. 113-118.

2.68 Hutchinson, J. N. Mass Movement. In The Encyclopedia of Geomorphology (Fairbridge, R. W., ed.), Reinhold Book Corp., New York, 1968, pp. 688-696.

2.69 Hutchinson, J. N. Field Meeting on the Coastal Landslides of Kent, 1-3 July, 1966. Proc., Geologists' Association, Vol. 79, Part 2, 1968, pp. 227-237.

2.70 Hutchinson, J. N. A Coastal Mudflow on the London Clay Cliffs at Beltinge, North Kent. Geotechnique, Vol. 20, No. 4, 1970, pp. 412-438.

2.71 Hutchinson, J. N. The Response of London Clay Cliffs to Differing Rates of Toe Erosion. Geologia Applicata e Idrogeologia, Vol. 8, Part 1, 1973, pp. 221-239.

2.72 Hutchinson, J. N., and Bhandari, R. K. Undrained Loading: A Fundamental Mechanism of Mudflows and Other Mass Movements. Geotechnique, Vol. 21, No. 4, 1971, pp. 353-358.

2.73 Jahn, A. Slopes Morphological Features Resulting From Gravitation. Zeitschrift fuer Geomorphologie, Supp. Band 5, 1964, pp. 59-72.

2.74 Jennings, J. E., and Robertson, A. M. The Stability of Slopes Cut Into Natural Rock. Proc., 7th International Conference on Soil Mechanics and Foundations Engineering, Mexico City, Vol. 2, 1969, pp. 585-590.

2.75 Johnson, A. M. Physical Processes in Geology. Freeman, Cooper, and Co., San Francisco, 1970, 577 pp.

2.76 Johnson, A. M., and Rahn, P. H. Mobilization of Debris Flows. Zeitschrift fuer Geomorphologie, Supp. Band 9, 1970, pp. 168-186.

2.77 Johnson, N. M., and Ragel, R. H. Analysis of Flow Characteristics of Allen II Slide From Aerial Photographs. In The Great Alaska Earthquake of 1964: Hydrology, National Academy of Sciences, Washington, D.C., 1968, pp. 369-373.

2.78 Jones, F. O. Landslides of Rio de Janeiro and the Serra das Araras Escarpment, Brazil. U.S. Geological Survey, Professional Paper 697, 1973, 42 pp.

2.79 Jones, F. O., Embody, D. R., and Peterson, W. L. Landslides Along the Columbia River Valley, Northeastern Washington. U.S. Geological Survey, Professional Paper 367, 1961, 98 pp.

2.80 Jordan, R. H. A Florida Landslide. Journal of Geology, Vol. 57, No. 4, 1949, pp. 418-419.

2.81 Kenney, T. C., and Drury, P. Case Record of the Slope Failure That Initiated the Retrogressive Quick-Clay Landslide at Ullensaker, Norway. Geotechnique, Vol. 23, No. 1, 1973, pp. 33-47.

2.82 Kent, P. E. The Transport Mechanism in Catastrophic Rock Falls. Journal of Geology, Vol. 74, No. 1, 1966, pp. 79-83.

2.83 Kesseli, J. E. Disintegrating Soil Slips of the Coast Ranges of Central California. Journal of Geology, Vol. 51, No. 5, 1943, pp. 342-352.

2.84 Kjellman, W. Mechanics of Large Swedish Landslips. Geotechnique, Vol. 5, No. 1, 1955, pp. 74-78.

2.85 Klengel, K. J., and Pašek, J. Zur Terminologie von Hangbewegungen. Zeitschrift fuer Angewandte Geologie, Vol. 20, No. 3, 1974, pp. 128-132.

2.86 Knapp, G. L., ed. Avalanches, Including Debris Avalanches: A Bibliography. Water Resources Scientific Information Center, U.S. Department of the Interior, WRSIC 72-216, 1972, 87 pp.

2.87 Kojan, E., Foggin, G. T., III, and Rice, R. M. Prediction and Analysis of Debris Slide Incidence by Photogrammetry: Santa-Ynez-San Rafael Mountains, California. Proc., 24th International Geological Congress, Montreal, Section 13—Engineering Geology, 1972, pp. 124-131.

2.88 Komarnitskii, N. I. Zones and Planes of Weakness in Rocks and Slope Stability. Consultants Bureau, New York, 1968, 108 pp. (translated from Russian).

2.89 Koppejan, A. W., Van Wamelon, B. M., and Weinberg, L. J. H. Coastal Flow Slides in the Dutch Province of Zeeland. Proc., 2nd International Conference on Soil Mechanics and Foundations Engineering, Rotterdam, Vol. 5, 1948, pp. 89-96.

2.90 Krinitzsky, E. L., and Kolb, C. R. Geological Influences on the Stability of Clay Shale Slopes. Proc., 7th Symposium on Engineering Geology and Soils Engineering, Moscow, Idaho, Idaho Department of Highways, Univ. of Idaho, and Idaho State Univ., 1969, pp. 160-175.

2.91 Kurdin, R. D. Classification of Mudflows: Soviet Hydrology, Vol. 12, No. 4, 1973, pp. 310-316.

2.92 Ladd, G. E. Landslides, Subsidences, and Rockfalls. Proc., American Railway Engineering Association, Vol. 36, 1935, pp. 1091-1162.

2.93 Lane, K. S. Stability of Reservoir Slopes. In Failure and Breakage of Rock (Fairhurst, C., ed.), Proc., 8th Symposium on Rock Mechanics, American Institute of Mining, Metallurgy and Petroleum Engineers, New York, 1967, pp. 321-336.

2.94 Laverdière, C. Quick Clay Flow-Slides in Southern Quebec. Revue de Geographie de Montreal, Vol. 26, No. 2, 1972, pp. 193-198 (French-English vocabulary).

2.95 Lawson, A. C. The California Earthquake of April 18, 1906: Landslides. Carnegie Institute of Washington, Washington, D.C., Vol. 1, Part 1, 1908, reprinted 1969, pp. 384-401.

2.96 Legget, R. F. Geology and Engineering. McGraw-Hill, New York, 2nd Ed., 1962, 884 pp.

2.97 Leighton, F. B. Landslides and Hillside Development. In Engineering Geology in Southern California, Association of Engineering Geologists, Special Publ., 1966, pp. 149-207.

2.98 Lemke, R. W. Effects of the Earthquake of March 27, 1964, at Seward, Alaska. U.S. Geological Survey, Professional Paper 542-E, 1966, 43 pp.

2.99 Liebling, R. S., and Kerr, P. F. Observations on Quick Clay. Geological Society of America Bulletin, Vol. 76, No. 8, 1965, pp. 853-878.

2.100 Lo, K. Y. An Approach to the Problem of Progressive Failure. Canadian Geotechnical Journal, Vol. 9, No. 4, 1972, pp. 407-429.

2.101 Mamulea, M. A. Suffusion and Slaking: Physical Processes Prompting the Mass Movements. Bulletin, International Association of Engineering Geologists, No. 9, 1974, pp. 63-68 (in French with English summary).

2.102 Matheson, D. S., and Thomson, S. Geological Implications of Valley Rebound. Canadian Journal of Earth Sciences, Vol. 10, No. 6, 1973, pp. 961-978.

2.103 McConnell, R. G., and Brock, R. W. The Great Landslide at Frank, Alberta. In Annual Report of the Canada Department of the Interior for the year 1902-03, Sessional Paper 25, 1904, pp. 1-17.

2.104 McRoberts, E. C., and Morgenstern, N. R. Stability of Thawing Slopes. Canadian Geotechnical Journal, Vol. 11, No. 4, 1974, pp. 447-469.

2.105 McRoberts, E. C., and Morgenstern, N. R. The Stability of Slopes in Frozen Soil: Mackenzie Valley, N.W.T. Canadian Geotechnical Journal, Vol. 11, No. 4, 1974, pp. 554-573.

2.106 Mencl, V. Engineering-Geological Importance and Possible Origin of the Stress Relief of the Rocks of the Cordillera Blanca, Peru. Bulletin, International Association of Engineering Geologists, No. 9, 1974, pp. 69-74.

2.107 Miller, W. J. The Landslide at Point Fermin, California. Scientific Monthly, Vol. 32, No. 5, 1931, pp. 464-469.

2.108 Mitchell, R. J., and Markell, A. R. Flowsliding in Sensitive Soils. Canadian Geotechnical Journal, Vol. 11, No. 1, 1974, pp. 11-31.

2.109 Moore, J. G., and Krivoy, H. L. The 1962 Flank Eruption of Kilauea Volcano and Structure of the East Rift Zone. Journal of Geophysical Research, Vol. 69, No. 10, 1964, pp. 2033-2045.

2.110 Morton, D. M. Seismically Triggered Landslides in the Area Above the San Fernando Valley. In The San Fernando, California, Earthquake of February 9, 1971, U.S. Geological Survey, Professional Paper 733, 1971, pp. 99-104.

2.111 National Research Council, Canada, and National Research Council, U.S. North American Contribution to 2nd International Conference on Permafrost, Yakutsk, USSR, July 1973. National Academy of Sciences, Washington, D.C. 1973, 783 pp.

2.112 Müller, L. New Considerations on the Vaiont Slide. Felsmechanik und Ingenieur Geologie, Vol. 6, No. 1-2, 1968, pp. 1-91.

2.113 Nemčok, A. The Development of Landslides on the Boundaries of Geological Formations. Sborník Geologických věd, Rada HIG, No. 5, 1966, pp. 87-105 (in Slovak with English summary).

2.114 Nemčok, A. Gravitational Slope Deformation in High Mountains. Proc., 24th International Geological Congress, Montreal, Section 13—Engineering Geology, 1972, pp. 132-141.

2.115 Nemčok, A. Gravitational Slope Deformations in the High Mountains of the Slovak Carpathians. Sborník Geologických věd, Rada HIG, No. 10, 1972, pp. 31-37.

2.116 Nemčok, A., Pasek, J., and Rybář, J. Classification of Landslides and Other Mass Movements. Rock Mechanics, Vol. 4, No. 2, 1972, pp. 71-78.

2.117 Newmark, N. M. Effects of Earthquakes on Dams and Embankments. Geotechnique, Vol. 15, No. 2, 1965, pp. 139-159.

2.118 Nossin, J. J. Landsliding in the Crati Basin, Calabria, Italy, Geologie en Mijnbouw, Vol. 51, No. 6, 1972, pp. 591-607.

2.119 Ostaficzuk, S. Large-Scale Landslides in Northwestern Libya. Acta Geologica Polonica, Vol. 23, No. 2, 1973, pp. 231-244.

2.120 Pain, C. F. Characteristics and Geomorphic Effects of Earthquake-Initiated Landslides in the Adelbert Range, Papua New Guinea. Engineering Geology, Vol. 6, No. 4, 1972, pp. 261-274.

2.121 Patton, F. D. Significant Geologic Factors in Rock Slope Stability. In Planning Open Pit Mines (Van Rensburg, P. W. J., ed.), Proc., Open Pit Mining Symposium, Johannesburg, South African Institute of Mining and Metallurgy, 1970, pp. 143-151.

2.122 Pautre, A., Sabarly, F., and Schneider, B. L'effet d'echelle dans les écroulements de falaise. Proc., 3rd Congress, International Society of Rock Mechanics, Denver, Vol. 2, Part B, 1974, pp. 859-864.

2.123 Peck, R. B. Stability of Natural Slopes. Journal of Soil Mechanics and Foundations Division, American Society of Civil Engineers, New York, Vol. 93, SM4, 1967, pp. 403-417.

2.124 Piteau, D. R. Geological Factors Significant to the Stability of Slopes Cut in Rock. In Planning Open Pit Mines (Van Rensburg, P. W. J., ed.), Proc., Open Pit Mining Symposium, Johannesburg, South African Institute of Mining and Metallurgy, 1970, pp. 33-53.

2.125 Piteau, D. R., Parkes, D. R., McLeod, B. C., and Lou, J. K. Overturning Rock Slope Failure at Hell's Gate, British Columbia. In Geology and Mechanics of Rockslides and Avalanches (Voight, B., ed.), Elsevier, New York (in press).

2.126 Plafker, G., Ericksen, G. E., and Fernandez Concha, J. Geological Aspects of the May 31, 1970, Peru Earthquake. Seismological Society of America Bulletin, Vol. 61, No. 3, 1971, pp. 543-578.

2.127 Popov, I. V. A Scheme for the Natural Classification of Landslides. Doklady of the USSR Academy of Sciences, Vol. 54, No. 2, 1946, pp. 157-159.

2.128 Prior, D. B., Stephens, N., and Archer, D. R. Composite Mudflows on the Antrim Coast of Northeast Ireland. Geografiska Annaler, Vol. 50A, No. 2, 1968, pp. 65-78.

2.129 Prior, D. B., Stephens, N., and Douglas, G. R. Some Examples of Modern Debris Flows in Northeast Ireland. Zeitschrift fuer Geomorphologie, Vol. 14, No. 3, 1970, pp. 276-288.

2.130 Radbruch-Hall, D. H. Large-Scale Gravitational Creep of Rock Masses on Slopes. In Geology and Mechanics of Rockslides and Avalanches (Voight, B., ed.), Elsevier, New York (in press).

2.131 Rapp, A. Recent Development of Mountain Slopes in Kärkevagge and Surroundings, Northern Scandinavia. Geografiska Annaler, Vol. 42, No. 2-3, 1960, pp. 71-199.

2.132 Rapp, A. The Debris Slides at Ulvådal, Western Norway: An Example of Catastrophic Slope Processes in Scandinavia. Nachrichten der Akademie der Wissenschaften Gottingen, Mathematisch-Physikalische Klasse, Vol. 13, 1963, pp. 195-210.

2.133 Reiche, P. The Toreva-Block: A Distinctive Landslide Type. Journal of Geology, Vol. 45, No. 5, 1937, pp. 538-548.

2.134 Rice, R. M., Corbett, E. S., and Bailey, R. G. Soil Slips Related to Vegetation, Topography, and Soil in Southern California. Water Resources Research, Vol. 5, No. 3, 1969, pp. 647-659.

2.135 Romani, F., Lovell, C. W., Jr., and Harr, M. E. Influence of Progressive Failure on Slope Stability. Journal of Soil Mechanics and Foundations Division, American Society of Civil Engineers, New York, Vol. 98, No. SM11, 1972, pp. 1209-1223.

2.136 Rybář, J. Slope Deformations in Brown Coal Basins Under the Influence of Tectonics. Rock Mechanics, Vol. 3, No. 3, 1971, pp. 139-158 (in German with English summary).

2.137 Rybář, J., and Dobr, J. Fold Deformations in the North-Bohemian Coal Basins. Sborník geologických věd, Rada HIG, No. 5, 1966, pp. 133-140.

2.138 St. John, B. J., Sowers, G. F., and Weaver, C. E. Slickensides in Residual Soils and Their Engineering Significance. Proc., 7th International Conference on Soil Mechanics and Foundation Engineering, Mexico City, Vol. 2, 1969, pp. 591-597.

2.139 Sangrey, D. A., and Paul, M. J. A Regional Study of Landsliding Near Ottawa. Canadian Geotechnical Journal, Vol. 8, No. 2, 1971, pp. 315-335.

2.140 Savarensky, F. P. Experimental Construction of a Landslide Classification. Geologo-Razvedochnyi Institut (TSNIGRI), 1935, pp. 29-37 (in Russian).

2.141 Scott, K. M. Origin and Sedimentology of 1969 Debris Flows Near Glendora, California. U.S. Geological Survey, Professional Paper 750-C, 1971, pp. C242-C247.

2.142 Scott, R. C., Jr. The Geomorphic Significance of Debris Avalanching in the Appalachian Blue Ridge Mountains. Univ. of Georgia, PhD dissertation, 1972; University Microfilms, Ann Arbor, Mich.

2.143 Seed, H. B. Landslides During Earthquakes Due to Soil Liquefaction. Journal of Soil Mechanics and Foundations Division, American Society of Civil Engineers, New York, Vol. 94, No. SM5, 1968, pp. 1053-1122.

2.144 Seed, H. B., and Wilson, S. D. The Turnagain Heights Landslide, Anchorage, Alaska. Journal of Soil Mechanics and Foundations Division, American Society of Civil Engineers, Vol. 93, No. SM4, 1967, pp. 325-353.

2.145 Sharp, R. P. Mass Movements on Mars. In Geology, Seismicity, and Environmental Impact, Association of Engineering Geologists, Special Publ., 1973, pp. 115-122.

2.146 Sharpe, C. F. S. Landslides and Related Phenomena: A Study of Mass Movements of Soil and Rock. Columbia Univ. Press, New York, 1938, 137 pp.

2.147 Shelton, J. S. Geology Illustrated. W. H. Freeman and Co., San Francisco, 1966, 434 pp.

2.148 Shreve, R. L. The Blackhawk Landslide. Geological Society of America, Memoir 108, 1968, 47 pp.

2.149 Shreve, R. L. Sherman Landslide. In The Great Alaska Earthquake of 1964: Hydrology. National Academy of Sciences, 1968, pp. 395-401.

2.150 Shroder, J. F. Landslides of Utah. Utah Geological and Mineralogical Survey Bulletin, No. 90, 1971, 51 pp.

2.151 Simonett, D. S. Landslide Distribution and Earthquakes in the Bewani and Toricelli Mountains, New Guinea: Statistical Analysis. In Landform Studies From Australia and New Guinea (Jennings, J. N., and Mabbutt, J. A., eds.), Cambridge Univ. Press, 1967, pp. 64-84.

2.152 Skempton, A. W. Soil Mechanics in Relation to Geology. Proc., Yorkshire Geological Society, Vol. 29, Part 1, 1953, pp. 33-62.

2.153 Skempton, A. W. Long-Term Stability of Clay Slopes. Geotechnique, Vol. 14, No. 2, 1964, pp. 77-101.

2.154 Skempton, A. W., and Hutchinson, J. N. Stability of Natural Slopes and Embankment Foundations. Proc., 7th International Conference on Soil Mechanics and Foundation Engineering, Mexico City, State-of-the Art Vol., 1969, pp. 291-340.

2.155 Skempton, A. W., and Petley, D. J. The Strength Along Structural Discontinuities in Stiff Clays. Proc., Geotechnical Conference on Shear Strength Properties of Natural Soils and Rocks, Norwegian Geotechnical Institute, Oslo, Vol. 2, 1967, pp. 29-46.

2.156 Smalley, I. J. Boundary Conditions for Flowslides in Fine-Particle Mine Waste Tips. Trans., Institution of Mining and Metallurgy, London, Vol. 81, Sec. A, 1972, pp. A31-A37.

2.157 Snopko, L. Study of Deformation Elements Developed in the Handlová Landslide. Geologické Prace, Vol. 28, No. 28, 1963, pp. 169-183.

2.158 So, C. L. Mass Movements Associated With the Rainstorm of June 1966 in Hong Kong. Trans., Institute of British Geographers, No. 53, 1971, pp. 55-65.

2.159 Sokolov, N. I. Types of Displacement in Hard Fractured Rocks on Slopes. In The Stability of Slopes (Popov, I. V., and Kotlov, F. V., eds.), Trans., F. P. Savarenskii Hydrogeology Laboratory, Vol. 35, 1963, pp. 69-83; Consultants Bureau, New York (translated from Russian).

2.160 Solonenko, V. P. Seismogenic Destruction of Mountain Slopes. Proc., 24th International Geological Congress, Montreal, Section 13—Engineering Geology, 1972, pp. 284-290.

2.161 Sowers, G. B., and Sowers, G. F. Introductory Soil Mechanics and Foundations. Macmillan, New York, 3rd Ed., 1970, 556 pp.

2.162 Sowers, G. F. Landslides in Weathered Volcanics in Puerto Rico. Proc., 4th Pan-American Conference on Soil Mechanics and Foundation Engineering, San Juan, American Society of Civil Engineers, New York, 1971, pp. 105-115.

2.163 Špůrek, M. Retrospective Analysis of the Climatic Sliding Agent. Sborník Geologických věd, Rada HIG, No. 7, 1970, pp. 61-79 (in Czech with English summary).

2.164 Suklje, L. A Landslide Due to Long-Term Creep. Proc., 5th International Conference on Soil Mechanics and Foundation Engineering, Paris, Vol. 2, 1961, pp. 727-735.

2.165 Swanston, D. N. Slope Stability Problems Associated With Timber Harvesting in Mountainous Regions of the Western United States. U.S. Forest Service, General Technical Rept. PNW-21, 1974, 14 pp.

2.166 Tabor, R. W. Origin of Ridge-Top Depressions by Large-Scale Creep in the Olympic Mountains, Washington. Geological Society of America Bulletin, Vol. 82, 1971, pp. 1811-1822.

2.167 Takada, Y. On the Landslide Mechanism of the Tertiary Type Landslide in the Thaw Time. Bulletin of the Disaster Prevention Research Institute, Kyoto Univ., Japan, Vol. 14, Part 1, 1964, pp. 11-21.

2.168 Tavenas, F., Chagnon, J. Y., and La Rochelle, P. The Saint-Jean-Vianney Landslide: Observations and Eyewitnesses Accounts. Canadian Geotechnical Journal, Vol. 8, No. 3, 1971, pp. 463-478.

2.169 Temple, P. H., and Rapp, A. Landslides in the Mgeta Area, Western Uluguru Mountains, Tanzania. Geografiska Annaler, Vol. 54A, No. 3-4, 1972, pp. 157-193.

2.170 Ter-Stepanian, G. On the Long-Term Stability of Slopes. Norwegian Geotechnical Institute, Publ. 52, 1963, 14 pp.

2.171 Ter-Stepanian, G. The Use of Observations of Slope Deformation for Analysis of Mechanism of Landslides: Problems of Geomechanics. Trans., Department of Geomechanics, Armenian SSR Academy of Sciences, No. 1, 1967, pp. 32-51.

2.172 Ter-Stepanian, G. Depth Creep of Slopes. Bulletin, International Society of Engineering Geologists, No. 9, 1974, pp. 97-102.

2.173 Ter-Stepanian, G., and Goldstein, M. N. Multi-Storied Landslides and Strength of Soft Clays. Proc., 7th International Conference on Soil Mechanics and Foundation Engineering, Mexico City, Vol. 2, 1969, pp. 693-700.

2.174 Terzaghi, K. Earth Slips and Subsidences From Underground Erosion. Engineering News-Record, Vol. 107, July 16, 1931, pp. 90-92.

2.175 Terzaghi, K. Mechanism of Landslides. In Application of Geology to Engineering Practice (Paige, S., ed.),

Geological Society of America, Berkey Vol., 1950, pp. 83-123.

2.176 Terzaghi, K., and Peck, R. B. Soil Mechanics in Engineering Practice. Wiley, New York, 2nd Ed., 1967, 729 pp.

2.177 Thomson, S., and Hayley, D. W. The Little Smoky Landslide. Canadian Geotechnical Journal, Vol. 12, No. 3, 1975, pp. 379-392.

2.178 Tilling, R. I., Koyanagi, R. Y., and Holcomb, R. T. Rockfall-Seismicity: Correlation With Field Observations, Makaopuhi Crater, Kilauea Volcano, Hawaii. U.S. Geological Survey, Journal of Research, Vol. 3, No. 3, 1975, pp. 345-361.

2.179 Torrance, J. K. On the Role of Chemistry in the Development and Behavior of the Sensitive Marine Clays of Canada and Scandinavia. Canadian Geotechnical Journal, Vol. 12, No. 3, 1975, pp. 326-335.

2.180 Trollope, D. H. Sequential Failure in Strain-Softening Soils. Proc., 8th International Conference on Soil Mechanics and Foundation Engineering, Moscow, Vol. 2, Part 2, 1973, pp. 227-232.

2.181 Van Rensburg, P. W. J., ed. Planning Open Pit Mines. Proc., Open Pit Mining Symposium, Johannesburg, South African Institute of Mining and Metallurgy, 1971, 388 pp.

2.182 Varnes, D. J. Landslide Types and Processes. In Landslides and Engineering Practice (Eckel, E. B., ed.), HRB, Special Rept. 29, 1958, pp. 20-47.

2.183 Varnes, H. D. Landslide Problems of Southwestern Colorado. U.S. Geological Survey, Circular 31, 1949, 13 pp.

2.184 Veder, C. Phenomena of the Contact of Soil Mechanics. International Symposium on Landslide Control, Kyoto and Tokyo, Japan Society of Landslide, 1972, pp. 143-162 (in English and Japanese).

2.185 Voight, B. Architecture and Mechanics of the Heart Mountain and South Fork Rockslides. In Rock Mechanics, The American Northwest (Voight, B., ed.), Expedition Guide, 3rd Congress of the International Society of Rock Mechanics, Pennsylvania State Univ., 1974, pp. 26-36.

2.186 Ward, W. H. The Stability of Natural Slopes. Geographical Journal, Vol. 105, No. 5-6, 1945, pp. 170-191.

2.187 Washburn, A. L. Periglacial Processes and Environments. St. Martin's Press, New York, 1973, 320 pp.

2.188 Williams, G. P. and Guy, H. P. Erosional and Depositional Aspects of Hurricane Camille in Virginia, 1969. U.S. Geological Survey, Professional Paper 804, 1973, 80 pp.

2.189 Wood, A. M. Engineering Aspects of Coastal Landslides. Proc., Institution of Civil Engineers, London, Vol. 50, 1971, pp. 257-276.

2.190 Yemel'ianova, Ye. P. Fundamental Regularities of Landslide Processes. Nedra, Moscow, 1972, 308 pp. (excerpts translated by D. B. Vitaliano for U.S. Geological Survey).

2.191 Youd, T. L. Liquefaction, Flow, and Associated Ground Failure. U.S. Geological Survey, Circular 688, 1973, 12 pp.

2.192 Záruba, Q. Periglacial Phenomena in the Turnov Region. Sborník Ústředního Ústavu Geologického, Vol. 19, 1952, pp. 157-168 (in Czech with English and Russian summaries).

2.193 Záruba, Q., and Mencl, V. Landslides and Their Control. Elsevier, New York, and Academia, Prague, 1969, 205 pp.

2.194 Zischinsky, U. On the Deformation of High Slopes. Proc., 1st Congress, International Society of Rock Mechanics, Lisbon, Vol. 2, 1966, pp. 179-185.

2.195 Zischinsky, U. Über Bergzerreissung und Talzuschub. Geologische Rundschau, Vol. 58, No. 3, 1969, pp. 974-983.

2.196 Zolotarev, G. S. Geological Regularities of the Development of Landslides and Rockfalls as the Basis for the Theory of Their Study and Prognosis. Géologie de l'Ingénieur, Sociéte Géologique de Beligique, Liège, 1974, pp. 211-235.

BIBLIOGRAPHIES

2.1 Collins, T. Bibliography of Recent Publications on Slope Stability. Landslide, The Slope Stability Review, Vol. 1, No. 1, 1973, pp. 28-37.

2.2 Fisher, C. P., Leith, C. J., and Deal, C. S. An Annotated Bibliography on Slope Stability and Related Phenomena. North Carolina State Highway Commission and U.S. Bureau of Public Roads, 1965, 89 pp; NTIS, Springfield, Va., PB 173 029.

2.3 Hoek, E. Bibliography on Slope Stability. In Planning Open Pit Mines (Van Rensburg, P. W. J., ed.), Proc., Open Pit Mining Symposium, Johannesburg, South African Institute of Mining and Metallurgy, 1971, pp. 365-388.

2.4 Holtz, W. G. Bibliography on Landslides and Mudslides. Building Research Advisory Board, National Academy of Sciences, Washington, D.C., 1973.

2.5 Knapp, G. L., ed. Avalanches, Including Debris Avalanches: A Bibliography. Water Resources Scientific Information Center, U.S. Department of the Interior, WRSIC 72-216, 1972, 87 pp.

2.6 Larew, H. G., and others. Bibliography on Earth Movement. Research Laboratory for Engineering Science, Univ. of Virginia, Charlottesville, 1964, 239 pp.; NTIS, Springfield, Va., AD 641 716.

2.7 Špůrek, M. Historical Catalogue of Slide Phenomena. Institute of Geography, Czechoslovak Academy of Sciences, Brno, Studia Geographica 19, 1972, 178 pp.

2.8 Tompkin, J. M., and Britt, S. H. Landslides: A Selected Annotated Bibliography. HRB, Bibliography 10, 1951, 47 pp.

2.9 Záruba, Q., and Mencl, V. Landslides and Their Control: Bibliography. Elsevier, New York, and Academia, Prague, 1969, pp. 194-202.

PHOTOGRAPH AND DRAWING CREDITS

Figure 2.3   Courtesy of U.S. Bureau of Reclamation
Figure 2.7   F. O. Jones, U.S. Geological Survey
Figure 2.8   F. O. Jones, U.S. Geological Survey
Figure 2.9   Courtesy of Tennessee Department of Transportation
Figure 2.11   Courtesy of Tennessee Department of Transportation
Figure 2.15   John S. Shelton
Figure 2.17   Courtesy of New Hampshire Department of Public Works and Highways
Figure 2.19   Courtesy of Journal of Geology
Figure 2.20   F. O. Jones, U.S. Geological Survey
Figure 2.21   F. O. Jones, U.S. Geological Survey
Figure 2.22   G. K. Gilbert, U.S. Geological Survey
Figure 2.23   John S. Shelton
Figure 2.24   F. O. Jones, U.S. Geological Survey
Figure 2.25   F. O. Jones, U.S. Geological Survey
Figure 2.26   F. O. Jones, U.S. Geological Survey
Figure 2.28   From *Geology Illustrated* by John S. Shelton (*2.147*), W. H. Freeman and Company, copyright 1966
Figure 2.29   From the *Encyclopedia of Geomorphology*, edited by R. W. Fairbridge, copyright 1968 by Litton Educational Publishing, Inc., reprinted by permission of Van Nostrand Reinhold Company
Figure 2.33   From *Geology Illustrated* by John S. Shelton (*2.147*), W. H. Freeman and Company, copyright 1966

Chapter 3

Recognition and Identification

Harold T. Rib and Ta Liang

The adage that half the solution to a problem is the recognition that a problem exists is especially appropriate to landslides. The recognition of the presence or the potential development of slope movement and the identification of the type and causes of the movement are important in the development of procedures for the prevention or correction of a slide.

Several basic guidelines developed through years of experience in investigating landslides form the basis for the present-day approach to landslide investigations.

1. Most landslides or potential failures can be predicted if proper investigations are performed in time.
2. The cost of preventing landslides is less than the cost of correcting them except for small slides that can be handled by normal maintenance procedures.
3. Massive slides that may cost many times the cost of the original facility should be avoided in the first place.
4. The occurrence of the initial slope movement can lead to additional unstable conditions and movements.

This chapter discusses techniques for recognizing the presence or potential development of landslides and the observable features that aid in identifying the types of slope movements and their probable causes. The techniques discussed include (a) a review of topographic maps and geologic, pedologic, and engineering reports and maps; (b) the analysis of aerial photography and other forms of aerial images; and (c) preliminary field reconnaissance surveys. These three techniques complement one another and together form the basis of the preliminary analysis in a landslide investigation. The follow-up detailed field investigations required to accurately delineate the landslides or landslide-prone areas, to identify the causative factors, and to determine the physical and chemical properties required for the design of corrective measures are discussed in subsequent chapters.

The use of aerial techniques for evaluating landslides is emphasized in this chapter because of their proven value and the unique advantages they offer. No other technique can provide a three-dimensional overview of the terrain from which the interrelations existing among slope, drainage, surface cover, rock type and sequence, and human activities on the landscape can be viewed and evaluated. In addition, the availability of new types of aerial imagery, including satellite, infrared, radar, and microwave radiometry, extend the advantages of this technique. (In this chapter, the term imagery is used to describe those data collected from sensor systems other than cameras.)

TERRAIN EVALUATION FOR LANDSLIDE INVESTIGATIONS

Basic Factors

Recognition and identification of landslides are as complex as are materials and processes that cause them. The basic causes of sliding movements are given in Chapter 2. Because, as Chapter 2 states, a landslide can seldom be attributed to a single definite cause, the overall terrain must be analyzed and the individual factors and the interrelations among those factors distinguished before a potential slope movement can be recognized and identified. For this purpose, the factors contributing to slides are more conveniently grouped into categories relating to features that can be delineated on maps and photographs or that can be measured and quantified, rather than into the mechanisms of failure listed in Chapter 2. The basic factors considered in evaluating the terrain and the major elements included within each are given in Table 1 and discussed below.

Geologic Factors

The present-day landscape—including its topography, geologic structure, and composition—is the result of millions of years of development and modification. The topographic features exposed on the surface are relatively young in age, but the geologic structures and composition from which the features were carved can be quite old. For example, the structural features that characterize the Rocky Mountains culminated at the close of the Cretaceous period (approximately 60 million to 70 million years ago), but little of the topography in that area dates beyond the Pliocene (several million years), and the present canyons and details of relief are of Pleistocene or Recent age (less than 1 million years old) (3.37). Thus, to properly evaluate the present-day landscape requires an appreciation of the geologic and climatic changes that occurred during recent geologic periods (a determinant of topography) as well as an understanding of the development of the geologic structure and composition. The science that deals with landscape development is geomorphology.

The basic unit of the landscape distinguished in geomorphology is the landform. The distinctive development of landforms depends on three factors: (a) initial composition and structure; (b) processes that act to modify the initial composition and structure; and (c) stage of development. A change in any one of these factors will produce a uniquely different landform.

To evaluate landforms, one must understand their geologic composition and structure including mineralogy and lithology; physical hardness of the constituent materials and their susceptibility to weathering; mode of deposition and subsequent stress history; structural attitude and presence of structural discontinuities and weaknesses such as joints, bedding planes, faults, and folds; and permeability of constituent materials or layers.

The many physical and chemical ways by which the original composition and structure are modified are called geomorphic processes. The most important of these processes are associated with changes due to (a) actions of water, wind, or ice; (b) weathering, mass wasting, and erosion; and (c) diastrophism and vulcanism.

The modification and eventual destruction of the landforms are considered to occur in stages that geomorphologists generally designate as youth, maturity, and old age. Qualifying adjectives such as early and late are often used to designate substages. Chronological age is not inferred, but rather a relative stage of development.

The concept of classifying distinct landforms is of prime importance in recognizing and identifying landslides. Experience has demonstrated that certain landforms are more susceptible to slope movement because of the nature of their development and the stage of their evolution.

Environmental Factors

The development of landforms and the occurrence of slope movements are greatly affected by environmental factors. Significantly different landscapes develop from the same geologic materials in different climatic zones. For example, a limestone bedrock will usually be found as a cliff or ridge in an arid area, but will commonly form a low undulating plain in a humid area. The former situation would be much more susceptible to landslides than the latter. Similarly, landslides occurring in arid regions are usually distinct and easily recognized because of lack of cover, while those occuring in wet tropical climates are weathered, more subdued, covered by vegetation, and difficult to discern.

Variations in microclimate, such as differences in altitude, exposure to moisture-bearing winds, and exposure to sunlight, can cause significant differences in the geomorphic processes. Various observers have reported that some south-facing slopes of east-west valleys in the northern hemisphere are less steep than adjacent north-facing slopes. North-facing slopes have snow cover longer, experience fewer days of freeze and thaw, retain their soil moisture longer, and probably have a better vegetative cover, all of which result in less active erosion and steeper slopes (3.37).

Numerous investigators have demonstrated that there is a relation between slope movement and precipitation. Záruba and Mencl (3.40) evaluated rainfall records for several different sites in Czechoslovakia covering periods of 50 to more than 75 years. They were able to correlate the years of heaviest precipitation with the most active periods of slope movements.

Sudden changes in the landscape, often represented by slope movement, have resulted from catastrophic occurrences, such as earthquakes, hurricanes, and floods. Figure 3.1 shows numerous slides that occurred in western Virginia along the path of hurricane Camille in 1969. These slides were rapidly documented from the analysis of aerial photographs taken after the storm (3.39). Other case histories documenting the effect of catastrophic occurrences are referenced in Chapters 1 and 2. Undercutting by stream and wave action and the erosion and mass movement of slopes by seepage, wetting and drying, and frost action are also well-known environmental phenomena resulting in slope movement.

Table 3.1. Basic factors considered in evaluating terrain.

Factor	Element	Examples
Geologic	Landform	Geomorphic history; stage of development
	Composition	Lithology; stratigraphy; weathering products
	Structure	Spacing and attitude of faults, joints, foliation, and bedding surfaces
Environmental	Climate and hydrology	Rainfall; stream, current, and wave actions; groundwater flow; slope exposure; wetting and drying; frost action
	Catastrophes	Earthquakes; volcanic eruptions; hurricanes, typhoons, and tsunamis; flooding; subsidence
Human	Human activity	Construction; quarrying and mining; stripping of surface cover; over loading, vibrations
Temporal[a]		

[a]Common to all categories and factors.

Human Factors

Human activities, such as construction, quarrying, and mining, have caused drastic landscape changes. Much has been said about the aesthetic damage (wastelands and scarred landscapes resulting from strip mining and open-pit quarrying) that human activities have caused, but far more damaging to life and property have been the instabilities occasioned by those activities. Oversteepening of slopes, removal of support, removal of protective cover, overloading, blockage of drainage, increasing moisture levels in the ground, and vibrations are some of the human activities that have caused slope movements.

Temporal Factor

Time is common to all other factors. It is a basic condition of nature to wear away the high areas by weathering, erosion, and mass wasting and to fill in the low areas in a process of leveling. Even the occurrence of landslides by catastrophic events normally can be related to conditions that have been building during a period of time. The time period of importance is the stage of development rather than an absolute time period. For example, for a given geologic material, evolution from youth to old age proceeds more rapidly under tropical wet conditions than under arid conditions. Thus, in the same elapsed time, the landform unit in a tropical wet area will be much further advanced than a similar unit in an arid area.

The stage of development achieved in a given area is expressed by its topographic characteristics. Thornbury (3.37) lists typical landscape characteristics for each stage of development for various idealized geomorphic cycles. The following major topographic features are typical of the respective stages of development in the common fluvial geomorphic cycle.

1. In youth, topography is relatively undissected, and only a few streams exist. Interstream tracts are extensive and poorly drained. Valleys have V-shaped cross sections, and their depths depend on the altitude of the region. Waterfalls or rapids exist where streams cross resistant rock.

2. In maturity, a well-integrated drainage system develops. Topography consists mostly of hillsides and valleysides. Drainage divides are sharp, and the maximum possible relief exists. Vertical cutting ceases, and lateral destruction becomes important.

3. In old age, valleys are extremely broad and gently sloping both laterally and longitudinally. Development of floodplains is considerable, and stream meandering prevails. Interstream areas have been reduced in height, and stream divides are not so sharp as in maturity.

According to Sharpe (3.29) and others, landslides are most prevalent during youth and the transition into maturity when valley walls are steepest and down-cutting the most active.

Regional Approach to Landslide Investigations

Analyzing the regional geology and terrain is the appropriate way to initiate preliminary landslide investigations. Geomorphologists have divided various regional areas into physiographic regions, that is, regional areas within which the method of deposition of earth materials and soils is approximately the same, the landforms are similar, and the climate is approximately identical.

Early efforts to rate the landslide severity of the various physiographic regions of the United States were reported by Baker and Chieruzzi (3.3). Their ratings of landslide severity were based on information gathered for the various physiographic regions with regard to frequency of occurrence, size of moving mass, and dollars expended per year. This information was obtained from responses to a questionnaire by state highway organizations and from various case histories and published records. Baker and Chieruzzi further noted in their development of the correlation of landslide severity to physiographic regions that specific geologic formations were usually associated with the landslides in particular regions. A listing of some of the more common landslide-susceptible formations is included in their report.

A map by Radbruch-Hall and others (3.24) is a recent effort at rating the severity of landslides in the United States. This map, a portion of which is illustrated in Figure 3.2, shows areas of relative incidence of landslides and areas susceptible to landslides. The accompanying text discusses the slope stability characteristics of the physiographic regions of the United States and the geologic formations and geologic conditions that favor landsliding in the various provinces.

Figure 3.2, together with the information provided by Radbruch-Hall and others and by Baker and Chieruzzi, are offered only as a guide for use in preliminary evaluation of landslide potential from a regional concept. They are not intended to depict precise boundaries and conditions. The delineation of areas of low incidence means not that extensive landslides do not occur in those areas but that they are negligible in comparison with occurrences and magnitudes of those in the other areas. It should also be recognized that, even though as noted in these reports certain geologic formations are normally associated with landslides, no formation has developed slides throughout the entire extent of its outcrop area. A more detailed investigation is required to pinpoint the actual slides or vulnerable locations.

Landforms Susceptible to Landslides

The designation of a specific landform connotes both a genetic classification and a type of landscape. For example, a sand dune landform denotes deposits formed by wind movement and sorting, which form unconsolidated, smooth, flowing hills and ridges. An appreciation of the genetic aspects of landforms enables one to estimate their potential susceptibility for movement. The type of landscape of each landform provides a basis for separating the various landforms and thus recognizing those most prone to sliding. Describing the genetic characteristics of landforms is beyond the scope of this presentation; however, excellent descriptions can be found in textbooks on geomorphology (3.14, 3.37). In this chapter, landscape characteristics of landforms are used as the basis for recognizing landslides and landslide-prone areas.

Figure 3.1. Catastrophic debris avalanches in Nelson County, Virginia, after torrential rains associated with hurricane Camille dumped 69 cm (27 in) of water August 19-20, 1969. Light-toned scars on tree-covered hillsides indicate locations of debris avalanches. Light-toned bands along stream floors are veneers of rubble and debris slide material.

Figure 3.2 Portion of USGS preliminary landslide overview map of the conterminous United States (3.24).

	AREA OF HIGH LANDSLIDE INCIDENCE		AREA OF HIGH LANDSLIDE SUSCEPTIBILITY
	AREA OF MODERATE LANDSLIDE INCIDENCE		AREA OF MODERATE LANDSLIDE SUSCEPTIBILITY
	AREA OF LOW LANDSLIDE INCIDENCE		

Landslides can occur in almost any landform if the conditions are right (e.g., steep slopes, high moisture level, no vegetative cover). Conversely, landslides may not occur on the most landslide-susceptible terrain if certain conditions are not present (e.g., clay shales on flat slopes with low moisture levels). Experience in observing and working with various landforms, however, has demonstrated that landslides are common in some landforms and rare in others. Table 3.2 provides a key to landforms and their susceptibility to landslides. The subdivisions are based on topographic expression and, in the case of hilly terrains, also on drainage patterns. This table gives only those landforms in which landslides are most common and is not meant to be all inclusive. Illustrations of some of these landforms and a brief description of their landscape characteristics are included later in this chapter in the section on landforms susceptible to landslides. Almost all landforms rated as highly susceptible to landslides are composed of alternate layers of pervious and impervious materials (rock or soil), a fact that needs to be specifically recognized.

Vulnerable Locations

The discussion of landslide-susceptible terrain to this point has progressed (a) from a regional concept in which the severities of landslide occurrence for various physiographic regions in the United States were indicated, (b) through the listing of geologic formations in the United States where landslides are common, and (c) to the rating of individual landforms as to their susceptibility to landslides. This progression from the general, regional overview to specific landforms can be carried one step further. Within the susceptible landforms are certain natural, vulnerable locations that are conducive to sliding. Typical vulnerable locations include areas of steep slopes, cliffs or banks being undercut by stream or wave action, areas of drainage concentration and seepage zones, areas of hummocky ground, and areas of fracture and fault concentrations. Special attention should be directed to those locations when maps or aerial photographs are examined and field studies are performed. In addition, areas that have recently slid require immediate and close scrutiny because additional movement may occur.

Steep Slopes

If slopes are steep enough, movement can occur on any landform. However, on landforms highly susceptible to landslides, other factors being equal, the steepest slopes are the most vulnerable locations. Only slopes of similar materials should be compared. For example, a slope cut in earth or talus should not be compared with a rock cliff in an adjacent landform, and slopes in bedrock generally are more stable, even though steeper, than slopes in adjacent soil areas.

The most common cause of the large number of slides that occur on steep slopes is residual or colluvial soils sliding on a bedrock surface. The loose, unconsolidated soils cannot maintain as steep a slope as the underlying rock surface and are, consequently, in a delicate balance. Any of

several factors, such as a sudden heavy rainfall (Figure 3.1) or an excavation at the toe of the slope (Figure 3.3) may result in sliding of the overlying soil mass. A study of cut-slope failures in North Carolina (3.12) reported that about two-thirds of the slides occurred in weathered soil materials and one-third occurred in rock slopes.

Cliffs and Banks Undercut by Streams or Waves

Landslides are common in cliffs or banks that are subject to attack by streams or waves. If the banks are made up of soil or other unconsolidated materials, the weakest (and hence the most favorable) slide position is often located at the point of maximum curvature of the stream. At this point, the bank receives the greatest impact from the water. In areas of rock outcrops, on the other hand, the exposure at and near the point of maximum curvature is often hard rock, and the weak spots are to be found upstream and downstream of this point. These conditions are shown in Plate 3.1 and Figure 3.23.

Many landslides occur along the edges of oceans and lakes because of undercutting by waves. Locating the point of maximum water impact is more complex and difficult

Table 3.2. Key to landforms and their susceptibility to landslides.

Topography	Landform or Geologic Materials	Landslide Potential[a]
I. Level terrain		
A. Not elevated	Floodplain	3
B. Elevated		
1. Uniform tones	Terrace, lake bed	2
2. Surface irregularities, sharp cliff	Basaltic plateau	1
3. Interbedded—porous over impervious layers	Lake bed, coastal plain, sedimentary plateau	1
II. Hilly terrain		
A. Surface drainage not well integrated		
1. Disconnected drainage	Limestone	3
2. Deranged drainage, overlapping hills, associated with lakes and swamps (glaciated areas only)	Moraine	2
B. Surface drainage well integrated		
1. Parallel ridges		
a. Parallel drainage, dark tones	Basaltic hills	1
b. Trellis drainage, ridge-and-valley topography, banded hills	Tilted sedimentary rocks	1
c. Pinnate drainage, vertical-sided gullies	Loess	2
2. Branching ridges, hilltops at common elevation		
a. Pinnate drainage, vertical-sided gullies	Loess	2
b. Dendritic drainage		
(1) Banding on slope	Flat-lying sedimentary rocks	2
(2) No banding on slope		
(a) Moderately to highly dissected ridges, uniform slopes	Clay shale	1
(b) Low ridges, associated with coastal features	Dissected coastal plain	1
(c) Winding ridges connecting conical hills, sparse vegetation	Serpentinite	1
3. Random ridges or hills		
a. Dendritic drainage		
(1) Low, rounded hills, meandering streams	Clay shale	1
(2) Winding ridges connecting conical hills, sparse vegetation	Serpentinite	1
(3) Massive, uniform, rounded to A-shaped hills	Granite	2
(4) Bumpy topography (glaciated areas only)	Moraine	2
III. Level to hilly, transitional terrain		
A. Steep slopes	Talus, colluvium	1
B. Moderate to flat slopes	Fan, delta	3
C. Hummocky slopes with scarp at head	Old slide	1

Note: This table updates Table 2 in the book on landslides published in 1958 by the Highway Research Board (3.10, p. 91).

[a]1 = susceptible to landslides; 2 = susceptible to landslides under certain conditions; and 3 = not susceptible to landslides except in vulnerable locations.

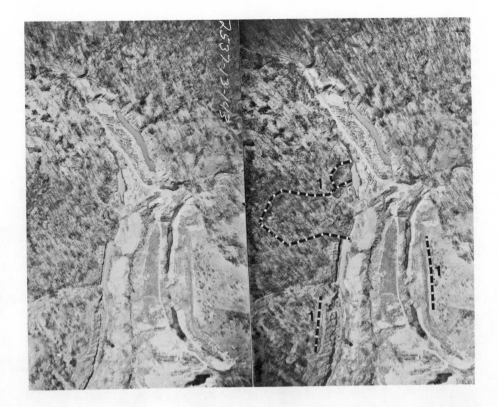

Figure 3.3. Slides (1) caused by undercutting of slopes by coal stripping operations in Washington County, Ohio (3.22).

along lake and ocean shores than along stream banks. Factors to be considered include shape and slope of the shoreline, direction of wave action, and frequency and magnitude of storms producing large waves. Data obtained at different periods of time are often of value in the analysis of these factors.

Areas of Drainage Concentration and Seepage

A survey conducted by the Federal Highway Administration of major landslides on the federal-aid highway system in the United States revealed that "water is the controlling or a major contributing factor in about 95 percent of all landslides" (3.5). Thus, careful study of the drainage network and areas of concentration or outfall of water is extremely important. Close scrutiny of existing slide scars often indicates that a line connecting the scars points to drainage channels on higher ground. Such drainage may appear as seepage water, which is responsible for the damage. An example of this condition is shown in Figure 3.4

Seepage with subsequent sliding is likely to occur in areas below ponded depressions, reservoirs, irrigation canals, and diverted surface channels (Figure 3.4). Such circumstances are sometimes overlooked on the ground because the water sources may be far above the landslide itself, but they become obvious in aerial photographs. The importance of recognizing the potential danger in areas below diverted surface drainage, especially in fractured and porous rocks, needs particular emphasis. Extensive field experience has proved repeatedly that within an unstable area one of the most dangerous sections is the lower part of an interstream divide through which surface water seeps from the higher to the lower stream bed. The recognition of seepage is

sometimes aided by the identification of near-surface channels, wet areas, tall vegetation on the slope, and displaced or broken roads adjacent to the slope. Delineating the drainage network, especially the presence of seeps or springs, is extremely important for planning new construction. Many

Figure 3.4. Oblique aerial view of extensive seepage zones (1) on scarp face of massive landslide in Kittitas County, Washington. Seepage was major cause of slide in these unconsolidated sediments. Remnant of dry sand flow (2) is seen above scarp face and is result of seepage at higher level. Landslide disrupted highway at base of slope and water canal carried through slope.

Figure 3.5. Landslide in highway fill section of Ohio-22 in Jefferson County. Fill slid en masse as block slide with fill above block slumping into resultant hole as graben. Ohio Department of Transportation staff report attributed failure to saturation of colluvial materials on side slopes beneath fill by springs outcropping on hillsides. Few small drainageways commencing part way down slope (1) and small slide (2) on natural hillside indicate presence of seepage zones and instability of natural slopes.

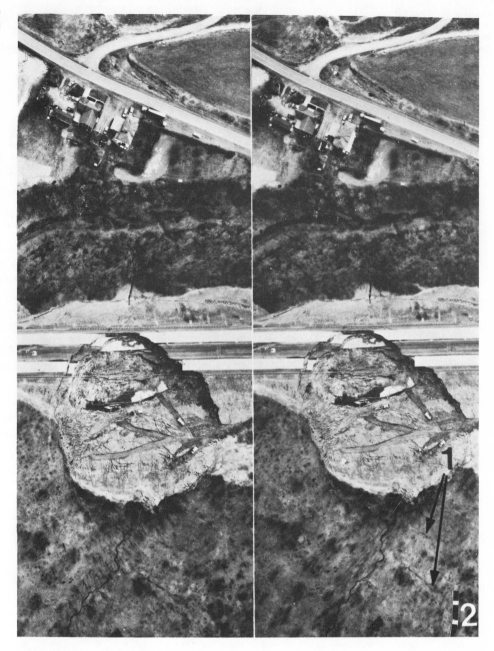

highway fills have failed because the natural drainage was blocked by the fill and allowance was not made for drainage. Figure 3.5 shows such a situation.

Areas of Hummocky Ground

The presence of hummocky ground whose characteristics are inconsistent with those of the general regional slopes and the presence of a scarp surface (sometimes not very distinct) at a higher elevation are often indications of an existing landslide. The older the landslide is, the more established the drainage and vegetation become on the slide mass. The drainage and vegetation thus help in determining the relative age and stability of the slide.

Once an old landslide is found, it serves as a warning that the general area has been unstable in the past and that new disturbances may start new slides. However, such a warn-

ing should not discourage construction unconditionally, because the unstable condition of the past may not necessarily exist today. In some parts of the western United States, for example, railroads built in extensive old landslide areas have been stable for a long time. Nevertheless, special care should be taken in construction on old slides. Figure 3.6 shows aerial photographs of the construction of a road on an existing slide and the consequences.

Areas of Concentration of Fractures and Bedding Planes

Movements of slopes may be structurally controlled by surfaces or planes of weaknesses, such as faults, joints, bedding planes, and foliation. These structural features can divide a rock mass into a number of individual units, which may act independently of one another. The result can be an incor-

Figure 3.6. Before (top) and after (bottom) construction of Ohio-78 in Monroe County on unstable ground (3.22). Before construction, presence of hummocky ground (1) indicates unstable nature of terrain. Relocation of side road to meet new grade of highway placed side road on slide-prone slope, which resulted in small slide of embankment (2). Crack crossing road diagonally, seen as faint dark line (3), is site of possible future movement.

rect slope design because the designer considered the rock to be one continuous mass rather than a series of individual blocks. These planes of weakness also provide egress for water and vegetation, which further weaken the individual units by wedging action, frost heave, and reduction of sliding friction. A careful search should be made to locate areas with close spacing of faults and joints, especially where they cross and divide the rock mass into smaller blocks.

Recent Landslides

The occurrence of a landslide does not mean that final adjustments to the unstable conditions have occurred and no further movement will occur. In many cases, especially in unconsolidated deposits, the materials present in the scarp face remain in an unstable condition because they are on a

very steep slope. The scarp face rapidly retrogrades uphill by continued slumping until a more stable condition occurs. Thus, a new landslide should be investigated as soon as possible not only to determine corrective measures but also to look for evidence of possible continued movement. The most significant sign of possible further instability is the presence of cracks on the crown of the slide. Figure 3.7 (top) shows some telltale signs at a recent slide. Figure 3.7 (bottom) shows the slide area 5 months later. Additional movement occurred in the area where the telltale signs were evident in the earlier photograph.

Procedure for Preliminary Investigations of Landslides

Not only must the presence or potential development of landslides be recognized, but the types and causes of

Figure 3.7. Sequential photography for use in evaluation of recent slide development along US-95 in Idaho County, Idaho, and investigation of possible further movements. Aerial photography (top) taken in February 1974 shortly after occurrence of slide (1) shows features present at crown of slide that forewarn of further movement: fissures above scarp face (2), loose colluvial materials on steep slopes at crown (3), and evidence of water feeding into these loose materials (4). Photography obtained 5 months later in July 1974 (bottom) shows that additional slide movement (5) occurred in areas indicated as potentially unstable in photograph above.

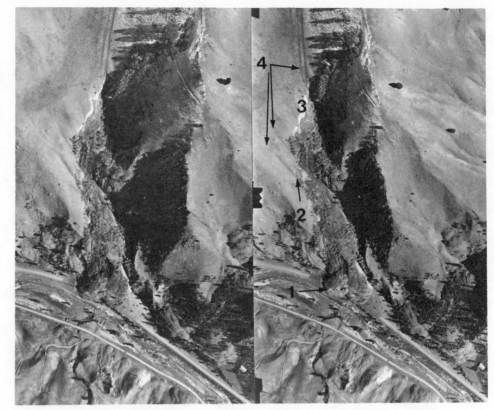

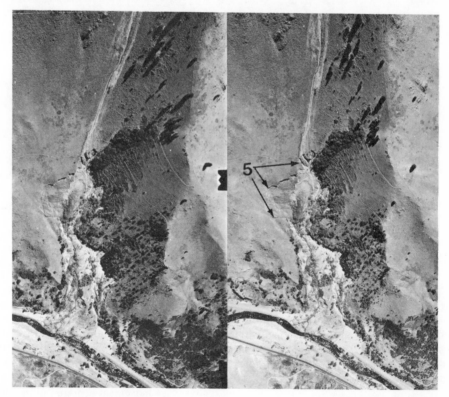

movement must be identified so that preventive or corrective action can be taken. Chapter 2, and specifically Figure 2.1, gives the classification of the various types of landslides based on the dominant types of movement and the types of materials involved. In this chapter, the emphasis is on how those various types of slides appear on the various data sources used in terrain investigations.

A typical procedure, which has been developed over the years, for performing terrain investigations is used for a variety of programs, such as engineering soil surveys, location of construction materials, and landslide investigations. The following steps are generally performed. Some excellent case studies demonstrating the application of this step-by-step procedure are reported by Mintzer and Struble (3.20).

1. Obtain aerial photography and other special coverage. Small-scale photography for regional overview and large-scale coverage for detailed study are obtained from available sources. If landslides are already present, photographs taken both before and after the occurrence should be obtained to aid in locating causative factors that may have been obliterated by the slide. Special types of photography should be acquired as needed (the section on use of aerial photography gives details on types and scales of photography needed).

2. Review literature and maps. A review is made of existing topographic maps, geologic maps and reports, water well logs, agricultural soil survey reports, and other literature to develop an areal concept of the area under investigation.

3. Analyze the photography and other special coverage. The patterns on the photographs are analyzed, and landforms are identified and related to the areal concept developed from the literature review. A careful examination is made of all vulnerable locations. Existing and potential landslide sites are delineated, and a three-dimensional concept of the terrain is developed. In this step, a mosaic is usually prepared and landform and drainage maps are developed. Sites that offer the best opportunities for confirming or extending the information developed are selected for field verification.

4. Perform field reconnaissance. A field reconnaissance of the area is performed to verify the three-dimensional concept developed in the earlier steps, to fill in information in questionable areas, and to observe the surface features and details that could not be determined from other data sources. The type of landslide movement is classified from the available data and from a study of the surface features and crack patterns observed in the field.

5. Conduct final analysis and plan field investigations. A final analysis of the photography is performed based on the results obtained from the field reconnaissance. A determination is made as to what additional information is needed to fully define the three-dimensional model of the site and what samples and test data are required to design the corrective procedures. Based on those needs, the field investigation program is planned (this process is described fully in Chapter 4).

MAP TECHNIQUES FOR LANDSLIDE DETECTION

The acquisition and the analysis of various types of maps constitute one of the first steps in landslide investigations. Maps depicting topography, geology, agricultural soils, and other special terrain and cultural features are available for many areas of the world. The maps represent, in a two-dimensional format, the authors' interpretations of terrain and cultural features determined from field investigations, a study of available literature and photography, and map-making procedures. The nature and quality of the information relating to the presence of landslides or potential for landslides that can be derived from existing maps depend on the purpose, type, scale, and detail used in preparing the map. From most maps, only a general indication of landslide susceptibility can be derived. In recent large-scale maps, more detail on locating existing landslides and rating the terrain as to landslide susceptibility is included.

Maps have certain inherent advantages for landslide studies. They are prepared at a uniform scale and, therefore, are subject to direct quantitative measurements. Furthermore, planimetric information is often included, thereby minimizing uncertainties in superimposing other information. Their disadvantage is that they are outdated as soon as they are published. Unless the maps are updated periodically or no changes occur, the maps will not show the most recent terrain and cultural features.

The various maps discussed in this section are published mainly by governmental and public agencies and may be obtained free or at nominal costs. The published maps can also be viewed at many governmental and university libraries and in some large public libraries. Included in the following discussions of the various types of maps is information on map sources.

Topographic Maps

Topographic maps show the size, shape, and distribution of features of the surface of the earth. These maps depict the features of relief, drainage, vegetation, and culture, usually in a color format. Some of the more recent topographic maps prepared in the United States use orthophotographs as the base for depicting drainage and culture as they actually appear and superimpose contours on the orthophotographs to depict the relief; these are published as orthophotomaps.

Major landslide areas that are clearly evident are sometimes labeled on topographic maps. On some maps, the boundaries of the slide and arrows pointing toward the direction of movement are also shown. Small slides, the types more commonly encountered in highway and other engineering works, are not usually labeled on such maps. Identification of these smaller slides or unlabeled larger slides can be accomplished by noting the following features on topographic maps (Figure 3.8):

1. Topographic expression observable, for example, steep slope (closely spaced contours) at scarp head of slide, hummocky topography in slide mass (irregular nonsymmetrical contour patterns with shallow depressions), and presence of detached mass and flow characteristics at the lower end;

2. Wavy contour lines, uneven or broken local roads, and other artificial lineaments such as transmission lines; and

3. Minor movements or irregularities at "vulnerable locations," as discussed in the section in this chapter on vulnerable locations.

The potential for identifying landslides on topographic maps is essentially limited by the scale and contour interval of the maps. On small-scale maps, even moderate-sized landslides significant to engineering works may be such microfeatures that they are not identifiable. The U.S. Geological Survey (USGS) publishes topographic maps at various scales from 1:20 000 to 1:1 000 000. Indexes to the status of topographic mapping for the United States are available free from USGS. Information on status of mapping in other countries can be obtained from similar mapping organizations in the respective countries.

Figure 3.8. Portion of USGS topographic map of Laguna Quadrangle, New Mexico, on which typical landslide characteristics described in text are evident (1). Figure 3.31, which shows landscape within same region but not same area, illustrates appearance of this landslide topography on aerial photograph.

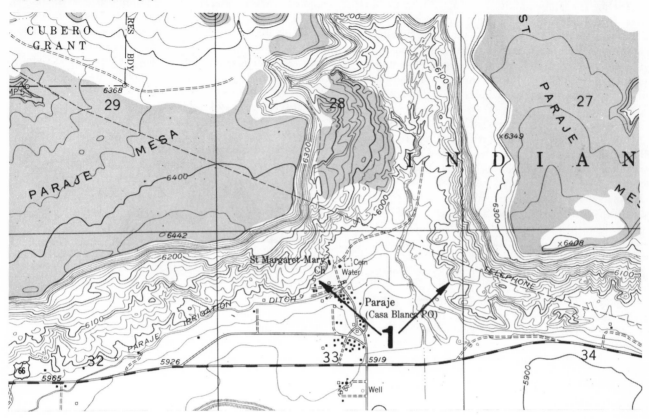

Geologic Maps and Reports

The principal sources of geologic information in the United States are the maps and publications of USGS and the state geologic surveys. Government geologic maps range from general, small-scale maps covering entire continents to large-scale maps covering counties and quadrangles. Not all the geologic literature published annually, however, is found in government publications. Numerous articles are published by geologic organizations, whose publications are referenced in two periodicals: *Bibliography of North American Geology*, published by USGS, for areas in North America, the Hawaiian Islands, and Guam; and *Bibliography and Index of Geology Exclusive of North America*, published by the Geological Society of America, for other areas. In addition, USGS publishes indexes of geologic maps and listings of maps and reports for each of the states.

The most extensive form of geologic maps published in the United States is the quadrangle map series, which includes surficial geology maps, bedrock geology maps, and standard geologic maps depicting both surface and bedrock units present in the map area. Although the recent quadrangle maps show landslide deposits, only relatively large slides are shown because of the small scale and large contour intervals. An example of this type of map is shown in Figure 3.9. The surficial geology maps show only the surface materials present and are usually limited to areas extensively covered by complex surficial deposits, such as in glaciated areas. These maps are useful for landslide studies,

particularly when surface and near-surface formations are of significance. Bedrock geology maps show only the bedrock and do not indicate surficial deposits. These maps must be used with care to avoid erroneous assumptions about surface conditions. For landslide investigations, bedrock geology maps are useful when little or no soil overlies the bedrock or when slides are deep-seated.

Other maps of special importance for landslide investigations are the USGS Miscellaneous Investigation Series and Miscellaneous Field Studies Series. Maps are now being produced that directly indicate landslides and rate the various terrain features as to their susceptibility to sliding. Figure 3.10 shows a portion of this type of map. Other maps in these series indicate items or features closely related to landslides, such as engineering geology, water resources, and groundwater. These maps are also of some value in landslide investigations. Unfortunately, most of the maps in these miscellaneous series are not well catalogued and are difficult to locate.

Even though geologic maps and reports may not specifically indicate the presence of landslides or landslide-susceptible terrain, this information can be extracted from the geologic literature by noting the physiographic regions, the geologic formations mapped, the topographic and stratigraphic setting, and the landforms present. As discussed in the earlier section on regional approach to landslide investigations, certain physiographic regions and geologic formations are noted for landslides. The presence of these susceptible formations can be determined from the geologic litera-

Figure 3.9. Black-and-white reproduction of a portion of USGS geologic quadrangle map GQ-970, Tyler Peak Quadrangle, Washington. Slide areas are identified by map symbol Ql and are outlined by dashed line. On original color map, surficial deposits including slide areas are shown in yellow and are more easily distinguishable.

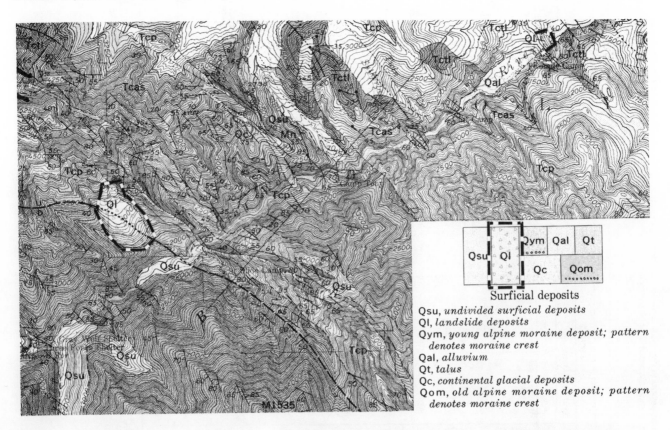

Surficial deposits

Qsu, *undivided surficial deposits*
Ql, *landslide deposits*
Qym, *young alpine moraine deposit; pattern denotes moraine crest*
Qal, *alluvium*
Qt, *talus*
Qc, *continental glacial deposits*
Qom, *old alpine moraine deposit; pattern denotes moraine crest*

Figure 3.10. Portion of USGS Miscellaneous Field Studies Map MF-685B, Susceptibility to Landsliding, Allegheny County, Pennsylvania.

SLOPES WITH MODERATE TO SERVERE SUSCEPTIBILITY TO LANDSLIDING

SLOPES WITH SLIGHT TO MODERATE, LOCALLY SEVERE, SUSCEPTIBILITY TO LANDSLIDING

GROUND WITH LITTLE SUSCEPTIBILITY TO LANDSLIDING

PREHISTORIC LANDSLIDES

STEEP SLOPES MOST SUSCEPTIBLE TO ROCK FALL

MAN-MADE FILL

GENERAL DIRECTIONS OF DIP OF ROCK LAYERING

45

Figure 3.11. Portion of soil survey report for Chilton County, Alabama (3.18). Information on susceptibility to sliding in situ is indicated in column, Soil Features Affecting Highway Location. Presence of possible seepage zones is also designated in this column. Some indication of slide susceptibility of soils placed in fill or embankment section is indicated in columns, Road Fill Material or Farm Ponds, Embankments.

Soil series and map symbols	Suitability as a source of—			Highway location	Farm ponds		Soil features affecting—		Terraces and diversions	Waterways
	Topsoil (surface layer)	Sand and gravel	Road fill material		Reservoir areas	Embankments	Agricultural drainage	Irrigation		
Luverne: LvB, LvB2, LvC, LvC2, LvD2, LwC2, LwD2, LwF. For Boswell part of LwC2, LwD2, and LwF, refer to Boswell series.	Fair to good	Poor: no sand or gravel available.	Fair to poor; high clay content; very plastic.	Susceptible to sliding; fair traffic-supporting capacity; slopes.	Moderate to slow seepage.	Low to moderate shear strength; poor stability.	No drainage needed	Slopes of 2 to 15 percent; moderate intake; medium available water capacity.	High erodibility	High erodibility.
McLaurin: McB, McC, McD.	Fair	Fair to poor: sand at a depth of 50 inches; poor: no gravel available.	Good	No limitations except slopes.	Rapid seepage	Moderate shear strength; fair stability.	No drainage needed.	Slopes of 2 to 15 percent; rapid intake; low to medium available water capacity.	Slight erodibility	Moderate erodibility.
Madison: MdB2, MdC2, MdD2.	Fair to good	Poor: no sand or gravel available.	Fair to good	Bedrock at a depth of 20 to 50 inches.	Slow to moderate seepage.	Moderate shear strength; fair stability.	No drainage needed.	Moderate intake; moderate permeability; medium available water capacity.	Slopes of 2 to 15 percent; moderate to high erodibility.	High erodibility.
Maseda: MsA.	Good to fair	Poor: no sand or gravel available.	Fair to good	Occasional flooding	Slow seepage	Fair stability	No drainage needed; bedrock at a depth of 30 to 50 inches.	Slopes of 0 to 2 percent; moderate intake; moderate available water capacity.	No terraces or diversions needed.	No waterways needed.
*Myatt: Mt, MyA. For Bibb part of MyA, refer to Bibb series.	Fair	Poor: no sand or gravel available.	Fair; high water table; flooding.	High water table.	Slow seepage	Fair to good stability.	High water table; moderate to moderately slow permeability; outlets difficult to establish.	Slopes of 0 to 2 percent; moderate to slow intake; poor soil for crops.	No terraces or diversions needed.	No waterways needed.
Ora: OrA, OrB, OrB2, OrC, OrC2.	Good	Poor: no sand or gravel available.	Good	Seepage in cuts along the fragipan layer.	Moderate seepage	Fair shear strength; fair stability.	No drainage needed.	Slopes of 0 to 10 inches; moderate intake; moderately slow permeability; medium available water capacity.	Moderate erodibility.	Moderate erodibility; fragipan.
Rock land: Ro. No interpretations; properties too variable.										
*Ruston: RsA, RsB, RsC, RsC2, RsD2, RtE, RtF. For Shubuta and Troup parts of RtE and RtF, refer to Shubuta and Troup series.	Good	Poor: no sand or gravel available.	Good	No limitations except slopes.	Moderate to rapid seepage.	Moderate shear strength; poor to fair stability.	No drainage needed.	Slopes of 0 to 15 inches; moderate intake; medium available water capacity.	Moderate erodibility.	Moderate erodibility.
Saffell: SaB, SaC, SaD.	Fair to poor: gravelly.	Fair to poor: underlain by sand and gravel at a depth of 4 to 6 feet; gravel good for road construction material.	Good	No limitations except slopes.	Rapid seepage	Poor to fair stability	No drainage needed.	Slopes of 2 to 15 inches; moderate to rapid intake; medium to low available water capacity.	Moderate erodibility.	Gravelly; little moisture.
Shubuta: ShD2.	Fair to good	Poor: no sand or gravel available.	Fair; high clay content; very plastic.	Susceptible to sliding; seepage in cuts; moderate shrink-swell potential.	Slow seepage	Low shear strength; fair to poor stability.	No drainage needed.	Slopes of 2 to 15 inches; moderate to slow intake; medium available water capacity.	High erodibility	High erodibility.

46

ture. In addition, as noted in the sections on landforms susceptible to landslides and on vulnerable locations, certain landforms and topographic settings are especially susceptible to landslides. For example, one of the most favorable settings for landslides is the presence of permeable or soluble beds overlying or interbedded with relatively impervious beds in an elevated topographic setting. The beds could be either rock or soil. Conditions such as this can be determined from the geologic literature.

Agricultural Soil Survey Reports

The Soil Conservation Service (SCS), U.S. Department of Agriculture, in cooperation with state agricultural experiment stations and other federal and state agencies, has performed and published soil surveys since 1899. The agricultural areas in the United States are, therefore, covered by fairly detailed soil surveys, and much of the remaining area is mapped in less detail. Similarly, agricultural areas in other countries are covered by soil surveys with varying degrees of detail and are available in published form.

Although SCS soil survey reports are compiled primarily for agricultural purposes, those published since 1957 are of special value for landslide investigations. Most of these soil surveys contain information on the engineering uses of the soils. In addition, most of the soil maps published in these reports are printed on a photomosaic base at a scale of 1:15 840 or 1:20 000, and the soil information is superimposed on the mosaic. This provides for easier identification of natural and cultural features.

Soil surveys provide a three-dimensional concept of the land and contain information about the areal extent and vertical profile for each soil unit. The insight given of each soil unit and its relation to adjacent soil units enables one to estimate the environmental setting and potential stability.

The engineering sections included in the recent reports contain engineering test data for the major soils mapped in the counties; a table of estimated engineering properties of the soils, including depth to water, depth to bedrock, typical profile, and grain-size distribution; a table of engineering interpretation of the soils; and accompanying text discussing the special engineering features of the soils. Estimates of susceptibility to landslides for the various soils mapped can be made from the data furnished. In some cases, these estimates are indicated in the table on engineering interpretations. Figure 3.11 shows a portion of such a table from the Chilton County, Alabama, soil survey report (3.18). Having identified from this table the soils susceptible to sliding, one can then locate those soil units on the photomosaic and thus delineate the most susceptible areas.

Other county soil surveys and generalized reconnaissance surveys can also provide useful information for landslide investigations, even though engineering sections are not included. Those surveys provide information on parent materials and their geologic origin, climate, physiographic setting, and soil profiles.

The status of availability of agricultural soil survey reports in the United States is published annually by SCS in *List of Published Soil Surveys*. In addition, those state soil conservation services that publish their own soil surveys provide lists for the respective states. For copies of the published reports, the unpublished field sheets, and the most up-to-date information about surveys in progress, one should check with the local soil conservation service office in the county of interest. The Canada Department of Agriculture provides lists of Canadian soil survey maps and reports. Sources of information on soil surveys in foreign countries include World Soil Map Office, SCS; libraries in major agricultural colleges and universities; and departments of agriculture in other countries.

Special Maps and Reports

The susceptibility of the terrain to landslides is also discussed in special maps and reports prepared by various organizations. Included in this group are engineering soil maps, engineering soil surveys, geologic reconnaissance surveys, and engineering guides to agricultural soil series.

Engineering soil maps depict the relation of the landform and engineering soils. The landform, which constitutes the foundation of this mapping, is subdivided on the basis of its engineering characteristics, such as soil texture, drainage conditions, and slope category. Although these reports generally do not indicate landslide conditions, estimates of landslide susceptibility can be determined by evaluating the landform types, soil characteristics, and slope conditions. In the United States, engineering soil mapping is performed at a variety of regional levels. For example, New Jersey and Rhode Island are completely covered; Indiana and Tennessee are preparing maps on a county basis; and Kansas, Louisiana, and New Mexico prepare maps on a project-by-project basis.

For most engineering projects, preliminary engineering soil surveys and geologic reconnaissance surveys are performed. In many of these surveys, landslides or landslide-susceptible areas are depicted prior to construction. Figure 3.12 shows an example of such an investigation performed for a route location in Kansas. The results of the survey influenced the final location of the highway (3.31).

Since large parts of the United States are covered by agricultural soil surveys and extensive data are available for the basic soil units mapped (the soil series), several state highway organizations have developed information on the

Figure 3.12. Delineation of landslide terrain during preliminary route location in Lincoln and Ellsworth counties, Kansas, aided in minimizing damage due to landslides (3.31).

47

engineering characteristics of the soil series mapped in their states. The information generally provided for each soil series includes origin, topographic setting, soil profile, drainage characteristics, and engineering significance for items such as location, slope stability, foundation and pavement support, and source of construction materials. Several states, such as Indiana, Michigan, New York, and Ohio, have summarized this information in report form. In many areas, and particularly in densely developed areas, existing reports for geologic and engineering investigations may be available and should be acquired for reference. Detailed boring logs, well logs, and similar valuable soil and geologic information can be obtained at little cost.

REMOTE-SENSING TECHNIQUES FOR LANDSLIDE DETECTION

Remote sensing is the collecting of information about an object or phenomenon by the use of sensing devices not in physical or intimate contact with the subject under investigation. The distance of separation might be as close as a few millimeters, as in the case of nuclear moisture-density devices, or as far as 800 km (500 miles) or more, as in the case of satellites. The devices range from cameras to various scanners and radiometers.

Most remote-sensing devices collect information from the electromagnetic spectrum (Figure 3.13), which extends through many decades of wavelength from very long radio waves to extremely short gamma and cosmic radiation waves. As shown in Figure 3.13, various types of sensors are available for obtaining information in the different regions of the spectrum. The *Manual of Remote Sensing* (*3.25*) gives greater detail on the electromagnetic spectrum and sensor systems.

The attractiveness of these systems to various investigators is that they greatly expand the scope of possibilities for uniquely and more accurately identifying subjects of interest. In addition to the sensors that provide information on the spectral reflectance phenomena of objects are those that provide information on the thermal-emission properties of objects in the infrared region, the backscatter of radar energy in the microwave region, and the reflectance and fluorescence phenomena in the ultraviolet region. Analyzing spectral regions individually provides special data about the properties of objects in that region (e.g., thermal characteristics or reflectance properties).

Collecting data simultaneously in several spectral regions and comparing the various responses in those regions, however, provide more data than can be obtained in each region individually and increase the accuracy of interpretation. This approach is referred to as multisensor analysis and is based on the following principle: Although two or more objects may have similar responses in one region, they will not respond similarly in all regions. Thus, evaluating the response of various objects in different regions of the spectrum increases the ability to uniquely identify each of the objects.

In almost all engineering evaluations of various sensor systems, the conclusion has been that photography is the best sensor system and provides the most information. Although other sensor systems provide some useful data that cannot be obtained from photography alone, except for

special applications, the additional data are not cost-effective. That being the case, the ensuing discussion on the use of remote-sensing techniques to recognize and identify landslides deals mostly with the use of aerial photography. The special applications in which other forms of remote-sensing data are useful are also discussed. Some comments are included on systems that have potential value but have yet to be demonstrated for this application.

Use of Aerial Photography

The material in this section expands and updates Chapter 5 in the book on landslides published by the Highway Research Board in 1958 (*3.10*).

The interpretation of aerial photography has proved to be an effective technique for recognizing and delineating landslides. No other technique can provide a three-dimensional overview of the terrain from which the interrelations of topography, drainage, surface cover, geologic materials, and human activities on the landscape can be viewed and evaluated. Aerial photographs at suitable scales are available for almost all of the United States and a large part of the world. New photography of various types and scales can readily be obtained. In addition, new techniques in production and interpretation processes have continued to extend the advantages of aerial photography.

The value of aerial photography in landslide investigations has been reported by many investigators (*3.13, 3.16, 3.19, 3.28, 3.31*). The effectiveness of this technique was well demonstrated by Maruyasu and others (*3.16*) in analyzing a mountainous region where landslides were common in Japan. Using photo-interpretation techniques, they identified 365 landslides. After the region was field checked, an additional 68 landslides were identified, resulting in an 84 percent accuracy in overall identification. An accuracy of 96 percent was reported in identification of landslides in areas such as paddy fields not covered by trees and shrubs.

The major advantages of using aerial photography in landslide investigations include the following:

1. Photographs present an overall perspective of a large area (when examined with a pocket or mirror stereoscope, overlapping aerial photographs provide a three-dimensional view);
2. Boundaries of existing slides can readily be delineated on aerial photographs;
3. Surface and near-surface drainage channels can be traced;
4. Important relations in drainage, topography, and other natural and man-made elements that seldom are correlated properly on the ground become obvious on the photographs;
5. A moderate vegetative cover seldom blankets details to the photo interpreter as it does to the ground observer;
6. Soil and rock formations can be seen and evaluated in their undisturbed state;
7. Continuity or repetition of features is emphasized;
8. Routes for field investigations and programs for surface and subsurface exploration can be effectively planned;
9. Recent photographs can be compared with old ones to examine the progressive development of slides;

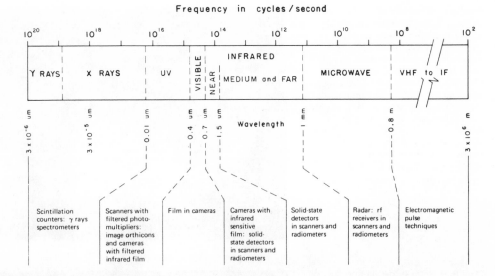

Figure 3.13. Electromagnetic spectrum and types of remote-sensing systems used for collecting data in each major spectral region.

10. Aerial photographs can be studied at any time, in any place, and by any person; and

11. Through the use of aerial photographs, information about slides can be transmitted to others with a minimum of ambiguous description.

Factors Influencing Aerial-Photograph Interpretation

Success in the application of aerial-photograph interpretation techniques for landslide investigations depends on four major items: (a) qualifications of the interpreter; (b) photographic parameters; (c) natural factors; and (d) equipment and analysis techniques used. Analysis by competent interpreters with suitable equipment will certainly be an important factor in the successful application of this technique. However, a proper understanding of the photographic parameters and natural factors that can influence the data is important to the proper use of aerial photography; these two items are discussed below.

Photographic Parameters

Photographic parameters include types of photography, format, scale, coverage, quality, and availability. General information concerning these parameters is given in many standard references, such as those published by the American Society of Photogrammetry (*3.7, 3.25, 3.30*). The following discussion covers only those items pertinent to landslide investigations.

Aerial photographs typically used for landslide studies include panchromatic and infrared black-and-white films and natural color and color infrared films. Panchromatic black-and-white film is the most common type used because it is low cost, convenient to handle, and readily available, but natural color and color infrared films are now being more extensively used. Color photography is especially valuable for outlining differences in moisture, drainage, vegetation conditions, and soil and rock contacts. Plates 3.1 through 3.6 show the uses and advantages of color films. The stratification in exposed soils and rocks is most easily recognized on natural color film. Color infrared films are most helpful for delineating the presence of water on the surface and for giving clues to subsurface water conditions by showing the vigor of the surface vegetative cover. This has made color infrared film especially valuable for locating the presence of seepage zones at or near the surface.

The format commonly used for landslide investigations is 23 x 23-cm (9 x 9-in) vertical photographs taken with an aerial mapping camera. Consecutive photographs having an average overlap of 60 percent are obtained to provide stereoscopic coverage. In addition to their use for interpretation purposes, these photographs can be used for preparing maps of the slide areas by photogrammetric methods. Details on flight planning and mapping are given in the *Manual of Photogrammetry* (*3.35*). Oblique photographs, such as those shown in Figures 3.1 and 3.4, are also obtained for landslide investigations. The view they provide of the unstable slopes and valley walls is more unobstructed than that provided in vertical photographs. In recent years, oblique views of landslide areas are being obtained with small format, nonmapping cameras in low-flying planes and helicopters; 35 and 70-mm cameras are commonly used.

Scale of photography can be critical to the interpreter's ability to identify landslides in that it influences his or her ability to denote topographic features indicative of landslides. In some cases, the landslides are so massive [sometimes involving more than $4 km^3$ ($1 mile^3$)] of material that they are virtually impossible to detect on the ground or on large-scale photographs (*3.9*). Small-scale photographs (1:48 000 or smaller) are required in order to view stereoscopically the full extent of massive slides. However, most landslides encountered in engineering construction are small. Cleaves (*3.6*) stated: "Within the scope of the author's experience, more than 90 percent of all landslides requiring control or correction have been small, averaging less than 500 feet in width from flank to flank, and 100 to 200 feet in length from crown to toe." Hence, many small landslides would not be readily discernible on normal-scale photography (1:15 000 to 1:30 000) and would require large-scale photography (1:4800 or larger) for identification. At these large scales, however, the area covered in each photograph is limited, and the overall perspective is more difficult to grasp. Whenever practicable, it is desirable to examine both small-

Plate 3.1. Slides in Douglas County, Oregon, caused by undercutting of slopes by stream. Slides largely composed of weathered colluvial materials are evident at almost every bend of stream where it undercuts naturally steep slopes (1). Advantages provided by use of color film include distinction of potentially more unstable colluvial material from more stable bedrock surfaces; ease of distinguishing seepage areas; and indication of vegetation under stress due to insufficient moisture, which could occur because of slope movement.

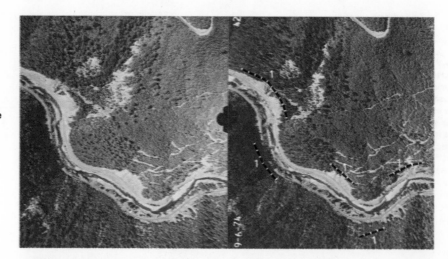

Plate 3.2. Color infrared photography of landslide terrain in Skamania County, Washington, best shows subtle features indicative of slide or slide-prone areas: presence of seepage areas as indicated by dark-toned streaks or "tails" (1) or by concentration of lush vegetation (2); vegetation under stress due to moisture deficiency as indicated by lighter reddish tones and by sparser leaf coverage (3); and presence of leaning and fallen trees (4).

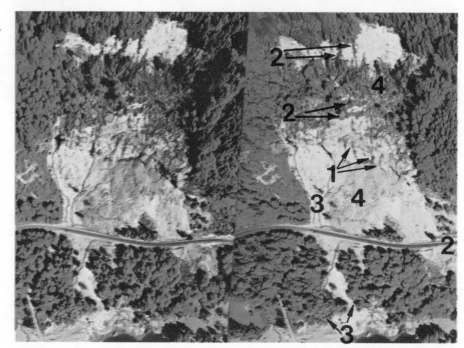

Plate 3.3. Oblique view of area shown in Plate 3.2 allows some of subtle features to be more easily identified.

Plate 3.4. Black-and-white photographs taken October 25, 1954, of delta deposits overlying lake bed in Tompkins County, New York. Many landslides in glaciated areas occur in lake-bed materials overlain by granular deposits. Slides occur on both sides of creek. Benching and other slope-stabilizing practices were incorporated to stop sliding. Large landslide is easily identified (1), but evidence of seepage and boundary between delta and lake-bed deposits is subtle and difficult to distinguish on this film type.

Plate 3.5. Natural color photographs taken April 9, 1976, of area shown in Plate 3.4. Presence of seepage at contact between delta and lake-bed deposits (2) and change in color tones of soils of these two materials make it easier to distinguish between these two deposits: Darker brown tones are seepage zones; bluish gray tones are lake bed; and lighter tan tones are delta deposits.

Plate 3.6. Color infrared photographs taken April 9, 1976, of area shown in Plate 3.4. Seepage zones and wet areas (2) are best seen on this film type. Dark tones are clearly distinct. It is too early for presence of luxurious vegetation, which could aid in delineating seepage areas. Boundary between two deposits is more distinct on this film than on black-and-white film, but contrast is not so sharp as on natural color film.

and large-scale photographs: the smaller scale (1:20 000 or smaller) to provide the regional setting and landform boundaries and the larger scale (1:8000 or larger) to provide the details.

Sufficient coverage is required in order to determine the full extent of the slide, the potential areas affected, and the changes occurring with time. In some instances the active slide is present only at the lower or toe end of a slope while unnoticed far up the mountainside are tension cracks that indicate the instability of the entire slope. With time, the slides may retrogress up the slope until the whole mountainside is affected. Clearly, the plans for stabilization will differ depending on whether one is dealing with a small confined slide area or with a whole mountainside. A situation of this type can be dealt with satisfactorily if two scales of photography are obtained, as previously recommended. Coverage at the larger scale should include the full height and depth (mountain top to valley bottom) of the area under investigation and at least three to five times the width of the present unstable area.

Many areas, particularly in the United States, have been photographed at various times since the 1930s, and this sequential photography is valuable for noting the changes in terrain conditions, such as drainage, slope stability, land use, and vegetation, that have occurred during a period of time. These changes provide a clue to the origin and rate of movement of a slide. Besides denoting changes, such photographs were probably taken at different seasons, times of day, and moisture conditions and will often show detailed features that might otherwise have been obscured. These older photographs can also be of value in planning new coverages.

The quality of the photography obtained has a direct influence on the amount of detail that an interpreter can derive from the photographs. The quality depends on the products used and the various processes the images pass through. Proper selection and control of the film-filter combinations require close cooperation between a knowledgeable interpreter who can specify the requirements of the end product needed and an experienced contractor who knows the capabilities and limitations of the various systems and can aid in selecting the optimum combination for the specific job.

Availability and access to existing photography are excellent in the United States and Canada and in much of the rest of the world. In the United States, most aerial photography has been taken for federal and state agencies. Many areas in the United States have been covered several times by one or more agencies. The National Cartographic Information Center (NCIC) of USGS publishes a catalog, *Aerial Photography Summary Record System Catalog*, of existing, in-progress, and planned aerial photography of the conterminous United States held by several federal agencies. The catalog includes a series of outline map indexes, indicating coverage within a 7.5 × 7.5-min geographic cell, and a computer listing on microfiche film that shows amount of cloud cover, camera used, specific scale, date of coverage, film type and format, and agency holding the photography. The NCIC address and other U.S. and Canadian agencies that have large holdings of aerial photography are given below. Several of the government agencies whose holdings are listed in the NCIC catalog also publish status maps of aerial photography held by their organizations. For example, the

Agricultural Stabilization and Conservation Service (ASCS), U.S. Department of Agriculture, periodically publishes the *ASCS Aerial Photography Status Maps* and the *Comprehensive Listing of Aerial Photography*, which indicate the coverage available from ASCS for each county in each of the 50 states. The National Archives in Washington, D.C., is the depository of all nitrate-base photography collected during the years 1936 to 1941. The EROS Data Center has USGS and National Aeronautics and Space Administration photography.

National Cartographic Information Center
U.S. Geological Survey
507 National Center
Reston, Virginia 22092

Aerial Photography Field Office
Agricultural Stabilization and Conservation Service
U.S. Department of Agriculture
2222 West, 2300 South
Post Office Box 30010
Salt Lake City, Utah 84125

Forest Service
U.S. Department of Agriculture
Washington, D.C. 20250

Headquarters, Defense Mapping Agency
U.S. Naval Observatory
Building 56
Washington, D.C. 20305

Coastal Mapping Division, C-3415
National Ocean Survey
National Oceanic and Atmospheric Administration
6001 Executive Boulevard
Rockville, Maryland 20852

EROS Data Center
U.S. Geological Survey
Sioux Falls, South Dakota 57198

Audiovisual Department
National Archives and Records Service
Eighth Street and Pennsylvania Avenue, NW
Washington, D.C. 20408

Maps and Surveys Branch
Tennessee Valley Authority
210 Haney Building
Chattanooga, Tennessee 37401

National Airphoto Library
Canada Department of Energy, Mines and Resources
601 Booth Street
Ottawa, Canada K1A0E8

Several state highway and transportation departments similarly have available for public review index maps that indicate the type and extent of coverage held by state organizations. Some state agencies publish lists of all the aerial coverage available for their states from all sources (*3.2*).

Using existing photography in a landslide investigation is of course less expensive than acquiring new coverage. Existing photographs, however, may be several years old, may not show the present terrain conditions, or may be at a scale too small for performing detailed analyses, and new coverage may thus be necessary. The existing photography can then be useful in the initial analysis and planning for the new coverage.

When the organization performing the investigation does not have the facilities for obtaining new photographic coverage, the best approach is to engage a reputable firm with a base of operation close to the study site. In some cases, a small plane can be hired locally, and a hand-held camera can be used to make oblique photographs that are economical and suitable for preliminary analysis and planning.

Natural Factors

Consideration of natural factors is important in planning aerial photography for landslide studies. Flight planning is certainly controlled by meteorological conditions, but consideration must also be given to time of year and time of day in which the photography is obtained.

Meteorological conditions include cloud cover, haze, turbulence, and solar angle, all of which affect the quality of definition of images on the film. Suitable conditions for color photography are when the sun angle exceeds 30° from the horizontal, cloud cover is less than 10 percent, visibility at aircraft altitude when one looks toward the sun exceeds 24 km (15 miles), and the atmosphere is free of turbulence (*3.30*, p. 43).

Time of year is also critical to the quality of aerial photography. Although no clear-cut recommendations can be made to fit all situations or localities, some of the desirable conditions are discussed below.

1. Optimum soil-moisture contrasts. The optimum contrast of soil moisture occurs under two different conditons: (a) when the soils are wet but not totally saturated or (b) when the moisture levels are low. The former occurs in the spring or early summer because of high water conditions. Wet areas and seepage zones present in unstable or potentially unstable areas are directly visible outcropping on hillsides or are evident by the presence of luxuriant vegetative growth over those wet areas. The latter occurs during dry periods in late summer or fall or over disrupted areas. During dry periods, vegetation in wet areas or seepage zones has more growth and color than that in areas that have normal moisture levels. Areas disrupted by cracks and fissures are better drained and thus drier, and vegetation located over those areas shows evidence of stress earlier. Contrasting vegetation and moisture conditions are best noted on color infrared photography, as illustrated in Plates 3.2 and 3.3.

2. Optimum ground cover. A minimal cover of snow and tree foliage is desirable. In humid, temperate climates, this usually occurs in early spring after the snow melts and before the deciduous trees leaf or in late fall after the leaves fall and before snow falls.

3. Optimum shadow conditions. Photographs taken when the sun angle is high and shadows on the hillsides and slopes are minimal are the best for interpretation. However, in some special cases, such as in areas of low topography, small slope failures are enhanced by the presence of long shadows produced when the photography is taken with a low sun angle (either early in the morning or late in the afternoon).

These optimum conditions are not always present at one time, and a compromise may be required to obtain as many of the ideal conditions as possible. Some detailed investigations may require photographs to be taken several times during the year.

Principles of Aerial-Photograph Interpretation

The interpretation of aerial photographs involves three major steps:

1. Examine the photographs to get a three-dimensional perception,
2. Identify ground conditions by observing certain elements appearing in the photographs, and
3. Using photo-interpretation techniques, analyze specific problems by the association of ground conditions with one's background experiences.

The quality and the reliability of any interpretation are, of course, enhanced in direct ratio to the interpreter's knowledge of the soils and geology of the area under study. The acquirement of such knowledge, either by field examination or by study of available maps and reports, should, therefore, be considered an essential part of any photo-interpretation job.

The technique used in interpreting aerial photographs is referred to as pattern analysis. This approach is based on the premise that the landforms, the basic units of the landscape, have distinct patterns in aerial photographs. Those landforms developed by the same geologic processes and in the same environmental setting will have similar patterns, while those developed by different processes will have different patterns. These patterns are composed of several major elements that are evaluated in the process of interpretation: topographic expression, drainage, erosion, soil tones, vegetation, and culture. These elements are discussed briefly in the following paragraphs; more thorough treatments are given by Lueder (*3.15*), Way (*3.38*), and Colwell (*3.7*).

Topographic Expression

The features of importance in describing topographic expression include topography, shape, and relative relief. Topography represents the three-dimensional aspect of the landscape, such as hill, basin, ridge, or mountain. Shape refers to a unique character of the landscape, such as conical hill, sinuous ridge, or A-shaped mountain, that aids in describing or separating different forms of topography. Relative relief refers to the comparative position of the landform in relation to other landforms in the vicinity, for example, the position (or relative relief) of the sloping fan-shaped plain lying at the foot of A-shaped hills and sloping downward to a level basin area. In this example, three different landforms are indicated based on their relative relief: the mountainous area, the sloping plain area, and the basin area. Topographic expression of landforms is easily determined by a stereoscopic examination of the photographs. In such an examination, various landforms can be separated and clues obtained as to the nature and stability of the materials constituting the landform.

Drainage and Erosion

The density and pattern of drainage channels in a given area directly reflect the nature of the underlying soil and rock. The drainage pattern is obvious in some cases, but more often the channels must be traced on a transparent overlay to allow the pattern to be successfully studied. Under otherwise comparable conditions, a closely spaced drainage system denotes relatively impervious underlying materials; widely spaced drainage, on the other hand, indicates that the underlying materials are pervious. In general, a treelike drainage pattern develops in flat-lying beds and relatively uniform material; a parallel stream pattern indicates the presence of a regional slope; rectangular and vinelike patterns composed of many angular drainageways are evidence of control by underlying bedrock; and a disordered pattern, interrupted by haphazard deposits, is characteristic of most glaciated areas. Other patterns have developed in response to special circumstances. A radial pattern, for instance, is found in areas where there is a domal structure, and a featherlike pattern is common in areas where there is severe erosion in rather uniform silty material, such as loess.

The shapes of gullies appearing in aerial photographs provide valuable information regarding the characteristics of surface and near-surface materials. Typically, long, smoothly rounded gullies indicate clays; U-shaped gullies, silts; and short, V-shaped gullies, sands and gravels.

Soil Tones

Soil tones are recognizable in photographs unless there is a heavy vegetative cover. On black-and-white photographs, the tones are merely shades of gray, ranging from black to white. Because gray tones are highly respondent to soil-moisture conditions on the ground, they are an important aerial-photographic element in landslide investigations.

A soil normally registers a dark tone if it has a high moisture content and a light tone if it has a low moisture content. The moisture condition is the result of the physical properties of the soil, its topographic position, and the position of the groundwater table. The degree of sharpness of the tonal boundary between dark and light soils aids in the determination of soil properties. Well-drained, coarse-textured soils show distinct tonal boundaries whereas poorly drained, fine-textured soils show irregular, fuzzy tonal boundaries.

Color photography greatly increases the number of tones that can be distinguished by the interpreter and improves the amount of detail and accuracy of the information derived. In addition, the use of special visual, photographic, or electronic enhancement techniques increases the effectiveness of analysis of the tonal pattern element.

Vegetation

Vegetative patterns reflect both regional and local climatic conditions. The patterns in different temperature and rainfall regions can be recognized in aerial photographs. Locally, a small difference in soil moisture condition is often detected by a corresponding change of vegetation. A detailed study of such local changes is helpful in landslide investigations. For instance, areas of wet vegetation, represented on the photographs by dark spots or "tails," are a clue to seepage on slopes. Cultivated fields, as well as natural growths, are good indicators of local soil conditions. Thus, an orchard is often found on well-drained soils; black ash, tamarack, and white elm are normally found on wet, fine-grained soils; and vegetation is sparse in nonproductive serpentine soils, where landslides are common.

The use of color photography, especially color infrared, greatly facilitates vegetation analysis and thus the overall interpretation. The vigor of vegetative growth and the presence of vegetation stress due to detrimental soil and moisture conditions are best noted on color infrared photography. Point 2 in Plates 3.2 and 3.3 shows areas of seepage, as indicated by lush vegetation, and point 3 indicates vegetation under stress on the unstable slide debris.

Culture

Cultural patterns reflect how humans adjust to the natural terrain on which they live and work. An understanding of these patterns and the reasons for their presence can be valuable indicators of soil, moisture, and other terrain conditions. For example, drainage ditches and tile fields indicate the presence of a high water table; steep, vertical cuts along highways denote the presence of bedrock; sloping cuts usually typify unconsolidated soils; and contour farming or an irregular wavy plowing pattern characterizes sloping terrain or erodible soils.

Interpretation of Landslides on Aerial Photographs

As previously described in the section of this chapter on landforms susceptible to landslides and the section on vulnerable locations, landslides are more prevalent in certain landforms and occur most frequently at certain vulnerable locations. The interpretation of aerial photographs by the pattern-analysis technique is a valuable tool for delineating landslide-susceptible landforms and pinpointing the vulnerable locations. The identifying pattern elements on the photographs and significant features about these landforms are summarized in the following discussion and illustrated in the accompanying figures.

Diagnostic Patterns of Landslides on Aerial Photographs

An investigator already familiar with the appearance of landslides on the ground should become oriented to the aerial view of landslides by an examination of photographs of some known examples. The difference between an aerial view and a ground view results chiefly from the fact that the former gives a three-dimensional perspective of the entire slide area, but at a rather small scale. Ground photographs, on the other hand, show only two dimensions, but on a larger scale.

The following features discernible on aerial photography are typical of landslides or landslide-susceptible terrain, but not all features are evident for each landslide. Most of these features are shown in Figure 3.14; those features not evident are shown in other figures in this chapter. The numbers in Figure 3.14 correspond to those in the list below.

Figure 3.14. Typical characteristic features of a landslide as shown in aerial photographs (*3.6*). Slide occurred in Yakima County, Washington. Numbered items are discussed in text.

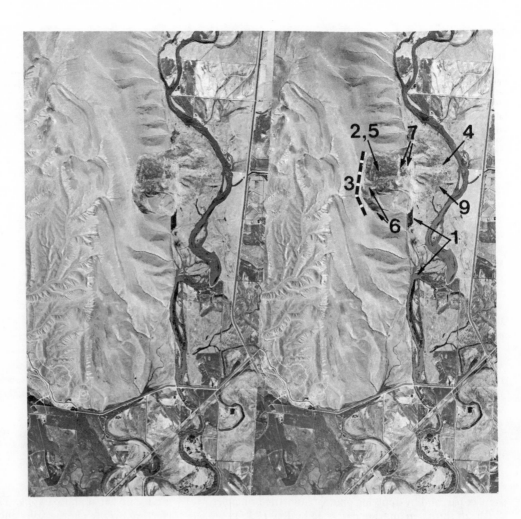

1. Land masses undercut by streams (Plate 3.1);
2. Steep slopes having large masses of loose soil and rock;
3. Sharp line of break at the scarp (head end) or presence of tension cracks or both;
4. Hummocky surface of the sliding mass below the scarp;
5. Unnatural topography, such as spoon-shaped trough in the terrain;
6. Seepage zones;
7. Elongated undrained depressions in the area;
8. Closely spaced drainage channels;
9. Accumulation of debris in drainage channels or valleys;
10. Appearance of light tones where vegetation and drainage have not been reestablished (Figure 3.1 and Plates 3.2 and 3.3);
11. Distinctive change in photograph tones from lighter to darker, the darker tones indicating higher moisture content;
12. Distinctive changes in vegetation indicative of changes in moisture (Plates 3.2 and 3.3); and
13. Inclined trees and displaced fences or walls due to creep (Plate 3.3).

In some cases the slide itself is not discernible, but indirect evidence of its existence is noted. For example, where a highway is built on unstable soil, the irregular outlines and nonuniform tonal patterns of broken or patched pavements are often visible (Figure 3.15).

Identification of Vulnerable Locations

Many slides are too small to be detected readily in photography at the scales normally available (i.e., 1:15 000 to 1:40 000). Consequently, the photographs should be closely examined for signs that indirectly indicate the presence of slides or, if signs are not visible, for the vulnerable locations where slides usually occur. Typical vulnerable locations include areas of steep slopes, cliffs or banks being undercut by stream or wave action, areas of drainage concentration, seepage zones, areas of hummocky ground, and areas of fracture and fault concentrations. The characteristics of these locations have been discussed in an earlier section of this chapter in which vulnerable locations are described.

Aerial photographs are valuable aids in identifying the vulnerable locations. The shape and slope of the terrain are readily discernible from the stereoscopic examination of the photographs. In fact, the vertical appearance of the terrain is exaggerated when viewed with a lens stereoscope. Moderate slopes appear steep, and steep slopes appear almost vertical, making them easier to delineate. In addition, the slopes can be measured on the photographs by using

55

Figure 3.15. Presence of limited sections of patched pavement (1) in generally unpatched roadway can be indicative of possible slope movement in area. Presence of fissures on hillside above this patched section (2) in Washington County, Ohio, confirms instability of slope.

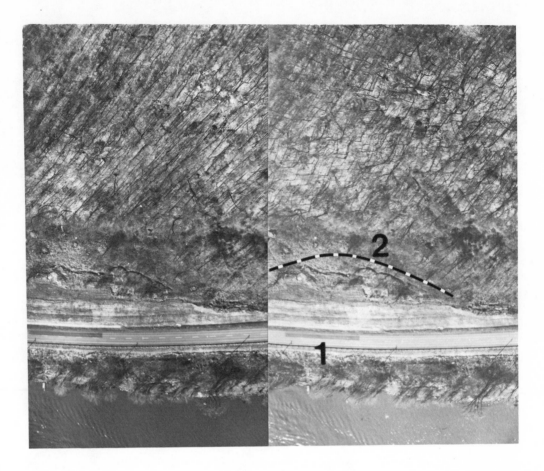

simple measuring devices, such as an engineer's scale or a parallax bar, and by applying photogrammetric principles. Details of this technique are described by Ray (3.23).

The presence of drainageways, seepage zones, fractures, and fault zones is also readily evident on aerial photographs. By means of stereoscopic examination, the complete drainage network can be mapped, including the intermittent streams and small gullies. The presence of wet zones or seepage areas is evidenced by darker tones caused by the higher moisture content in the soils or by a more luxuriant vegetative growth over the wet areas. Areas of drainage or water concentration above a slope should be closely examined because they are vulnerable locations. Subsurface seepage from these areas can lead to slope failures; Figure 3.4 shows such a case. Fracture and fault zones are indicated on the photographs by dark linear or curvilinear lines. The darker tones are usually due to the better growth of vegetation along the fracture zones where it is easier for the roots of plants to grow and where moisture levels are usually higher. In delineating fracture zones, care must be taken not to interpret man-made features, such as fence lines or field boundaries, as fracture zones. Generally, features having straight lines, right-angle intersections, and standard geometric patterns are man-made. Several of the above features are illustrated in succeeding figures that show landslides in various landforms.

Landforms/Geologic Materials Susceptible to Landslides

Landslides are rare in some landforms and common in others.

Most of the forms susceptible to landslides are readily recognizable on aerial photographs. The identifying elements on black-and-white photography and significant facts about the landforms are summarized and illustrated in the following sections. Table 3.2 gives the landslide potential of various landforms/geologic materials grouped on the basis of their topographic form. The order of presentation in this section follows a sequence based on origin and character of the materials.

Sedimentary Rocks and Their Residual Soils

The discussion of rocks and their residual soils is combined in this and the following two sections because the recognition of types of residual soils depends primarily on the recognition of the landform developed on the parent rocks. The determination of depth of residual soil requires considerable judgment. However, investigators working constantly in their own regions should have no difficulty in estimating the depth once they are familiar with local conditions. In general, rounded topography, intricate, smoothly curving drainage channels, and heavy vegetation are indicators of probable deep soils; in contrast, sharp, steep, resistant ridges and rock-controlled angular channels are commonly found in areas of shallow soils. The local climatic and erosion patterns should be considered in the interpretation.

Many landslides occur in residual soils and weathered rocks; they are usually in the form of slides or flows. For example, in a study in North Carolina (3.12), two-thirds of the slides involved in cut-slope failures occurred in the

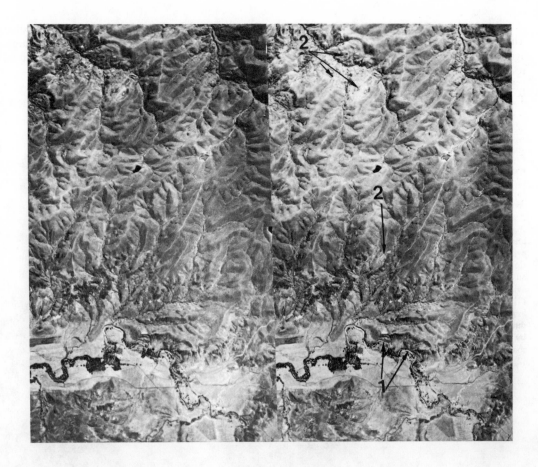

Figure 3.16. Pierre shale, Lyman County, South Dakota, a formation that is most highly susceptible to landslides (*3.4*). Numerous slides are evident throughout area, some due to undercutting by streams (1) and others to steep slopes, seepage, or stress release along joints (2).

weathered soil material. Of the sedimentary rocks, shales are the most susceptible to landslides. A typical example is shown in Figure 3.16. Shales are especially susceptible when interbedded with pervious sandstones or limestones or other pervious rock types. This geologic setting of alternate layers of pervious and impervious materials on a slope is conducive to sliding.

Landslides in massive horizontal sandstones or limestones are uncommon unless they are interbedded with shales or other soft rocks (Figure 3.17). In steeply tilted positions, any sedimentary rock may fail by sliding (Figure 3.18). Depending on the dip angle, joint system, and climate, slides may take one or a combination of forms, such as rock falls, rock slides, topples, block slides, slumps, debris falls, debris slides, and earth flows. River undercutting and artificial excavation are important factors in initiating landslides in both horizontal and tilted rocks (Figures 3.3 and 3.16).

Methods of identification of sedimentary rocks in aerial photographs are well established.

1. Hard sandstones are evidenced by their high relief, massive hills, angular and sparse drainage, light tones, and general lack of land use.
2. Clay shales are noted for their low rounded hills, well-integrated treelike drainage system, medium tones, and gullies of the gentle swale type. They are generally farmed in regions with sufficient rainfall.
3. Soluble limestones are characterized by low rolling topography with sinkhole development in temperate, humid areas, by rugged karst topography in some tropical regions, by a general lack of surface drainage, and by mottled light

and medium dark tones. They usually are heavily cultivated in humid areas.

4. Interbedded sedimentary rocks show a combination of the characteristics of their component beds. When horizontally bedded, they are recognized by their uniformly dissected topography, contourlike stratification lines, and treelike drainage; when tilted, parallel ridge-and-valley topography, inclined but parallel stratification lines, and trellis drainage are evident.

The identification of landform as a means of detecting associated landslides is important in the flat-lying sedimentary group because the slides there are often small and, therefore, not very obvious on aerial photographs. This is particularly true for slides in colluvial deposits formed from sedimentary rocks. These conditions are common since sedimentary rocks are the most widespread of all surface rocks.

Igneous Rocks and Their Residual Soils

Basaltic lava flows are among the most common extrusive igneous rocks, and they are readily identifiable on aerial photographs. Basalts and other volcanic flows are highly susceptible to different types of landslides (Figures 3.19 and 3.31). They often form the caprock on plateaus or mesas; their sharp, jagged (saw-toothed) cliff lines are clearly visible on photographs. Surface irregularities or flow marks, sparseness of surface drainage and vegetation, and dark tones are confirming characteristics.

If a basalt flow is underlain by or interbedded with soft layers, particularly if it occupies the position of a bold es-

57

carpment, a favorable condition for large slumps is present. The joints and cracks in basalt give rise to springs and seepage zones and greatly facilitate movement. Rock falls and rock slides along rim rock are usually favored by vertical jointing of basalt and by undercutting of basaltic cliffs. Talus accumulations of various magnitudes are found at the foot of cliffs. Disturbance of talus slopes during road construction has caused some large slides of talus materials. Old slides and fissures indicating incipient slides often can be seen on photographs. In areas of relatively deep weathering, the landscape is somewhat modified. A more rounded topography and heavier vegetation develop, although dark

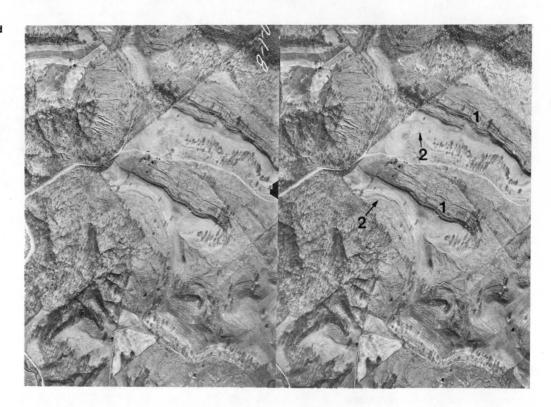

Figure 3.17. Interbedded sedimentary rocks (*3.22*). Hummocky topography, heavy erosion, and extensive drainage characterize shale material underlying sandstone cap rock in Morgan County, Ohio. Almost all hillsides show evidence of instability (1). Evidence of slides in capping sandstone due to undermining by sliding out of the shales is also apparent (2).

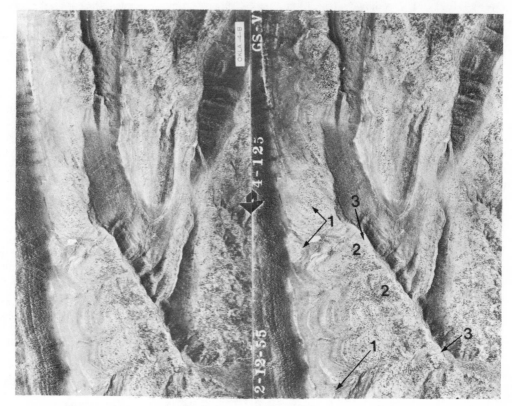

Figure 3.18. Synclinal structure composed of sandstone and shale bedrock provides steeply dipping slopes ideal for slides (*3.8*). No large slide masses are apparent in this Le Flore County, Oklahoma, area, but evidence of mass wasting is indicated by oversteepened slopes lacking vegetation and having light tones (1); bulging of lower slopes (2); and accumulation of loose scree materials at base of slopes (3).

tones still predominate. Slumps of both large and small size are common in basaltic soils.

Granite and related rocks constitute the most widely occurring intrusive igneous rock types. The landslide potential of granitic rocks varies widely, depending on the composition of the rock and its fracture pattern, the topography, and the moisture conditions. In granites that are highly resistant to weathering or are of low relief, there is generally no slide problem. In hilly country where the granite is deeply weathered, slumps in cut slopes, as well as in natural steep slopes, are common. Fractures in the rock and high moisture conditions undoubtedly are favorable factors in producing landslides. Figure 3.20 shows an example of an extensive slide in granitic materials. Granitic masses are identified on aerial photographs by the rounded (old) to A-shaped (young), massive hills and by the integrated tree-like drainage pattern with characteristic curved branches. The presence of criss-cross fracture patterns and light tones and the absence of stratification and foliation aid in confirming the material.

Metamorphic Rocks and Their Residual Soils

As reported by Leith and others (*3.12*) and other investigators, the frequency of landslides per unit area is greater in metamorphic rocks than in most other rock types. The presence and attitude of the foliation and joints in these rock types greatly influence the stability of the slopes. Although the characteristics of the major types of metamorphic rocks—gneiss, schist, slate, and serpentinite—have been identified on aerial photographs for specific areas, they are not sufficiently consistent to develop typical pattern elements. Nevertheless, the presence and attitude of the foliation and joints

can be distinguished on aerial photography by an indication of banding or lineation of the topography and drainage network, and landslides, when present, are clearly evident. Figure 3.21 shows an example of landslides in metamorphic terrain (mica schists). The attitude of the foliation seen at point 3 indicates that the dip of these layers may have been a significant factor in the presence of numerous slides (point 1) found just across the river where the layers are steeply dipping toward the river.

Within the metamorphic group, many slides are associated with serpentinite areas, which are identified on aerial photographs by sinuous ridges, smoothly rounded surfaces, short steep gullies, poor vegetative cover, and dull gray tones. Within a general area, local conditions, such as vegetation, moisture, and slope, may create special, favorable circumstances for landslides. In general, low relief and low rainfall are among the factors responsible for the stability of some of the serpentinite slopes.

Glacial Deposits

Landslides are common in some glacial and glaciofluvial deposits. Although most of the distinct glacial forms are easily identified on aerial photographs, there are complex areas that require a high degree of skill for their identification. Moraines are found in nearly all glaciated areas. They are identified on aerial photographs by jumbled, strongly rolling to hilly terrain. Moraines, particularly in semiarid areas, contain a large proportion of untilled land. Deranged drainage patterns, irregular fields, and winding roads are confirming clues. Minor slumps, debris slides, and earth flows are common in cut slopes in moraines as a result of undrained depressions and seepage zones in the mass. Because morainic hills are usually small, these slides are not exten-

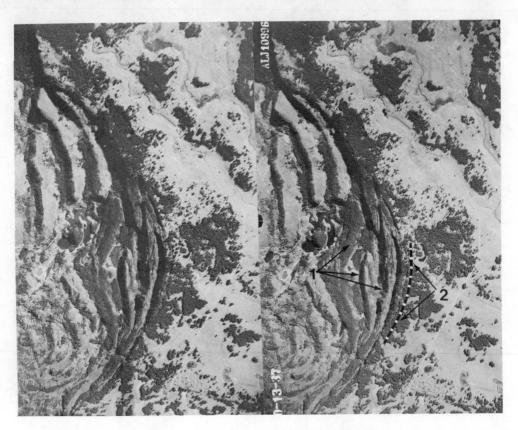

Figure 3.19. Basalt cap rock overlying impermeable shale deposits has resulted in massive landslides in Mesa County, Colorado. Large blocks of material have toppled and slumped off scarp face (1). Presence of incipient slide is indicated by cracks on crown of slide (2).

sive. They are, nevertheless, large enough to cause continual trouble to many maintenance engineers (Figure 3.22).

In areas subject in part to mountain glaciation, transportation routes must follow valleys formerly occupied by glaciers and be built on their deposits. Under these conditions slides in moraines and colluvium are common. Slides also occur in the shallow mantle overlying bedrock and take the form of slumps, debris slides, and debris falls; they often contribute to failures in artificial fills placed on them. The landslides most commonly occur along valley walls that

have been oversteepened by glaciation (Figure 3.23). The topography of such areas is basically that of the underlying bedrock with slight local modifications, depending on the thickness of the mantle.

Slides seldom occur in other glacial or glaciofluvial deposits, such as kames, deltas, eskers, outwash plains, and till plains. They have occurred, however, in places where these materials are underlain by fine-grained impervious materials, such as lake-bed deposits, when these deposits are present in elevated positions (Plates 3.4, 3.5, and 3.6).

Figure 3.20. Large debris slide (1) with remnant scar (2) in granodiorite rocks on east front of Sierra Nevada Mountains just east of Lake Tahoe in Washoe County, Nevada. An estimated 95 Mm^3 (125 million yd^3) was involved; portions of toe of slide are buried under younger alluvium (*3.36*).

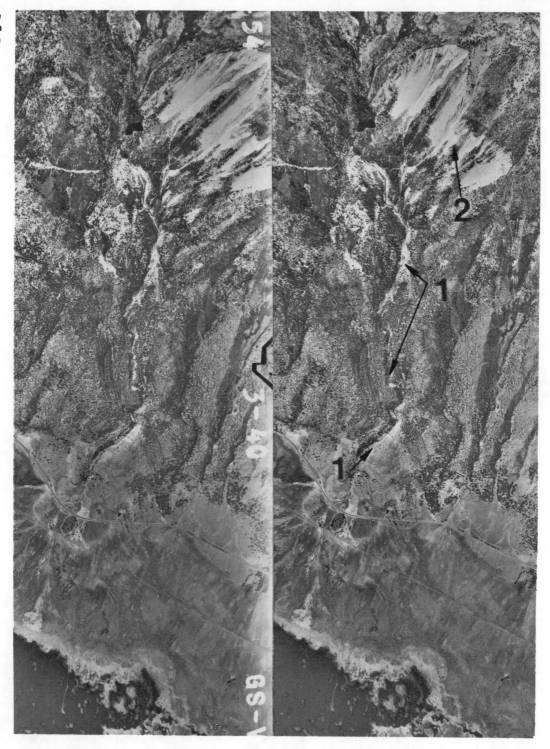

Water-Laid Deposits

Water-deposited landforms most susceptible to landslides include coastal plain deposits, river terraces, and lake beds.

Coastal plain deposits can be subdivided into three distinct subforms for analysis by aerial photographic methods: the dissected or upper coastal plain, the undissected or lower coastal plain, and the beach zone. Landslides are uncommon in the beach zone. The lower coastal plain is identified by its low, flat topography; its association with tidal flats, marshes, and swamps; and the presence of broad, shallow, tidal stream channels. The upper coastal plain is identified by its rolling to rugged topography; an integrated drainage system with wide, vegetation-filled main drainageways; and irregular land-use patterns. It is also associated with coastal features and appears on aerial photographs to be somewhat similar to areas underlain by consolidated sedimentary rocks. In the lower coastal plains, landslides offer a problem only in the construction of canals or similar structures that require deep excavation in flat lands. In the upper

Figure 3.21. Typical landslide topography in weathered mica schists evident on valley walls on both sides of Little Salmon River in Idaho County, Idaho. Old slide scars with hummocky terrain below can be seen (1). Slide debris has covered road many times (2) necessitating high maintenance costs and final relocation. Presence and attitude of foliation are also apparent (3).

Figure 3.22. Moraine and lake bed, Tompkins County, New York (3.10). New slumps and debris slides, as well as old slide scars, are common occurrences in this glaciated valley where postglacial erosion has dissected moraines and overlying lake deposits. Morainal deposits (1) are recognized by their irregular, jumbled topography. Overlying lake deposits have smoothed topographic appearance; their silty clay composition is characterized by smooth slopes, high degree of dissection, and gradual change of gray tones (2). Prominent old slump is visible (3). Almost all highway excavations experience slumps and debris slides (4). Although deep-seated and large-scale slides are not common in such an area, continual maintenance work in clearing sliding material and in protecting slopes from erosion has been necessary.

61

coastal plains, however, slumps in natural hill slopes, as well as in road cuts, are common (Figure 3.24). The stratified and unconsolidated nature of the sands, silts, and clays that characterize most coastal plains provides a favorable situation for landslides.

Terraces are easily recognized on aerial photographs as elevated flat land along major or minor valleys. Terraces of gravel and sand are usually stable, maintaining clean slopes on their faces. However, where terraces are composed of fine sands or interbedded silts and clays or where the natural equilibrium is disturbed by artificial installations, slides will occur. Slides in terraces naturally start on unsupported slopes facing the low land (Figure 3.25).

Lake-bed deposits generally display flat topography unless they are dissected. Although generally composed of clays, lake beds have little chance to slide except when exposed in valley walls or cuts (Figure 3.26). Slides of considerable magnitude have occurred in lake clays under each of the following circumstances: (a) where lake clays are interbedded with or, especially, are overlain by granular de-

Figure 3.23. Thin till deposit overlying bedrock in glaciated valley in Broome County, New York. Highway cuts in hillsides oversteepened by glaciation create slides (A and B). Slides generally extend to full depth of soil material until competent rock bed (in this case sandstone) is encountered. Between A and B, section of highway entirely in rock has not experienced any slides. In this respect, landslides serve as indicators of depth of soil to rock. On other side of river, site of future slide is indicated at C, an area of concentration of drainage, loose fill, and lack of bedrock control.

posits and (b) where lake clays overlie bedrock at shallow depth and the base level of erosion of the area is generally lowered. Examples of flow slides in these types of deposits are shown in Figures 3.27, 3.33, and 3.34. Undissected lake clays are easily identified in aerial photographs by their characteristic broad level tracts, dark gray tones, and artificial drainage practices. Dissected and complex lake-bed areas are relatively difficult to identify, particularly by those not familiar with the local geologic conditions. Again, the presence of existing slides is the most reliable warning signal.

Eolian or Wind Deposits

Loess, or wind-deposited silt, can be identified unmistakably in aerial photographs by its vertical-sided gullies, which are evenly spaced along wide, flat-bottomed tributaries to show a featherlike drainage pattern. Such a landform is confirmed by equal slopes on hills and valleys (an indication of uniform material), heavy vegetative cover in dissected areas, extensive farming in undissected areas, and soft gray tones (Figure 3.28). Earth flows and minor slumps, generally referred to as cat

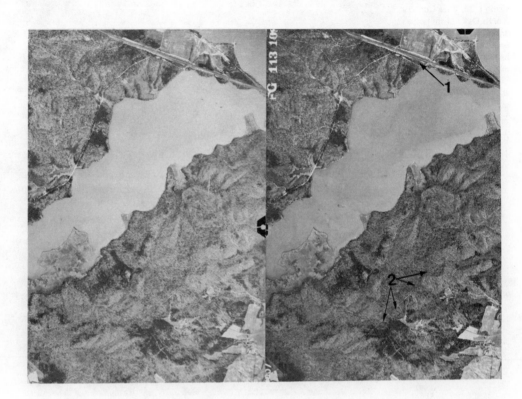

Figure 3.24. Dissected coastal plain in Prince Georges County, Maryland, is identified by low, soft hills and associated tidal channels (more evident when larger, adjacent areas are examined) (3.10). Most railroad (1) and highway cut slopes in this region have experienced slide problems. Cut slopes steeper than natural slopes are susceptible to slides unless adequate precautions are taken. For highway locations, grade line would better be set below critical clay layers if at all possible so that, even if slide occurs, foundation of road would not be affected. Presence of amphitheater scarp hillsides (2) with highly eroded and hummocky topography below indicates natural erodibility and instability of materials in this area.

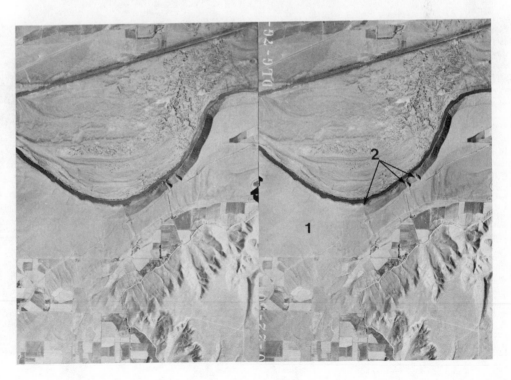

Figure 3.25. Sandy terrace (1) overlies basalt flow along Snake River in Elmore County, Idaho. Several dry sand flows (2) indicating sandy nature of materials are evident on terrace face.

steps, are commonly found in loess. The cat steps appear as fine, roughly parallel, light tone contours on the aerial photographs. Because of their small size, they are not always evident. The individual steps of these small slumps are commonly about a meter wide and several centimeters to a meter high. These subtle features are shown in Figure 3.29.

Gravity Deposits

Loose, unconsolidated talus and colluvial materials formed by weathering of parent soil and rock materials and moved downslope by gravity are found on steep slopes. These deposits are easily identified on aerial photographs as bare slopes in mountainous areas, but they are not so obvious on

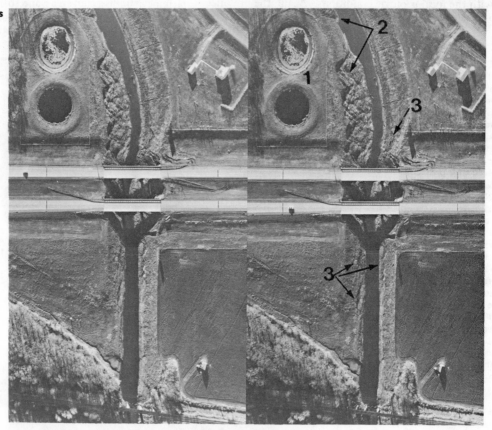

Figure 3.26. Sewage lagoons (1) at Harwood Rest Area along I-29, Cass County, North Dakota, may be endangered by slides occurring along stream bank just below (2). Lake-bed deposits in this area are highly susceptible to slides as shown by presence of other slides along stream bank (3).

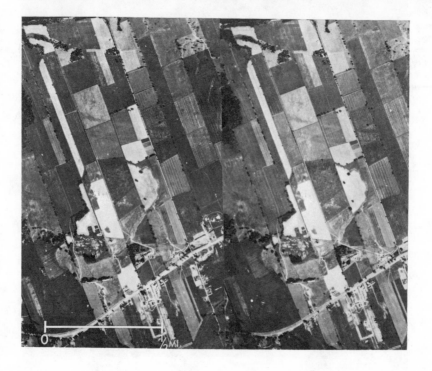

Figure 3.27. Example of earth flow in sensitive clays along Ottawa River east of Ottawa, Canada (3.17).

64

Figure 3.28. Highly dissected nature of the loess terrain in Lincoln County, Nebraska, illustrates great erodibility of these materials (3.10). Presence of earth flows (1) is indicated by broad amphitheater scarp faces at crown, steep side slopes of gullies, and light-toned deposits in gulley bottoms. Numerous little erosional scars and slumps can be seen on hillsides (2). Section (3) is shown enlarged in Figure 3.29.

Figure 3.29. Cat steps can be seen on this enlarged section of area shown in Figure 3.28. Because of their small size, they appear as thin, wavy contours on hillside.

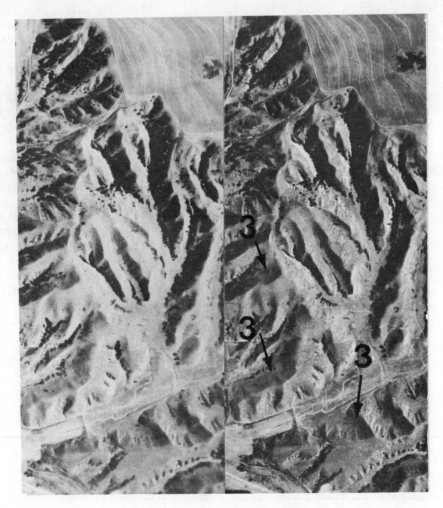

vegetated lower slopes. Excavation into these materials usually creates stability problems. Old landslide deposits are themselves part of this group (Figure 3.30).

Complex Forms

Most of the landforms previously described can be called simple forms because they consist predominantly of one type of material in each unit. In nature, however, complex or superimposed forms are numerous and of common occurrence, especially in glaciated areas, as mentioned previously. They are further emphasized here because of their significance in landslide studies. A change of material vertically or horizontally in complex areas often affects the internal drainage characteristics and creates slope stability problems. The common situation most favorable to slides is one in which pervious formations are underlain by relatively impervious beds.

Procedure for Landslide Investigation Using Aerial Photographs

The following procedure is recommended for photographic studies of landslides.

1. Lay out sites of planned facility on photographs.
2. Delineate areas that show consistent characteristics of topography, drainage, and other natural elements and classify into landform types. Large and obvious slides are identified at this stage.
3. Evaluate the general landslide potential of the landform types. Use Table 3.2 and the section on landforms susceptible to landslides as a guide.
4. Make a detailed study of cliffs or banks adjacent to river bends and all steep slopes. Compare slopes within the same landform type: For instance, slopes in bedrock landforms are more stable, even though steeper, than slopes in adjacent soil areas. Because slides are usually small in size on photographs, look carefully and inspect slopes in minute detail. Give particular attention to the following features:

Slide	Feature
Existing	Hillside scarps and hummocky topography
	Parallel spoon-shaped dark patches on hillsides
	Irregular outlines of highways and random cracks or patches on existing pavements.
Potential	Ponded depressions and diverted drainageways
	Seepage areas suggested by faint dark lines, which may mean near-surface channels, and fan-shaped dark patches, probably reflecting wet vegetation

Relatively new slides appear in light tones because vegetation and drainage are not well established. The parallel spoon-shaped dark patches on hillsides are likely to reflect vegetation in minor depressions. Lines drawn through the axes of the scarps in the slides often point to drainageways on higher ground that contribute to landslide movement.

5. Ground check all suspected slides.

Use of Other Remote-Sensing Systems

Multisensor investigations for terrain analysis and landslide investigations have been performed by several investigators, including Alföldi (3.1), Gagnon (3.11), Rib (3.26), Stallard and Myers (3.32), and Tanguay and Chagnon (3.33). The general consensus of those studies is that large-scale aerial photography (most preferred color) provides the most information on terrain conditions, specifically for landslide detection. Several of the reports did indicate, however, that satellite and infrared imagery offered some unique information that would prove useful for landslide investigations.

Satellite Imagery

Since July 1972, multispectral satellite coverage has been obtained for the United States, most of Canada, and many other areas of the world. Data are collected in four bands on an 18-d cycle:

Band	Color	Wavelength (μm)
4	Green	0.5 to 0.6
5	Red	0.6 to 0.7
6	Near infrared	0.7 to 0.8
7	Near infrared	0.8 to 1.1

A second satellite launched in 1975, as well as others scheduled for launching, provides data on a shorter time cycle. Each satellite scene covers an area of approximately 33 000 km^2 (10 000 miles^2) and has a resolution of about 80 m (250 ft). The satellite scene can be provided as a 70-mm film at a scale of 1:3 369 000; as a 185 × 185-mm (7.3 × 7.3-in) black-and-white print of each band; as a color composite at a scale of 1:1 000 000; and on computer-compatible tapes for computer processing. The scenes can also be provided on 35-mm slides.

In addition to the LANDSAT images provided by these two satellites, the National Aeronautics and Space Administration (NASA) has obtained photographic and image coverages of considerable areas of the United States and scattered areas of many foreign countries during the manned satellite SKYLAB missions (May 1973 through February 1974). SKYLAB data of special interest to landslide investigators are those collected by (a) the multispectral camera (S-190A) at a scale of 1:2 850 000 with 30 to 79-m (100 to 260-ft) resolution and (b) the earth terrain camera (S-190B) at a scale of 1:950 000 with 17 to 30-m (55 to 100-ft) resolution. Black and white, black-and-white infrared, color, and color infrared products are included. Detailed information about film types, spectral characteristics, and areal coverages has been published by NASA (3.21). Compared with LANDSAT products, SKYLAB products provide better resolution but lack universal coverage and repeated sequence.

At the small scale of satellite imagery, only extremely large landslides can be identified directly. Figure 3.31 shows a typical stereographic aerial photograph coverage of a large landslide zone. Figure 3.32 shows the satellite scene covering this landslide zone. The scalloped edges of the scarp slopes and the hummocky topography of this large slide are evident on the satellite scene.

Since most landslides are much smaller than the landslide shown in Figure 3.31, they are not directly identifiable on satellite imagery. However, the value of satellite imagery, as noted by Alföldi (3.1) and Gagnon (3.11), is that the

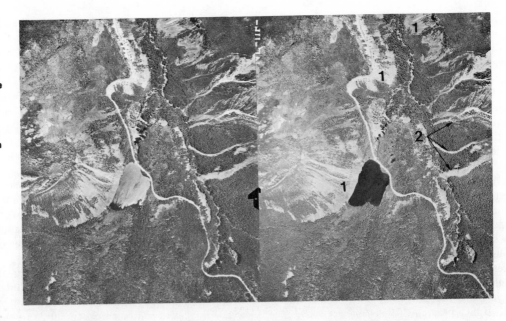

Figure 3.30. Unstable talus slopes in Yakima County, Washington, are indicated by steep slopes either bare of vegetation or with streaks of vegetation remnants (1). Some debris avalanches (2) are also evident on steep valley walls. Instability of slopes was recognized at time of location of road. Alternate location on opposite side of valley was considered worse because of numerous deep drainage channels in addition to talus slides.

landslide susceptibility of an area can be determined indirectly from some of the features that are identifiable at those scales. Regional physiography, geologic structure, and most landforms as well as land-use practices and distribution of vegetation are evident on the satellite imagery. These features in conjunction with the tonal patterns present on the imagery provide clues to the types of surface materials present, the surface moisture conditions, and the possible presence of buried valleys. Correlating these factors to geology and topography and using local experience in a region make it possible to rate the susceptibility of various areas to sliding. For example, Alföldi noted in his study of landsliding in eastern Ontario that on the satellite image the clay plains were easy to spot because they are almost 100 percent cultivated; the till plains were recognizable because they form a poorer agricultural area and field and forest sections are intermixed; and the elevated sand plains of the old Ottawa River delta (which overlay the clay plains) are kept mostly in forest.

An additional advantage noted by Alföldi for satellite imagery was the frequent coverage available. Seasonal changes in vegetative cover and moisture levels—as indicated by tonal changes—can be evaluated to increase the accuracy of interpretation of terrain conditions. Also, any changes noted during the year in the landslide-susceptible zones, such as urban expansion, clear-cutting of forest, forest fires, and draining of swamps, might presage renewed or new landslide activities. This could alert the interpreter to the necessity for a more detailed investigation in these areas.

Satellite photography and imagery for large parts of the world are available from the EROS Data Center of USGS. They can be ordered by providing the geographic coordinates (latitude and longitude) of the area of interest or by indicating particular frame numbers from the Single Landsat Coverage Map available from the above-noted organization. Canadian imagery can be obtained from the Canada Centre for Remote Sensing, Department of Energy, Mines and Resources. Addresses of these organizations are given earlier in this chapter. Experience has indicated that band 5 (0.6 to 0.7 μm) or the infrared color composite is the

most beneficial for landslide investigations.

Infrared Imagery

The infrared region extends from approximately 0.7 μm to 1 mm (Figure 3.13). Atmospheric interference, however, limits the areas within the infrared region available for investigation to only certain clear zones or "windows." Some of the common windows used in infrared surveys are in the following bands: 1.0 to 1.4 μm, 1.5 to 1.8 μm, 2.0 to 2.6 μm, 3.0 to 5.0 μm, and 8 to 14 μm. Daytime infrared surveys collecting data in the region below 3.0 to 3.5 μm record infrared reflectance phenomena from various objects in the scene. Daytime and nighttime surveys collecting data in the region above 3.0 to 3.5 μm record infrared heat-emission phenomena.

Since two basically different phenomena are being recorded in the infrared region (i.e., reflectance and heat emission), a distinction is made in the terminology used to indicate these phenomena. Images recorded in the bands below 3.0 to 3.5 μm are referred to as infrared reflectance and those recorded in the bands above as infrared imagery. The most useful window for terrain analysis is the 8 to 14-μm band. Further discussions on the characteristics of infrared surveys are given by Reeves (*3.25*) and by Rib (*3.26*).

Infrared imagery offers some unique information that cannot be obtained directly from the analysis of aerial photography. The combination of aerial photography and infrared imagery provides a more accurate and complete portrayal of terrain conditions than can be obtained from either system alone. Infrared imagery provides the following types of supplemental information that is valuable for evaluating existing landslide and landslide-susceptible terrain:

1. Surface and near-surface moisture and drainage conditions;
2. Indication of the presence of massive bedrock or bedrock at shallow depths;
3. Distinction between loose colluvial materials that are present on steep slopes and are susceptible to landslides

67

and the massive bedrock that is more stable on steep slopes; and

4. Diurnal temperature changes that occur in soil masses (these provide clues to the soil-water mass conditions).

Tanguay and Chagnon (*3.33*) demonstrated the value of infrared imagery and aerial photography for evaluating the moisture and drainage regime associated with a landslide. A flow slide had occurred within the crater of a former slide in clay lake beds in the vicinity of Saint-Jean-Vianney, Quebec. To plan a drilling program to evaluate the potential of further movements required that the areas of seepage, water runoff, and wet soils be identified. By means of photography alone, these features could not be uniquely separated from areas of standing water, topographic shadows, and dense vegetation (brush and forested zones) because they all produced similar dark tonal patterns. However, the combination of photography and daytime and nighttime imagery made it possible to separate these various features and identify the critical items for planning the drilling program.

Figures 3.33 and 3.34 show some of the results reported by Tanguay and Chagnon. Figure 3.33 shows daytime thermal infrared imagery obtained in the 8 to 14-μm band and black-and-white infrared photographs of a portion of the area shown in the imagery. Several of the seepage areas (points a and b) interpreted from this figure were drilled and indicated very deep and soft clay starting from the surface. The c points indicate both seepage and runoff in a farm field; the d points show standing water coming from snowmelt and surface drainage; and the e points depict the boundary of the recent slide. Point f indicates the uppermost boundary of an older slide surface. In this figure,

seepage areas, runoff, and standing water are dark on both images. On the predawn infrared imagery in Figure 3.34, standing water (points a and f) is warmer and has light tones while seepage zones and runoff (points c and d) are cooler and have darker tones. On the photography, all of these areas have dark tones. Vegetated areas occurring between the areas of standing water in this figure have a medium tone on the imagery and appear as white specks on a dark background on the photography.

Figure 3.35 shows a nighttime (predawn) infrared image of an area along a railroad line being investigated for locating potential areas for landslides. The railroad had been plagued for years with landslide problems. The circled darker areas were interpreted as zones of seepage and high moisture levels—potential for landslides. Based on this analysis, the circled areas were drained, and no further slides have occurred in those areas.

Another example of the value of infrared imagery is shown in Figure 3.36, which illustrates nighttime infrared imagery of an area of tilted sedimentary rocks. The massive bedrock areas (point 2) are indicated by light tones. The fractured rock zones and colluvial slopes, which are more susceptible to landslides, are indicated by medium dark tones (point 3). These types of data would be useful in conjunction with aerial photography for rating the landslide susceptibility of the terrain.

Other remote-sensing techniques, such as multispectral imagery and microwave radiometry, offer potential for indicating the presence of landslides. Multispectral systems obtain simultaneous coverage of data in several spectral regions—usually a portion of the ultraviolet, the visible, and both infrared reflectance and infrared emission regions. The data are collected on magnetic tape, which makes them

Figure 3.31. Old, massive landslide terrain in Rio Arriba County, New Mexico. Landslides (1) are of such magnitude that they are easily identifiable on common 1:20 000 aerial photographs as well as on smaller scaled photo-index sheets or even 1:1 000 000 satellite imagery (Figure 3.32). Most slides encountered in engineering projects are of smaller magnitude and difficult to identify at scales of 1:20 000.

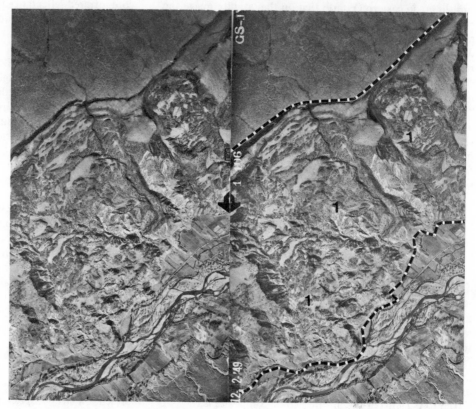

amenable to the use of computer analysis techniques. Microwave radiometers collect radiometric (temperature varying) data in the microwave or radar regions. The particular advantage of this system is that at the longer wavelengths information from subsurface layers is included in the data. Rib and others (*3.27*) have demonstrated that, under certain conditions, information on subsurface moisture conditions (i.e., presence of zones of high moisture level) can be determined by comparative evaluation of photography, nighttime infrared imagery, and microwave radiometry. These techniques are not described in detail because at this

time either further development is needed or the systems are too costly in comparison to the level of information furnished. However, they do offer some future potential for landslide investigations.

FIELD RECONNAISSANCE TECHNIQUES

The discussion in this section is based to a large extent on information in Chapter 4 in the book on landslides published by the Highway Research Board in 1958 (*3.10*).

Figure 3.32. LANDSAT-1 satellite scene 1657-17031, band 5, May 11, 1974, north central New Mexico. Arrow points to edge of small basalt remnant shown in Figure 3.31. Massive landslides are evident on satellite scene by irregular, scalloped edges. Hummocky appearance of slide material can be discerned by close examination of scarp edges.

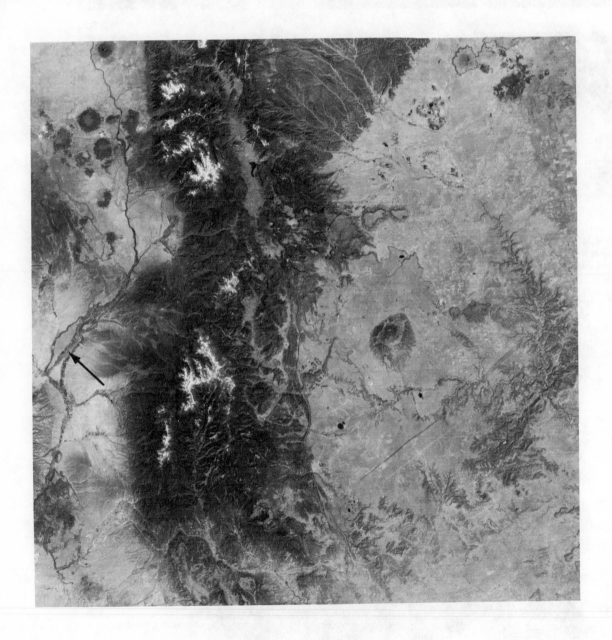

Figure 3.33. Daytime thermal infrared imagery in 8 to 14-μm band (above) and mosaic prepared from black-and-white infrared photography (below) taken in Ontario, Canada (3.33). Lines AB and A'B' represent equivalent coverages on the two image formats. Lettered areas are discussed in text.

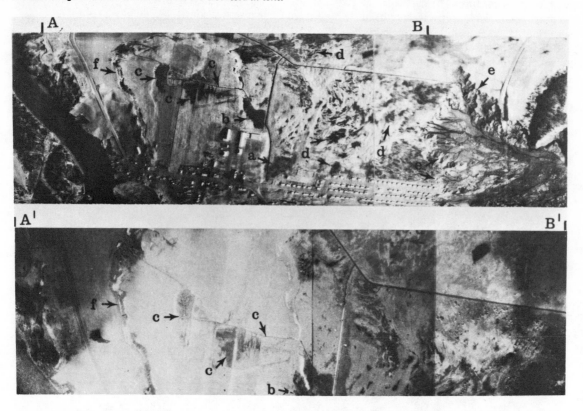

Figure 3.34. Predawn infrared imagery in 4 to 5-μm band (above) and mosaic prepared from black-and-white infrared photography (below) taken in Ontario, Canada (3.33). Line c-c represents common line on both images. Lettered areas are discussed in text.

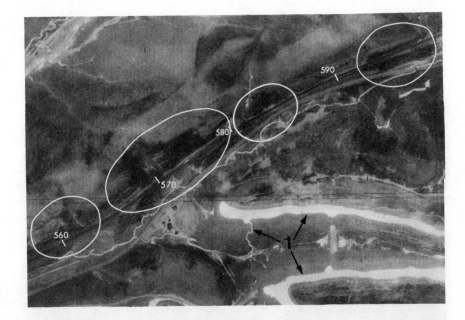

Figure 3.35. Nighttime infrared imagery, 0645 h, station 560 to 580, Muskingum Mine Railroad in Ohio. Dark-toned regions within circled areas are interpreted as zones of seepage and high moisture levels and have been associated with slides along railroad right-of-way. Light-toned areas (1) represent standing water on surface.

Figure 3.36. Nighttime infrared imagery of tilted sedimentary rocks in Schuylkill County, Pennsylvania. Lightest tones (1) are water or roads; next lightest (2) are shallow soils over continuous bedrock; medium gray tones (3) in bedrock area generally indicate more fractured rock and colluvial slopes.

An important phase in the preliminary investigation is field reconnaissance to verify the three-dimensional concept of the terrain developed from the literature review and the analysis of remote-sensing data. Many of the fine details and more subtle evidences of slope movement cannot be identified at the scales of photographs and maps analyzed and can be detected only in a field survey. Until that survey is made, part, or even most, of the concepts developed remains conjecture. Performing a field survey at the time corrective measures are being planned is especially important. Many landslides are complex and as time passes frequently change their physical characteristics and marks of identification. For instance, a landslide examined a year after its occurrence may have changed remarkably from the conditions immediately following the original movement. If a landslide developed as a slump slide and eventually turned into a flow, the original report on the slide would be invalid as a basis for planning a correction of the slide. The identification of type should be made at the same time the slide is to be corrected.

Field Evidence of Movement

The term landslide, by definition, implies that movement has taken place; hence, an analysis of the kind and amount of movement becomes a key to the nature of an active slide. It is imperative to learn to recognize the features that first indicate the onset of ground movement and to learn to recognize the kind of slide that already exists. Quite commonly the first visible sign of ground movement is recorded by settlement of the roadway or, depending on the location of the roadway within the moving mass, a bulge of the pave-

ment. Evidence may also be found of landslide movement that has not yet affected the highway but that may do so in time: Minor failure in an embankment, material that falls on the roadway from an upper slope, or even the progressive failure of the region below a fill may presage a larger landslide that will endanger the road itself. Other evidences of movement are to be found in broken pipe or power lines, spalling or other signs of distress in concrete structures, closure of expansion joints in bridge plates or rigid pavements (Figures 3.37 and 3.38), or loss of alignment of building foundations. In many cases, arcuate cracks and minor scarps in the soil give advance notice of serious failure.

Table 3.3 gives the chief evidence of movement in the various parts of each type of landslide. The following discussion elaborates on the significance of cracks in recognizing and classifying slides and on features that aid in identifying the various types of landslides.

The ability to recognize small cracks and displacements in the surface soils and to understand their meaning deserves cultivation because it can produce accurate knowledge of the cause and character of movement that is prerequisite to correction. Surface cracks are not necessarily normal to the direction of ground movement, as is commonly assumed by some. For example, cracks near the head of a slump are indeed normal to the direction of horizontal movement, but the cracks along its flank are nearly parallel to it.

Small en echelon cracks commonly develop in the surface soil before other signs of rupture take place; thus, they are particularly important in the recognition of potential or incipient slides. They result from a force couple in which the angle between the direction of motion and that of the cracks is a function of the location within the landslide area. Thus, in many cases a map of the en echelon cracks will delineate the slide accurately, even though no other visible movement has taken place (Figure 3.39).

In addition to indicating incipient or actual movement, cracks in surface soils are locally useful in helping to determine the type of slide with which one is dealing. For ex-

Figure 3.37. Early signs of impending debris slide along Clear Creek, Colorado (3.10). Displacement of fence, bulging of pavement, and distress in bridge abutment (Figure 3.38) all give early indications of movement at toe of incipient slide. In several places, now covered by patching, centerline stripe was offset along cracks.

Figure 3.38. Right wing wall shown in lower center of Figure 3.37 (3.10). Distress in bridge abutment indicates incipient slide. In addition to offset of wing wall shown here, rockers beneath bridge girders were tipped.

ample, in a slump the walls of cracks are slightly curved in the vertical plane and concave toward the direction of movement; if the rotating slump block has an appreciable vertical offset, the curved cracks wedge shut at depth. In block slides, on the other hand, the cracks are nearly equal in width from top to bottom and do not wedge out at depth because failure in a block slide begins with tension at the base of the block and progresses upward toward the surface. The presence of a few major breaks in the upper parts of a block slide (Figure 3.40) distinguish it from lateral spreading, which is characterized by a maze of intersecting cracks. The inclination of cracks is commonly almost vertical in block slides in cohesive soils, regardless of the dip of the slip plane, but depends on the joint systems in the rock in block slides of rock.

Distinguishing between incipient block slides and slumps is one of the most helpful applications of a study of cracks. If the outline of the crack pattern is horseshoe-shaped in plan, with or without concentric cracks within it, a slump is almost certainly indicated. If, on the other hand, most of the surface cracks are essentially parallel to the slope or cliff face, a block slide is probably in the making. In either case, additional cracks may develop as major movement gets under way, but these will generally conform to the earlier crack pattern (Figures 2.14 and 2.15).

Evidence of soil creep and of "stretching" of the ground surface should also be sought. Stretching indicates comparatively deep-seated movement whereas soil creep is of surficial origin. The phenomenon of stretching is most commonly observed in noncohesive materials that do not form or retain minor cracks readily. The best evidence of stretching consists of small cracks that surround or touch some rigid body, such as a root or boulder, in otherwise homogeneous materials; these cracks form because the tension forces tend to concentrate at or near the rigid bodies.

Field Identification of Slope-Movement Types

Once it has been established that slope movement has taken

Figure 3.39. Tension cracks that typically develop in slump slide in cohesive materials (3.10, 3.34).

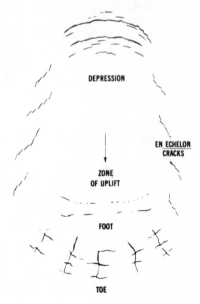

DEPRESSION

EN ECHELON CRACKS

ZONE OF UPLIFT

FOOT

TOE

place or is in progress, the next essential step is to identify the type of movement. Table 3.3 gives the surface features that aid in the field identification of slope movements. This table gives only the most common types and not all of the types shown in Figure 2.1. Further generalizations are given in the following paragraphs.

Falls and Topples

Although the initial mechanisms of falls and topples are different (Chapter 2), their final appearances after movement are similar. Falls and topples are best recognized by the accumulation of material that is not derived from the underlying slope and that is foreign to normal processes of erosion. In most cases, this material consists of blocks of rock, debris, or earth scattered over the surface or forming a talus slope. If undercutting by lake or stream waters has caused the fall, the rate of failure is proportional to the ability of the water to remove the fallen material. A fast-moving stream may remove material almost as fast as it falls, thus removing the evidence but encouraging continuing falls. On the other hand, a lake or some parts of an ocean shore must depend only on wave action to disintegrate and remove the fallen material; hence, the evidence tends to remain in sight but continued falls tend to be inhibited. Most of the material yielded by rock topples and rock falls is necessarily close to the steep slopes from which it came, but some may bound down the slope and come to rest far from its present source.

In active or very recent slides, the parent cliff is commonly marked by a fresh irregular scar that lacks the crescent shape that is characteristic of slumps. Instead, the irregularity of its surface is controlled by the joints and bedding planes of the parent material (Figures 2.3 and 3.41).

Some idea of the intensity and state of activity of a rock fall or rock topple can be inferred from the presence or absence of vegetation on the scarp and by the damage done to trees by the falling rocks. In active areas, the trees are scarred and debarked or show evidence of healed wounds. If the rock fall or rock topple is severe or of long duration, conifers and other long-lived trees are absent; their places may or may not be taken by aspen or other fast-growing species. Many rock falls follow chutes or dry canyons that can usually be differentiated from normal watercourses or paths cut by snow avalanches (Figure 3.42).

Some debris and earth falls and topples exhibit most of the characteristics of rock falls and rock topples. Others proceed by mere spalling of the surface, but even this activity, if long continued, can lead to removal of considerable quantities of material. Any fall or topple may, of course, presage major landslide movement in the near future.

Slides

Slides, as distinct from falls, topples, lateral spreads, and flows, are characterized by a host of features that are observable at the surface. These features are related to the kind of material in which the slide occurs and to the amount and direction of motion. Rotational slides are characterized by rotation of the block or blocks of which they are composed whereas translational slides are marked by lateral separation with little vertical displacement and by vertical,

rather than concave, cracks. Lateral spreading with few, if any, cracks, on the other hand, is characteristic of earth flows.

Translational rock slides are generally easy to recognize because they are composed wholly of rock, boulders, or rock fragments. Individual fragments may be very large and may move great distances from their source. Rock slides

Figure 3.40. Block slide in cohesive materials, near Portage, Montana (3.10). The slide, in alluvium, colluvium, and some wind-deposited silt, is moving out over surface of alluvium-filled stream channel with little or no rotation of block and without developing zone of uplift at foot. Parallel step scarps along main scarp and drag effects along flanks are both characteristic of block slides. Arrows indicate overbreak cracks that develop after main scarp is formed; possibly because of abrupt break in slope, these are more sharply curved than are those above most slump slides.

Figure 3.41. Rock slide and rock fall along Penn-105 (3.10). Beds are nearly flat; slide is controlled by joints dipping at steep angle toward road.

Table 3.3. Features that aid in recognition of common types of slope movements.

Type of Motion	Kind of Material	Parts Surrounding Slide		
		Crown	Main Scarp	Flanks
Falls, topples	Rock	Consists of loose rock; probably has cracks behind scarp; has irregular shape controlled by local joint system	Is usually almost vertical, irregular, bare, and fresh; usually consists of joint or fault surfaces	Are mostly bare edges of rock
	Soil	Has cracks behind scarp	Is nearly vertical, fresh, active, and spalling on surface	Are often nearly vertical
Slides Rotational *Slump*	Soil	Has numerous cracks that are mostly curved concave toward slide	Is steep, bare, concave toward slide, and commonly high; may show striae and furrows on surface running from crown to head; may be vertical in upper part	Have striae with strong vertical component near head and strong horizontal component near foot; have scarp height that decreases toward foot; may be higher than original ground surface between foot and toe; have en echelon cracks that outline slide in early stages
	Rock	Has cracks that tend to follow fracture pattern in original rock	Is steep, bare, concave toward slide, and commonly high; may show striae and furrows on surface running from crown to head; may be vertical in upper part	Have striae with strong vertical component near head and strong horizontal component near foot; have scarp height that decreases toward foot; may be higher than original ground surface between foot and toe; have en echelon cracks that outline slide in early stages
Translational *Block*	Rock or Soil	Has cracks most of which are nearly vertical and tend to follow contour of slope	Is nearly vertical in upper part and nearly plane and gently to steeply inclined in lower part	Have low scarps with vertical cracks that usually diverge downhill
Rock	Rock	Contains loose rock; has cracks between blocks	Is usually stepped according to spacing of joints or bedding planes; has irregular surface in upper part and is gently to steeply inclined in lower part; may be nearly planar or composed of rock chutes	Are irregular
Flows Dry	Rock	Consists of loose rock; probably has cracks behind scarp; has irregular shape controlled by local joint system	Is usually almost vertical, irregular, bare, and fresh; usually consists of joint or fault surfaces	Are mostly bare edges of rock
	Soil	Has no cracks	Is funnel shaped at angle of repose	Have continuous curve into main scarp
Wet *Debris avalanche* *Debris flow*	Soil	Has few cracks	Typically has serrated or V-shaped upper part; is long and narrow, bare, and commonly striated	Are steep and irregular in upper part; may have levees built up in lower parts
Earth flow	Soil	May have a few cracks	Is concave toward slide; in some types is nearly circular and slide issues through narrow orifice	Are curved; have steep sides
Sand flow *Silt flow*	Soil	Has few cracks	Is steep and concave toward slide; may have variety of shapes in outline: nearly straight, gentle arc, circular, or bottle shaped	Commonly diverge in direction of movement

Note: This table updates Table 1 in book on landslides published in 1958 by Highway Research Board (*3.10*, chap. 4 by A. M. Ritchie, pp. 56-57).

Head	Body	Foot	Toe
Is usually not well defined; consists of fallen material that forms heap of rock next to scarp	Falls: Has irregular surface of jumbled rock that slopes away from scarp and that, if large and if trees or material of contrasting colors are included, may show direction of movement radial from scarp; may contain depressions Topples: Consists of unit or units tilted away from crown	Is commonly buried; if visible, generally shows evidence of reason for failure, such as prominent joint or bedding surface, underlying weak rock, or banks undercut by water	Is irregular pile of debris or talus if slide is small; may have rounded outline and consist of broad, curved transverse ridge if slide is large
Is usually not well defined; consists of fallen material that forms heap of rock next to scarp	Is irregular	Is commonly buried; if visible, generally shows evidence of reason for failure, such as prominent joint or bedding surface, underlying weak rock, or banks undercut by water	Is irregular
Has remnants of land surface flatter than original slope or even tilted into hill, creating at base of main scarp depressions in which perimeter ponds form; has transverse cracks, minor scarps, grabens, fault blocks, bedding attitude different from surrounding area, and trees that lean uphill	Consists of original slump blocks generally broken into smaller masses; has longitudinal cracks, pressure ridges, and occasional overthrusting; commonly develops small pond just above foot	Commonly has transverse cracks developing over foot line and transverse pressure ridges developing below foot line; has zone of uplift, no large individual blocks, and trees that lean downhill	Is often a zone of earth flow of lobate form in which material is rolled over and buried; has trees that lie flat or at various angles and are mixed into toe material
Has remnants of land surface flatter than original slope or even tilted into hill, creating at base of main scarp depressions in which perimeter ponds form; has transverse cracks, minor scarps, grabens, fault blocks, bedding attitude different from surrounding area, and trees that lean uphill	Consists of original slump blocks somewhat broken up; has little plastic deformation; has longitudinal cracks, pressure ridges, and occasional overthrusting; commonly develops small pond just above foot	Commonly has transverse cracks developing over foot line and transverse pressure ridges developing below foot line; has zone of uplift, no large individual blocks, and trees that lean downhill	Has little or no earth flow; is often nearly straight and close to foot; may have steep front
Is relatively undisturbed and has no rotation	Is usually composed of single or few units; is undisturbed except for common tension cracks that show little or no vertical displacement	Has none and no zone of uplift	Plows or overrides ground surface
Has many blocks of rock	Has rough surface of many blocks, some of which may be in approximately their original attitude but lower if movement was slow translation	Usually has none	Consists of accumulation of rock fragments
Has none	Has irregular surface of jumbled rock fragments sloping down from source region and generally extending far out on valley floor; shows lobate transverse ridges and valleys	Has none	Composed of tongues; may override low ridges in valley
Usually has none	Is conical heap of soil, equal in volume to head region	Has none	
May have none	Consists of large blocks pushed along in a matrix of finer material; has flow lines; follows drainageways and can make sharp turns; is very long compared to breadth	Is absent or buried in debris	Spreads laterally in lobes; if dry, may have a steep front about a meter high
Commonly consists of a slump block	Is broken into many small pieces; shows flow structure	Has none	Is spreading and lobate; consists of material rolled over and buried; has trees that lie flat or at various angles and are mixed into toe material
Is generally under water	Spreads out on underwater floor	Has none	Is spreading and lobate

commonly occur only on steep slopes. The numerous small block units with random rotation are mixed in a matrix of finer grained material; the large rock fragments tend to float on or in the matrix. Rock slides have translational failure surfaces rather than the concave surfaces of slumps, and they do not move as unrotated multiple units, like block slides. Most rock slides are controlled by the spacing of joints and bedding planes in the original rock. There is particular danger of forming a rock slide of serious proportions if construction is undertaken in an area marked by a system of strongly developed joints or bedding planes that dip steeply outward toward the natural slope. This is especially true if the natural slope angle is steeper than the angle of repose of the broken rock. Water is not so important a factor in causing rock slides as in causing slides in soil. In some instances, however, water helps to weaken bedding or joint planes that otherwise would offer high frictional resistance. Any seepage that is apparent after a rock slide has taken place is most likely to be visible in the scarp region or, perhaps, in the slide material itself.

Many rock slides are thinner than either slumps or block slides because they commonly are restricted to the weathered zone in bedrock or to surficial talus. The shape of the shear zone, therefore, conforms to the unweathered bedrock surface and is not controlled to any large extent by joints or bedding planes in the bedrock. Many talus slopes produce rock slides by failure within the body of talus.

Rotational slumps rarely form from solid, hard rocks, although special combinations of factors have been known to produce them. Slumps are widespread, however, in sands, silts, and clays and in the weaker bedded rocks. There they can be identified readily from surface indications, though only after considerable movement has taken place. The head region of a slump is characterized by steep escarpments and by visible offsets between separate blocks of material (Figure 3.19). The highest escarpment is commonly just below the crown; because the crests of the flanks are lower than the crown, they can be recognized as flanks, even if the escarpment on one of them happens to be larger than the one below the crown. If the landslide is active or has been active recently, the scarp is bare of vegetation and may be marked by striations or grooves that indicate the direction of movement that has taken place. Striations at the head tend to reflect downward movement whereas striations along the flanks may be nearly horizontal. If the slump is compound, its several crescent-shaped scarps will appear as a scalloped edge in plain view (Figure 2.21).

Undrained depressions and perimeter lakes, bounded on the uphill side by the main scarp, characterize the head regions of many slumps; even if internal drainage prevents such ponds from holding water for long periods, their depressions may be evident. In humid regions, the head area may remain greener than surrounding areas because of the swampy conditions. In the San Juan region of southwestern Colorado, for instance, groves of aspen trees are commonly good indications of wet ground conditions and hence of slides and unstable ground. In northern West Virginia, the swampy areas in the head remain green during the winter whereas the well-drained low areas are brown.

The tension cracks near the head of a slump are generally concentric and parallel to the main scarp. Many such cracks are obscured by rubble or other noncohesive materials, but

Figure 3.42. Rock fall and rock slide near Skihist, British Columbia, on Canadian National Railroad (3.10). Note bare active slopes, closely spaced jointing of rocks, rock chutes, and absence of water. Wooden and concrete sheds bypass debris falling over tracks from above.

even so they may be indicated by evidence of surface stretching, by lines of rock fragments that have been displaced, or even by blades of grass that have been pulled down into the cracks by sand or other loose surface material as it sifted into them. The head region of a slump or an earth block slide with surficial slump characteristics may also be recognized by the presence of slump grabens that have experienced some rotation (Figures 3.19 and 3.43). These are depressed fault blocks of soil or rock, caused in part by decrease in curvature of the shear plane. This produces tension and ultimate failure in the main slump block because of lack of support on its uphill side.

The amount and direction of rotation that any slump block has undergone can usually be ascertained by determining the slope of its surface as compared with that of the original slope. Comparison of the dip and strike of bedded material within the slump block with the original attitude of the unslumped material is an even more exact and foolproof method of determining the amount of rotation, because erosional alternation of the surfaces may give an erroneous impression. In a general way, the amount of rotation is a measure of the amount of displacement.

The part just above the foot of a slump is a zone of compression. The slumped material is confined by the foot and by the flanks of the main scarp so that it is compressed by the load above into the bottom part of the bowl-shaped surface of rupture. In this region, therefore, there are no open cracks. The foot region is marked by a zone of tension and uplift because the slumped material is required to stretch over the foot before it passes farther downhill. This stretching destroys any remaining slump blocks because of the change in direction of the forces. The small blocks and fragments that result tend to weather to rounded forms, producing hummocky ground. Pavement uplifts and cracks, so disconcerting to the motorist and highway engineer, are also most likely to take place in the foot zone. Seeps, springs, and marshy conditions commonly mark the foot and toe of a slump. Moreover, trees tend to be tilted down-

Figure 3.43. Graben on slump slide (3.10). Rotation of slump block is uphill, resulting in flattening of original ground surface, but graben block, which breaks off from slump block, rotates downhill. Grabens do not form if slump block has sheared on surface that approaches arc of circle; instead, they form on slump blocks that slide over principal surface of rupture having marked decrease in its curvature, causing greater horizontal movement for each unit of vertical offset in center portion of slide than in head region.

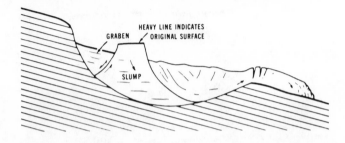

Figure 3.44. General orientation of trees on slump landslide (3.10). As a result of rotation, trees on blocks are bowed uphill because tree tops tend to grow vertically while stump portion changes with rotating land surface. Contrast head and toe regions and compare with Plates 3.2 and 3.3.

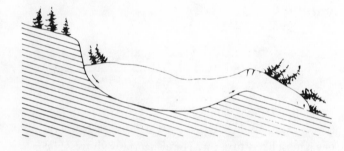

hill rather than uphill as they near the head (Figure 3.44 and Plate 3.3).

The approximate age of some slumps can be inferred by study of the bent trees. If older trees are bent but younger ones are straight, for instance, the slide probably has not moved during the life of the younger trees. On the other hand, the sizes of the tree trunks at the points where the bends occur give a running history of the rotation.

Just below the foot of a slump the ground is commonly marked by long transverse ridges that are separated from one another by open tension cracks. These cracks seldom remain open for long periods of time, and they do not form scarps or other evidence of displacement because the material is no longer confined but spreads out laterally and develops radial cracks at the toe.

Lateral Spreads and Flows

In lateral spreads, a coherent upper block is broken into strips or blocks and extended or stretched by plastic flow or liquefaction of the lower layer. The mechanism of failure can involve not only rotation and translation, but also flow, and thus surface characteristics will contain elements of both slides and flows. One form of lateral spread, in which a plastic layer is squeezed out by the weight of an overlying rigid layer, is here termed a piston slide. In this type, a part of the upper layer may drop vertically,

without rotation, into the space left by removal of the plastic layer.

Dry flows are not difficult to recognize after they have occurred, but it is virtually impossible to predict them in advance. They are commonly very rapid and short-lived. Dry flows are rarely composed of rock fragments; they usually consist of uniformly sized silt or sand. They exhibit no cracks above the main scarp and have poorly developed or nonexistent flow lines. Except for sand runs, they have no well-defined foot. Dry loess may become fluid and flow down a slope because of an earthquake or other external vibrations. Sand runs also behave somewhat as fluids, but in the latter part of their course the sand particles are more likely to slide than to flow.

Wet flows occur when fine-grained soils, with or without coarser debris, become mobilized by an excess amount of water. Most of them behave like wet concrete in a chute (differences are due to water content), but the flow of some wet silts and fine sands is triggered by shocks. "Quick" or sensitive clays may be liquefied by the leaching of salts or by other causes that are not completely understood.

Wet flows are generally characterized by their great lengths, generally even gradients and surfaces, absence of tension cracks, and lack of blocky units and minor scarps. If tension cracks are present, they are bowed in the direction of movement (Figure 3.45), showing the effects of movement of wetter materials at depth beneath the drier crust. An older flow that has had time to dry may show large shrinkage cracks or flow lines. In many cases, the main scarp area is emptied by the removal of all flow material and resembles a glacial cirque. In other cases, the gradation may be imperceptible downward from soil creep to mud flow.

The rate of flow is dependent on the total amount of material that feeds the slide; the accumulated debris imposes a load on the entire mass below it and tends to maintain a constant rate of movement. The flow is under pressure everywhere from the material above it; consequently, the mass shows few, if any, cracks over the foot. Flows can and do make sharp turns and move around firmly fixed obstacles that appear in their paths. A very wet flow moving at high velocity may leave debris marks high on trees or other objects and ridges of debris called torrent levees along its sides.

Hidden Landslides

Among the most difficult kinds of slides to recognize and guard against are old landslides that have been covered by glacial till or other more recent sediments. To predict, in detail, the existence of old buried slides or the effect that they may have on new construction work that exposes them is probably impossible. One who knows the recent geologic history of the region intimately, however, may be able to make some educated guesses as to the probable existence of such slides and even as to where they are most likely to be found. One example is shown in Figure 3.46; the indicated unstable body of soft shale and chalk is clearly related to a fault in the bedrock, but it also has the characteristics of a surface landslide. Since this slide took place, the area was covered by two or more layers of loess, which completely obscured the landslide until the rising

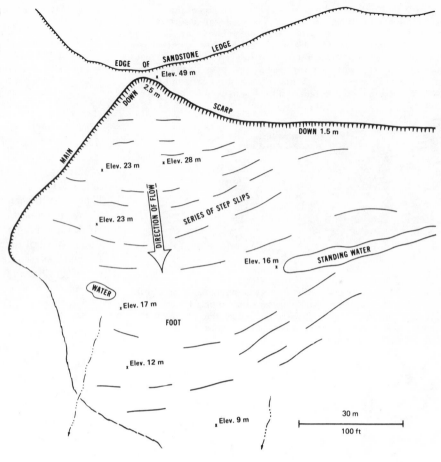

Figure 3.45. Crack pattern in slump on south side of Reservoir Hill, Dunbar, West Virginia, indicates flowage at depth beneath harder material at surface (*3.10*). Broken pipes from reservoir at top of hill dumped large amount of water into old slide and reactivated it. Horseshoe-shaped scarp is imperfect, differentiating it from that of true slump. Greatest movement is near center of slide, as indicated by arrangement of cracks and of standing water. Presence of convex cracks is indicative of flow movement at depth.

Figure 3.46. Hidden landslide exposed when overburden of loess was removed by bank-cutting along shore of Fort Randall Reservoir in South Dakota (*3.10*). Buried soil profiles indicate two periods of loess deposition after faulting and landsliding took place in underlying shale and chalk of Cretaceous age. All surface evidence of landsliding was obliterated by loess until lake waters cut new face.

waters of the lake undercut the bank and exposed the old slide.

CONCLUSIONS

All landslide investigations must start with either the recognition of a distressed condition on the natural or artificial slope or the determination of the vulnerable locations where one is most likely to occur. The evidence for distressed conditions that may be present or induced lies chiefly in evidence of movements, minor or major, that have already taken place or of geologic, soil, and hydrologic conditions that are likely to cause movement in the future. Once the fact of land movement, actual or potential, has been established, the next essential step is to identify the type of landslide. One would not apply the same corrective procedure to a rock fall and a block slide any more than one would attempt to prevent a slide without knowing the kind of slide expected. If maximum benefit is to be accrued from the preventive or corrective measures finally employed, it is imperative to learn to recognize the kind of slide that exists or is expected.

One final point needs emphasizing. Even if the preliminary examination of the general environment has indicated that no landslide movements have yet taken place or are imminent, the investigator must still determine whether the ground to be disturbed by the proposed construction will prove reasonably stable. No one is capable, nor is money available, of studying in detail and of guaranteeing the stability of all slopes for most large construction projects. As a general rule, the amount of investigation that is warranted is a function of the landslide susceptibility of the surrounding country and also of the degree of damage that might be expected to occur to persons or installations if a slide should occur. In other words, the more serious the consequences of a landslide are, the more detailed the search for potential slides should be.

78

REFERENCES

3.1 Alföldi, T. T. Regional Study of Landsliding in Eastern Ontario by Remote Sensing. Department of Civil Engineering, Univ. of Toronto, MS thesis, 1974, 79 pp.

3.2 Anderson, R. R., Hoyer, B. E., and Taranik, J. V. Guide to Aerial Imagery of Iowa. Iowa Geological Survey, Public Information Circular 8, Sept. 1974, 142 pp.

3.3 Baker, R. F., and Chieruzzi, R. Regional Concept of Landslide Occurrence. Highway Research Board, Bulletin 216, 1959, pp. 1-16.

3.4 Bruce, R. L., and Scully, J. Manual of Landslide Recognition in Pierre Shale, South Dakota. South Dakota Department of Highways, Pierre, Final Rept. of Research Project 615(64), Dec. 1966, 70 pp.

3.5 Chassie, R. G., and Goughnour, R. D. National Highway Landslide Experience. Highway Focus, Vol. 8, No. 1, Jan. 1976, pp. 1-9.

3.6 Cleaves, A. B. Landslide Investigations: A Field Handbook for Use in Highway Locations and Design. U.S. Bureau of Public Roads, 1961, 67 pp.

3.7 Colwell, R. N., ed. Manual of Photographic Interpretation. American Society of Photogrammetry, Falls Church, Va., 1960, 868 pp.

3.8 Denny, C. S., Warren, C. R., Dow, D. H., and Dale, W. J. A Descriptive Catalog of Selected Aerial Photographs of Geologic Features in the United States. U.S. Geological Survey, Professional Paper 590, 1968, 79 pp.

3.9 Dishaw, H. E. Massive Landslides. Photogrammetric Engineering, Vol. 32, No. 6, June 1967, pp. 603-609.

3.10 Eckel, E. B., ed. Landslides and Engineering Practice. Highway Research Board, Special Rept. 29, 1958, 232 pp.

3.11 Gagnon, H. Remote Sensing of Landslide Hazards on Quick Clays of Eastern Canada. Proc., 10th International Symposium of Remote Sensing of Environment, Environmental Research Institute of Michigan, Ann Arbor, 1975, pp. 803-810.

3.12 Leith, C. J., Fisher, C. P., Deal, C. S., Gupton, C. P., and Yorke, C. A. An Investigation of the Stability of Highway Cut Slopes in North Carolina. North Carolina State Univ., Raleigh, Project ERO-110-U, June 1964, 129 pp.

3.13 Liang, T. Landslides: An Aerial Photographic Study. Cornell Univ., Ithaca, N.Y., PhD thesis, 1952, 274 pp.

3.14 Lobeck, A. K. Geomorphology. McGraw-Hill, New York, 1939, 731 pp.

3.15 Lueder, D. R. Aerial Photographic Interpretation: Principles and Applications. McGraw-Hill, New York, 1959, 462 pp.

3.16 Maruyasu, T., Nakano, T., Nishio, M., Takeda, Y., Kawasaki, T., Kimata, K., Ozarki, Y., and Fuchimoto, M. Statistical Analysis of Landslides and Related Phenomena on Aerial Photographs. Journal of Japan Society of Photogrammetry, Special Vol. 1, June 1964, pp. 93-100.

3.17 Mathur, B. S., and Gartner, J. F. Principles of Photo Interpretation in Highway Engineering Practice. Materials and Testing Division, Ontario Department of Highways, 1968, 236 pp.

3.18 McNutt, R. B., and others. Soil Survey of Chilton County, Alabama. U.S. Soil Conservation Service, U.S. Forest Service, Alabama Agricultural Experiment Station, and Alabama Department of Agriculture and Industries, Oct. 1972.

3.19 Mintzer, O. W. Application of Photo Interpretation to Highway Engineering Design. Engineering Experiment Station, Ohio State Univ., Columbus, Final Rept. EES 196, Oct. 1966, 213 pp.

3.20 Mintzer, O. W., and Struble, R. A. Manual of Terrain Investigation Techniques for Highway Engineers. Engineering Experiment Station, Ohio State Univ., Columbus, Rept. 196-2 and Appendix I, Oct. 1965, 384 pp.

3.21 National Aeronautics and Space Administration. Skylab Earth Resources Data Catalog. National Aeronautics and Space Administration, Houston, 1974, 359 pp.

3.22 Norell, W. F. Air Photo Patterns of Landslides in Southeastern Ohio. Ohio Department of Highways, Columbus, Vols. 1 and 2, 1965.

3.23 Ray, R. G. Aerial Photographs in Geologic Interpretation and Mapping. U.S. Geological Survey, Professional Paper 373, 1960, 230 pp.

3.24 Radbruch-Hall, D. H., Colton, R. B., Davies, W. E., Skipp, B. A., Lucchitta, I., and Varnes, D. J. Preliminary Landslide Overview Map of the Conterminous United States. U.S. Geological Survey, Miscellaneous Field Studies Map MF-771, 1976.

3.25 Reeves, R. G., ed. Manual of Remote Sensing. American Society of Photogrammetry, Falls Church, Va., Vols. 1 and 2, 1975, 2144 pp.

3.26 Rib, H. T. An Optimum Multisensor Approach for Detailed Engineering Soils Mapping. Purdue Univ., Lafayette, Ind., Joint Highway Research Project 22, Vols. 1 and 2, 1966, 406 pp.

3.27 Rib, H. T., Spencer, J. M., Jr., Falls, C. P., and Koca, J. F. Evaluation of Aerial Remote Sensing Systems for Detecting Subsurface Cavities in Kansas. Federal Highway Administration, Rept. FHWA RD-75-119, 1977.

3.28 Scully, J. Landslides in Pierre Shale in Central South Dakota. South Dakota Department of Transportation, Pierre, Final Rept. of Project 635(67), Dec. 1973, 707 pp.; Federal Highway Administration.

3.29 Sharpe, C. F. S. Landslides and Related Phenomena: A Study of Mass Movements of Soil and Rock. Columbia Univ. Press, New York, 1938, 137 pp.

3.30 Smith, J. T., Jr., ed. Manual of Color Aerial Photography. American Society of Photogrammetry, Falls Church, Va., 1968, 550 pp.

3.31 Stallard, A. H., and Anschutz, G. Use of the Kelsh Plotter in Geoengineering and Allied Investigations in Kansas. Highway Research Board, Highway Research Record 19, 1963, pp. 53-107.

3.32 Stallard, A. H., and Myers, L. D. Soil Identification by Remote Sensing Techniques in Kansas: Part I. State Highway Commission of Kansas, Topeka, 1972, 107 pp.; Federal Highway Administration.

3.33 Tanguay, M. G., and Chagnon, J. Y. Thermal Infrared Imagery at the St. Jean-Vianney Landslide. Proc., 1st Canadian Symposium on Remote Sensing, Canada Center for Remote Sensing and Canadian Institute of Surveying, Ottawa, 1972, pp. 387-402.

3.34 Terzaghi, K., and Peck, R. B. Soil Mechanics in Engineering Practice. Wiley, New York, 1948, 566 pp.

3.35 Thomas, M. M., ed. Manual of Photogrammetry. American Society of Photogrammetry, Falls Church, Va., Vols. 1 and 2, 3rd Ed., 1966, 1199 pp.

3.36 Thompson, G. A., and White, D. E. Regional Geology of the Steamboat Springs Area, Washoe, Nevada. U.S. Geological Survey, Professional Paper 458 A, 1964.

3.37 Thornbury, W. D. Principles of Geomorphology. Wiley, New York, 1954, 618 pp.

3.38 Way, D. S. Terrain Analysis: A Guide to Site Selection Using Aerial Photographic Interpretation. Dowden, Hutchinson and Ross, Stroudsburg, Penn., 1973, 329 pp.

3.39 Williams, G. P., and Guy, H. P. Erosional and Depositional Aspects of Hurricane Camile in Virginia, 1969. U.S. Geological Survey, Professional Paper 804, 1973, 80 pp.

3.40 Záruba, Q., and Mencl, V. Landslides and Their Control. Elsevier, New York, and Academia, Prague, 1969, 205 pp.

PHOTOGRAPH CREDITS

Plate 3.1 Courtesy of Region 10, Federal Highway Administration

Plate 3.2 Courtesy of Washington State Highway Commission

Plate 3.3 Courtesy of Washington State Highway Commission

Plate 3.4 Courtesy of U.S. Department of Agriculture

Plate 3.5 Courtesy of National Aeronautics and Space Administration

Plate 3.6 Courtesy of National Aeronautics and Space Administration

Figure 3.1 Courtesy of Virginia Division of Mineral Resources

Chapter 4

Field Investigation

George F. Sowers and David L. Royster

The field investigation is the central and decisive part of a study of landslides and landslide-prone areas (*4.9*). The investigation serves two essential purposes: (a) to identify areas subject to sliding when future construction is being planned and (b) to define features of and environmental factors involved in an existing slide. Unstable areas prone to sliding usually exhibit symptoms of past movement and incipient failure; most of these can be identified in a detailed field investigation before design. Such investigations can show how to prevent or at least minimize future movements, and they can suggest alternate routes that are less likely to slide. Once a landslide has developed (either before the construction of a facility or after work is under way), the investigation is made to diagnose the factors responsible for the movement and to determine what corrective measures are appropriate to prevent or minimize continuing movements.

Because landslides are continually changing phenomena, the field investigation cannot be considered an isolated or easily defined activity; instead, it is iterative. New data generate new questions that require more data for answers. Although the investigation must continue for a period of time consistent with the shifting topography and changing environment, it is constrained by the timely needs for preventive or corrective design. Thus, field investigations should commence long before construction is anticipated and sometimes continue long after the area has been changed by the anticipated construction.

SCOPE OF FIELD INVESTIGATIONS

A number of features require study in a field investigation; these are enumerated below as a checklist for planning a study.

 I. Topography
     A. Contour map
        1. Land form
        2. Anomalous patterns (jumbled, scarps, bulges)
     B. Surface drainage
        1. Continuous
        2. Intermittent
     C. Profiles of slope
        1. Correlate with geology (II)
        2. Correlate with contour map (IA)
     D. Topographic changes
        1. Rate of change by time
        2. Correlate with groundwater (III), weather (IV), and vibration (V)
 II. Geology
     A. Formations at site
        1. Sequence of formations
        2. Colluvium
           a. Bedrock contact
           b. Residual soil
        3. Formations with bad experience
        4. Rock minerals susceptible to alteration
     B. Structure: three-dimensional geometry
        1. Stratification
        2. Folding
        3. Strike and dip of bedding or foliation
           a. Changes in strike and dip
           b. Relation to slope and slide
        4. Strike and dip of joints with relation to slope
        5. Faults, breccia, and shear zones with relation to slope and slide
     C. Weathering
        1. Character (chemical, mechanical, and solution)
        2. Depth (uniform or variable)
III. Groundwater
     A. Piezometric levels within slope
        1. Normal
        2. Perched levels, relation to formations and structure

3. Artesian pressures, relation to formations and structure
B. Variations in piezometric levels—correlate with weather (IV), vibration (V), and history of slope changes (VI)
  1. Response to rainfall
  2. Seasonal fluctuations
  3. Year-to-year changes
  4. Effect of snowmelt
C. Ground surface indications of subsurface water
  1. Springs
  2. Seeps and damp areas
  3. Vegetation differences
D. Effect of human activity on groundwater
  1. Groundwater utilization
  2. Groundwater flow restriction
  3. Impoundment and additions to groundwater
  4. Changes in ground cover and infiltration opportunity
  5. Surface water changes
E. Groundwater chemistry
  1. Dissolved salts and gases
  2. Changes in radioactive gases

IV. Weather
A. Precipitation
  1. Form (rain or snow)
  2. Hourly rates
  3. Daily rates
  4. Monthly rates
  5. Annual rates
B. Temperature
  1. Hourly and daily means
  2. Hourly and daily extremes
  3. Cumulative degree-day deficit (freezing index)
  4. Sudden thaws
C. Barometric changes

V. Vibration
A. Seismicity
  1. Seismic events
  2. Microseismic intensity
  3. Microseismic changes
B. Human induced
  1. Transport
  2. Blasting
  3. Heavy machinery

VI. History of slope changes
A. Natural process
  1. Long-term geologic changes
  2. Erosion
  3. Evidence of past movement
  4. Submergence and emergence
B. Human activity
  1. Cutting
  2. Filling
  3. Changes in surface water
  4. Changes in groundwater
  5. Changes in vegetation cover, clearing, excavation, cultivation, and paving
  6. Flooding and sudden drawdown of reservoirs
C. Rate of movement
  1. Visual accounts
  2. Evidence in vegetation

3. Evidence in topography
4. Photographs (see Chapter 3)
  a. Oblique
  b. Aerial
  c. Stereoptical data (photographic)
  d. Spectral changes
5. Instrumental data (see Chapter 5)
  a. Vertical changes, time history
  b. Horizontal changes, time history
  c. Internal strains and tilt, including time history
D. Correlations of movements
  1. Groundwater—correlate with groundwater (III)
  2. Weather—correlate with weather (IV)
  3. Vibration—correlate with vibration (V)
  4. Human activity—correlate with human-induced vibration (VB)

The techniques for obtaining the data will be discussed at greater length later in this chapter. The extent to which any one feature requires evaluation is difficult to determine in advance. Whether any more data are needed will depend on the amount of information already available. However, the process is seldom completed; instead, the work stops when time, money, and patience commensurate with risks and potential costs have been exhausted.

Topography

The topography or geometry of the ground surface is an overt clue to past landslide activity and potential instability. More detail than that shown on existing topographic or project design maps is usually required for landslide studies. Moreover, interpretations made by topographers who are not specifically looking for landslide features can obscure the special geometric or topographic forms that are diagnostic of landslides. Therefore, special mapping is usually necessary. Because the topography of a landslide is continually changing, the area must be mapped at several different times: (a) several years before construction (if data are available), (b) at the time a specific landslide investigation is initiated, (c) at appropriate intervals during the progress of the investigation, and (d) after remedial measures are undertaken. Ultimately, the effectiveness of corrective measures is expressed by whether the topography changes.

Because of the long time span involved, the location of the area should be referenced with some standard location system. Latitude and longitude may be sufficient for studies in remote areas or studies encompassing large sites. State survey coordinate systems and road stationing outside the area of study can provide an effective frame for location. In addition, nearby prominent topographic features should be referenced because man-made features often change and those who must use the data referenced to them may not be aware of those changes.

Geology

The geometry of the subsurface is the most important single factor in the analysis of a landslide. First, the various soil and rock units must be identified. Although it is desirable to associate the geologic units with the named geologic formations that have been defined or previously studied in the area, this activity frequently dissipates more time and effort than are relevant to the evaluation of the landslide. Because the name of a formation seldom reflects its engineering behavior, determining the structural, lithologic, and engineering properties of the formation is far more important than determining its exact age and identification. However, names do have value in comparing landslide activity in one area with that in another that includes the same formations, and relative age can help in describing structural features.

Geologic structure is frequently a major factor in landslides. Although this includes the major large-scale structural features such as folds and faults to which most importance is attached in regional geologic studies, the minor structural details, including joints, small faults, and local shear zones, may be even more important. Changes in the geologic conditions with location give important clues to areas of distress and to potential landslide activity. Offsets in strata, changes in joint orientation, and abrupt changes in dips and strikes are indicators of nonuniform geologic conditions. These must be identified both within the area of potential or observed movement and far enough beyond that area to predict the effect of any planned construction surrounding the landslide zone.

Water

Water is a major factor in most landslides. Concentrations of surface water from rainfall runoff, seeps, and springs are overt indications of topographic changes in landslide-prone areas. For example, unusually deep erosion gullies sometimes suggest that soils and rocks have been weakened by landslide activity. Moreover, surface water entering cracks and fissures caused by earth movement aggravates instability.

Seeps and springs that serve as groundwater exits are another dimension of the water problem. Sometimes, however, too much attention is paid to the exit and not enough to the source. If there is a groundwater exit, there also must be a path of seepage leading to it; thus, the exit of groundwater is a clue to internal water pressures elsewhere. Unfortunately, the layperson sometimes considers seeps and springs to be the cause of earth movements. Seeps uphill from the slide serve as a source of surface water that can infiltrate back into the soil within the slide, thus contributing to instability. However, within the slide zone and downhill from it, the opposite is usually true. Less pressure builds up when water is seeping out of the ground than when the exits for groundwater are blocked. For example, in one major slide area, landslide activity was always preceded by a stoppage of spring discharge near the slide toe; the cessation of movement was marked by an increase in spring discharge.

The various joints, fissures, and more pervious strata that conduct water underground must be identified. These features constitute a part of the subsurface geometry that is not always reflected by the strike and dip of the formations. Even the most insignificant water-bearing stratum can sometimes be the source of trouble and is too often overlooked in favor of the more obvious, but perhaps less dangerous, aquifers. Within each water-bearing formation or fissure is a definite (but changing) groundwater pressure

or piezometric level. The groundwater is sometimes perched, leaving the overlying and underlying strata only partially saturated. The formation or fissure can be carrying water under sufficient confined pressure to cause significant reduction in effective stresses within the soil or rock between blocks of impervious materials separated by fissures. To the geologist interested in rock formations and to the geotechnical engineer interested in soil samples, groundwater aquifer identification and piezometric pressure studies are often considered to be of secondary interest, but, in many cases, they constitute the most valuable data gained in the investigation.

Physical Properties

Evaluating the stability of a zone of questionable movement or determining the effectiveness of various corrective measures requires a knowledge of the physical properties of the soil and rock strata. The required data include the properties of both the intact formation and the formation after it has been subject to water pressure changes and to large strains.

The physical properties can be determined in three ways. The conventional approach, which is to test undisturbed samples in the laboratory, has the advantage that stresses and water pressures can be controlled to simulate future environmental changes produced by the anticipated construction and climate variations. However, the process of sampling, transportation to the laboratory, and preparation for testing causes definite changes in the character of the soil and rock before their laboratory testing. A second approach, which is to test the soil and rock in place in the ground, minimizes the disturbing effects of sampling. However, the range in environmental changes that can be induced in the field is severely limited. A third approach is to use indirect tests that measure some physical property related to the engineering parameters involved in the landslide. For example, the compression-wave velocity in a formation, as determined by a seismic refraction geophysical investigation, is related to the joint spacing, density, and rigidity of the soils and rocks through which the wave passes. A formation exhibiting a low compression-wave velocity is generally weaker and less rigid than one exhibiting a high velocity. Unfortunately, the indirect tests require subjective interpretation; therefore, their use in measuring the physical properties of the materials involved is limited by the experience and imagination of the interpreter.

Ecological Factors

A landslide, even one that occurred in the geologic past, has a distinct influence on life systems in the surrounding area. For example, a landslide in a steep, fractured rock mass can produce a loose, jumbled mass of colluvium that is more easily saturated by rain and therefore supports a more luxuriant growth of vegetation than does the undisturbed slope. Thus, the ecosystem of an area can provide a subtle hint of landslide activity. Conversely, the ecosystem effects changes in landslides. For example, deep masses of roots can provide sufficient reinforcement to distort the geometry of the failing soil mass; trees with deep tap roots may even curtail severe movement. A thick cover of vegetation minimizes rainwash and provides more uniform infiltration of moisture. In contrast, concentrated gullying occurs when an area has been exposed by excavation or when vegetation has been destroyed by fire. Therefore, the ecosystem, and particularly the vegetation within the landslide area, should be compared to that within the surrounding area. Since life systems are changing, a single study at a particular time may not be sufficient; continuing studies during an extended period usually are necessary in order to define the significance of the interrelation between the earth movements and the ecology of an area.

PLANNING INVESTIGATIONS

Area of Investigation

The area of an investigation is controlled by the size of the project and by the extent of the topographic and geologic features that are to be involved in the landslide activity. At sites where there is potential for movement that has not yet developed, the area that must be investigated cannot be easily defined in advance. The extent of the investigation can be better defined once a landslide has occurred. However, in either case, the area studied must be considerably larger than that comprising the suspected activity or known movement for two reasons: (a) The landslide or potential landslide must be referenced to the stable area surrounding it, and (b) most landslides enlarge with passage of time (moreover, many landslides are much larger than first suspected from the obvious overt indications of activity). As a crude rule of thumb, the area studied should be two to three times wider and longer than the area suspected. In some mountainous areas, it is necessary to investigate to the top of the slope or to some major change in lithology or slope angle. The lateral area must encompass sources of groundwater and geologic structures that are aligned with the area of instability.

The depth of the investigation is even more difficult to define in advance. Borings or other direct techniques should extend deep enough to identify those materials that have not been subject to past movement, but that could be involved in future movement, and the underlying formations that are likely to remain stable. The boring depth is sometimes revised hourly as field operations proceed. When instrumentation of a landslide yields data on the present depth of activity, planned depths are sometimes found to be insufficient and increases are necessary. The specifications should be flexible enough to allow additional depth of investigation when the data obtained suggest deeper movements. Longitudinal cross sections should be drawn through the center of the slide and depict possible toe bulges and uphill scarps; circular or elliptical failure surfaces sketched through these limits can suggest the maximum depth of movement. Continuous thick hard strata within the slope can limit the depth. However, at least one boring should extend far below the suspected depth of shear; sometimes deep slow movements are masked by the greater activity at shallower depths. For a second estimate, the depth of movement below the ground surface at the center of the slide is seldom greater than the width of the zone of surface motion.

Time Span

Since most landslides are influenced by climatic changes, a minimum period for investigation should include one seasonal cycle of weather–1 year in most parts of the world. However, because long-term climatic cycles that occur every 11 or 22 years are superimposed on the yearly changes, a landslide investigation could be necessary for more than 2 decades. Such a long investigation is almost impossible, however, because of the need to draw conclusions and take corrective action. Investigations made during a period in which the climatic conditions are less severe than the maximum will prove too optimistic, and those made during a period of bad climate may appear too pessimistic. The worst climatic conditions that develop during the life of the project control the risk to engineering construction. Experience has indicated that many false conclusions have been reached regarding the causes of landslides and the effectiveness of corrective measures because worsened climatic changes were not considered by the engineers and geologists concerned.

Stages of Investigation

An investigation of landslides is a continuing process but, from the practical point of view, may be divided into four stages.

1. The first stage is a preliminary investigation or reconnaissance in which a general overall view of the problem area is gained. The work begins with a review of the published geology in the area and accounts of past land instability. The field study is largely visual and the interpretation is highly subjective. The results of this preliminary evaluation are used as a guide in planning the more intensive, specific investigation during which the major part of the quantitative data are obtained.

2. The second stage, which is more intensive and detailed, includes boring, sampling, trenching, and other specific techniques designed to obtain the data needed to satisfy the objectives outlined in the above checklist. Because of climatic changes, the intensive investigation is preferably undertaken during the season that is least favorable for stability. For example, borings to determine the minimum soil strength or groundwater studies to determine maximum water pressures should be made during periods of snowmelt or following heavy rainfall. A similar investigation made during the driest and hottest period of the year could give completely misleading data.

3. The third stage is iterative. As new data are obtained, they will point to the need for additional data from specific locations. The investigative plan must provide for additional work that was not a part of the initial scope. Experience indicates that the additional work stemming from information obtained during the planned study will range from 30 to 50 percent more than that which was originally considered adequate for the intensive investigation.

4. The fourth stage involves continuing surveillance of any area where activity is suspected or where corrective action has been taken. The surveillance period is indeterminate but should extend through at least one cycle of annual climatic change and, to be most meaningful, should be longer term to include the worst climatic conditions. For example,

corrective work may be done during a number of consecutively dry years, and the area may not be subjected to its worst climatic test until 4 or 5 years later (assuming an 11-year cycle of climatic change). If the climatic changes occur in a 22-year cycle, the period of observation required could be correspondingly much longer. Rarely do those who finance such long-term surveillance have the wisdom and foresight to continue the data gathering, because the period required extends beyond the tenure of most public officials or private managers. However, public and professional interests are best served if surveillance is maintained long enough for the occurrence of the full range of environmental conditions that can reasonably be expected; otherwise, the extent of continuing risk may never be adequately defined.

SITE TOPOGRAPHY

As previously stated, the site topography (surface geometry) is the first clue to potential instability and the degree to which an area has undergone landslide activity. The topography is first determined by aerial surveys (photogrammetry), which provide an overall view of the site conditions. However, because of the detail required and the masking of the vegetation and the landslide itself, detailed ground surveys must be included as a major tool in landslide investigation.

Aerial Survey

Existing Sources of Photography

All of the United States and most of the remainder of the world have been covered by some form of aerial photography or remote sensing, including LANDSAT; the principal sources are given in Chapter 3. Generally, these existing photographs have been made at moderately high altitudes (typically at a scale of 1:10 000 to 1:40 000). Probably the most widely used scales of photography by these agencies are 1:20 000 (1 cm = 200 m; 1 in = 1667 ft) or 1: 24 000 (1 cm = 240 m; 1 in = 2000 ft). At this scale, most landslides appear as minor topographic anomalies and their detailed features, such as scarps, toe bulges, and seeps, can seldom be identified, even from enlargements. However, these photographs are useful in identifying areas that may be prone to landsliding because they may show the typical undulating disturbed topography or arcuate scarps that are associated with past landslides. Such existing photography is frequently a good starting place for evaluating the surface geometry of the site, but it is seldom of sufficient accuracy to provide useful data for detailed landslide studies.

Smaller scale photography can be obtained by the more sophisticated remote-sensing systems of high-altitude photography (1:60 000 and 1:120 000) provided by the National Aeronautics and Space Administration. Although the imagery obtained from LANDSAT with a resolution of 56 × 79 m (185 × 260 ft) is not sufficient to identify the topography of most landslides, it does provide clues to the major geologic structures that sometimes influence landslide activity. However, such high-altitude, small-scale data should be considered as supplementary information that provides clues of regional significance rather than as the primary source of topographic data on landslide-prone areas. One distinct advantage of existing photographic and remote-sensing imagery is that it provides information on ground surface conditions before the particular site becomes of interest for construction. Some aerial photographs of the U.S. Department of Agriculture date from the late 1930s, and imagery of LANDSAT and the Earth Resources Technology satellite dates from the early 1970s.

Color and Color Infrared Photography

For mapping purposes, black-and-white panchromatic photography generally provides the best detail at the lowest cost. However, in the study of landslides, certain features become more obvious in color (4.15). Color photography emphasizes differences in vegetation and wet areas through changes in soil color and vegetation vigor. Stratification in exposed soil and rock is often more easily recognized and mapped in color. Color infrared photography is particularly helpful because of two properties. First, most water appears blue in such photography; seeps issuing from bare ground frequently will have a blueish tinge that is far more obvious in the color infrared photography than in either black-and-white or color photography. Second, color infrared photography is useful in mapping hidden cracks and fissures through their influence on growth of vegetation. Fissures above the water table act as drains; the color or image depicts dull red colors of inhibited growth, and springs and sheared zones that hold water sometimes feature more vigorous growth by brilliant reds. However, such inferences should be verified on the ground before they are included on maps.

Thermal Sensing

Thermal sensing involves scanning the ground with a detector that is sensitive to selected portions of infrared radiation. The image produced by the shorter wave (2 to 5 μm) is comparable to the image produced by black-and-white or color infrared photography. In the second range (approximately 8 to 14 μm), the radiation is comparable to the heat radiated from the ground. Thus, for any given texture of ground surface, the level of radiation measured is proportional to the ground temperature. This range of thermal sensing is highly sensitive to small differences in ground temperature. For example, if the imagery is obtained during winter, seeps and springs issuing from the warmer, deep ground will be depicted as warm points contrasting with the cold ground; similar imagery obtained during summer nights will show the seeps and springs as colder than the surrounding ground surface. To be most effective, this sensing must be done during the night or early dawn when there is insignificant infrared reflectance from the sun.

Because of changing angles associated with aircraft movement, the thermal infrared image is seldom a scale representation of the ground feature. Instead, it must be used in conjunction with aerial photographs or accurate maps so that the distorted geometry of the infrared image can be related to the actual geometry of the ground surface and the points depicted on the imagery can be located exactly. The resolution of the imagery is not great because the sensing generally encompasses 0.002 to 0.005 rad (0.115° to 0.286°) of angle from the aircraft or 2 to 5-m (7 to 16-ft) resolution for 1000-m (3300-ft) height. At high altitude, the heat radiation from significant areas is averaged and small features cannot be detected. Therefore, for locating seeps, springs, small water courses, and other features of particular significance in landslide investigation, the thermal infrared imagery must be flown at low levels, generally 300 to 1000 m (980 to 3300 ft) above the ground surface. In hilly country, this presents problems for aircraft navigation.

In planning for thermal infrared imagery, one must be sure that the exact needs or objectives of the work are transmitted to those who obtain the imagery so that the proper altitudes and sensing range can be selected. Thermal infrared imagery detects temperature differences as small as 1°C (1.8°F), and those differences, in turn, reflect variations in soil moisture and ground heat storage.

Use of Aerial Survey Data

Aerial survey data are used in two ways. First, and most important, when used in coordination with ground surveys (including reference points established on the ground surface), they provide detailed topographic maps of the area through photogrammetric processes. This is the procedure with which organizations obtaining aerial survey data are most familiar and that requires the least guidance. A second use of aerial photography and remote-sensing data is in interpretation or terrain evaluation (described in Chapter 3). Because the technical quality of photographs and imagery for accurate mapping is more demanding, the specifications for photography should be directed by topographic requirements. However, the special needs for interpretation should also be considered so that one photographic mission can serve both purposes.

Terrestrial Photogrammetry

Photogrammetric measurements of ground geometry can be made from oblique photographs obtained at the ground surface. For example, two or more permanent photographic sites that overlook a slide area can be used to monitor slide movement by successive sets of simultaneous photographs. Such observations have been used to measure changing dam deflections. Although the data reduction is more complex than that for photogrammetric mapping, the technique is useful for determining movement of any selected points, provided they can be seen in the photograph.

Specifying Scale and Coverage Quality

The scale of photography and thermal imagery must be determined on the basis of the size of the features to be identified. In the mapping of springs, seeps, and fissures, a resolution circle 0.5 to 1 m (1.5 to 3 ft) in diameter is desirable. A suitable scale for the aerial photography for landslide studies is of the same order that is required for final highway alignment studies (from 1:3000 to 1:6000 on contact prints). Maps are drawn to a scale of 1:1000 (1 cm = 10 m; 1 in = 83 ft) to 1:5000 (1 cm = 50 m; 1 in = 417 ft) and with contour intervals of 0.5 m (1 to 2 ft) for most slides. Good quality aerial photographs for mapping and interpretation are necessary tools at the beginning of an investigation, but additional coverage is desirable after episodes of

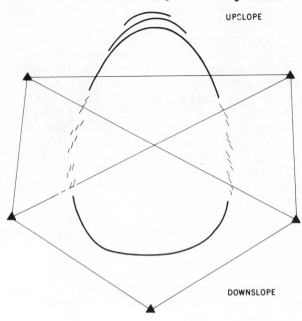

Figure 4.1. Bench marks and triangulation leveling network.

enhanced movement. Thermal imagery is useful after periods of unusually dry or wet weather. However, the various forms of thermal imagery are most revealing when the ground surface can be observed (that is, when deciduous vegetation is free from leaves). Aerial photographs taken at intervals for several years after corrective measures are taken provide an excellent tool for assessing the overall benefits achieved by the measures and for obtaining a preliminary view of the effect of the landslide and associated corrective measures on the ecosystem of the area. Since satisfying all those needs is not always possible, it may not be feasible to bring all of the tools of aerial photography and remote sensing into play in any single investigation.

Ground Surveys

On-the-ground surveys are necessary to (a) establish the ground control for photogrammetric mapping and instrumentation, (b) obtain topographic details where the ground surface is obscured by vegetation (this is particularly important because of the accuracy required in mapping landslides), and (c) establish a frame of reference against which movements of the ground surface can be compared.

Ground Control for Instrumentation

The first requirement is a system of local bench marks that will remain stable during the course of the investigation and as far into the future as movements will be observed. These must be located far enough outside the suspected zone of sliding that they will not be affected by any movements. Ultimately, the bench marks should be referenced to control monuments of federal and state survey systems. However, for convenience, a subsystem of local bench marks should be established close enough to the zone of movement that they can be used as ready references for continuing surveys. At least two monuments of position and elevation should be established on each side of the zone of sliding or suspected movement; as indicated in Figure 4.1, these should be as close as possible to the movement zone, but not influenced by future enlargement of the slide. Experience suggests that the distance from a bench mark to the closest point of known movement should be at least 25 percent of the width of the slide zone. In areas of previous landslides, the minimum distance may be greater. In mountainous areas, adequate outcrops of bedrock can sometimes be found uphill or downhill from the landslide; in areas of thick soil, deep-seated bench marks may be necessary.

Control Network

The bench marks should be tied together by triangulation and precise leveling loops. If there are enough bench marks, the movement of any one can be detected by changes in the control network. Intermediate or temporary bench marks are sometimes established closer to the zone of movement for use in the more frequent surveys of the landslide area. However, these should be checked against the permanent monument grid each time they are used.

Topography

As previously stated, topography obtained from aerial photography may not be sufficiently accurate or detailed for landslide studies because vegetation obscures the ground surface. Therefore, detailed on-site mapping is necessary. Major features, such as scarps, bulges, and areas of jumbled topography, should be defined (Figure 4.2). Because of the changing nature of landslides, the surface surveys should be conducted at the same time the photography is taken; otherwise, the movements of the landslide will confuse the topographic map. Even then, it may not be possible to obtain a precise correlation between the surface topography determined on the ground over a period of days or weeks and that obtained at a single instant from aerial surveys. Differences should be expected, and these should be noted on the topographic maps that are produced.

Cracks, Seeps, and Bulges

Although many cracks, seeps, and bulges, as well as other minor topographic details, can be identified in aerial photography, their full extent can seldom be determined unless the photographs are taken with an unusually high degree of resolution in vegetation-free areas. Therefore, independent crack and bulge surveys should be made by surface methods. Developing cracks are often obscured by grass, leaves, and root mats, particularly at their ends; these cracks should be

Figure 4.2. Cracks, bulges, scarps, and springs.

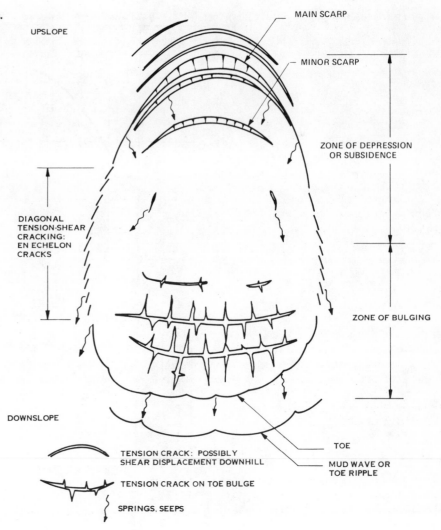

87

carefully uncovered so their total extent can be mapped. Hidden cracks can be identified by subtle changes in leaf mold patterns, tearing of shrubs, and distortion of trees and tree root systems. Boulder alignments or sliding trajectories should also be noted. Cracks should be staked on both sides, as well as referenced to the movement system, because the entire crack system shifts with continuing landslide movement.

Seeps and springs are the ultimate exits for water-bearing strata and cracks and thus are clues to the water paths that influence soil and rock stability. Because seeps often follow cracks that have been opened by soil or rock movement, they can sometimes be traced to sources uphill. The points of disappearance of surface runoff into cracks and fissures should also be mapped. Seeps, springs, and points of water loss change with rainfall, snowmelt, and ground movement. Thus, meaningful data on their location and shifts cannot be obtained by a single survey or regular intervals of observation. Instead, they should be located during and shortly after periods of intense rainfall or snowmelt and after episodes of significant movement.

Movement Grids and Traverses

The continuing movement of a landslide can be measured by a system of traverses or grids across the landslide area (Figures 4.3 and 4.4). Typically, a series of lines more or less perpendicular to the axis of the landslide, spaced 15 to 30 m (50 to 100 ft) apart with stakes at intervals of 15 to 30 m (50 to 100 ft), should be maintained and referenced to the control bench marks. Grids should be laid out so that the reference points are aligned with trajectories of maximum slope or apparent movement if sliding is continuing. In addition, where soil and rock weaknesses cause secondary movements that are skewed to the major slide, intermediate points should be established. For small slides or widely spaced areas of suspected movements, single traverse lines of reference (Figure 4.4) are often used.

Appropriate reference wands or flags should be placed nearby so that the staked points can be found despite severe movement. The elevation and coordinates of each point on the traverse or reference grids should be determined by periodic surveys. In areas where highly irregular topography suggests rapid differences in movement from one point to another, reference points should be spaced more closely, regardless of any predetermined grid pattern. Such closely spaced stakes help to define the lateral limits of the landslide, as well as the direction of movement of localized tongues within the slide. This is particularly important in the later stages of movement if secondary flows develop

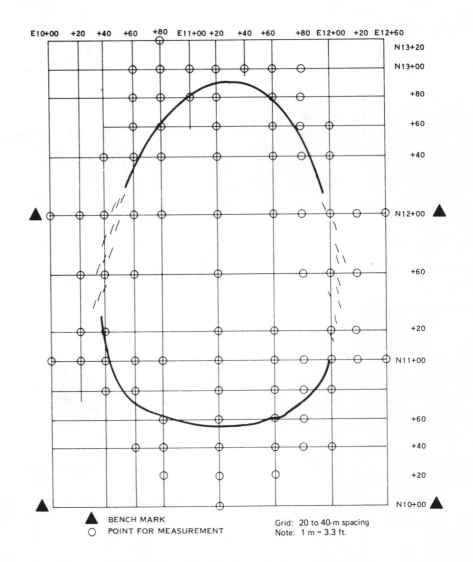

Figure 4.3. Observation grid.

▲ BENCH MARK
○ POINT FOR MEASUREMENT

Grid: 20 to 40-m spacing
Note: 1 m = 3.3 ft.

from the weakening of the soil by sliding. Depending on the rate of movement, these grid points should be checked at intervals ranging from a few days to several months. In addition, they should be observed after periods of unusual weather changes, such as snowmelt, high rainfall, or sharp temperature changes. In this way, any relation between landslide movement and climatic changes can be established.

Crack Measurement

Most earth movements are accompanied by cracking of the ground (Figure 4.2). The principal scarp is most prominent; it is paralleled by developing scarps (arc- or crescent-shaped cracks with the points of the crescents pointing downhill from the scarps) and secondary or older scarps downhill. Along the lateral limits of the slide, cracks are formed by differential shear between the moving mass and the intact soil beyond. The shear often generates parallel diagonal tension cracks (termed en echelon) in this zone. In the bulge zone near the toe of the slide, there are frequently short tension cracks parallel to the direction of movement as well as crescent-shaped tension cracks with the points of the crescents pointing uphill. If the movement extends below the toe of the slope, there will be bulge and shear cracks or subtle ripples in the soil well beyond the slide toe. Survey points should be set on the more prominent of these features and beyond them, provided they are not close to the grid or traverse points.

Instrumentation of landslides is discussed in detail in Chapter 5; however, certain supplementary measurements should be a part of the survey program. For example, crack width changes should be measured directly by taping across stakes set on each side of the crack; vertical offsets on cracks and scarps should be obtained by direct measurement. These direct measurements serve as a check to the more sophisticated systems (discussed in Chapter 5) that determine the in-depth movements. In addition, the surface location and elevations of reference points of the instrumentation systems must be determined at each time of observation.

Representation of Topographic Data

The photogrammetric data are correlated with the ground survey controls and detailed topography, and they are used to establish two or more maps of the landslide area. The first encompasses the landslide (or suspected landslide) plus the surrounding area; the topography extends uphill and downhill beyond major changes in slope or lithology. Topography should be developed on each side for a distance of approximately twice the width of the sliding area (or more when the zone of potential movement is not well defined). Typical scales for such mapping of large slides may be 1:2500 to 1:5000 (1 cm = 25 m to 1 cm = 50 m; 1 in = 208 ft to 1 in = 417 ft).

The second topographic map is more detailed and encom-

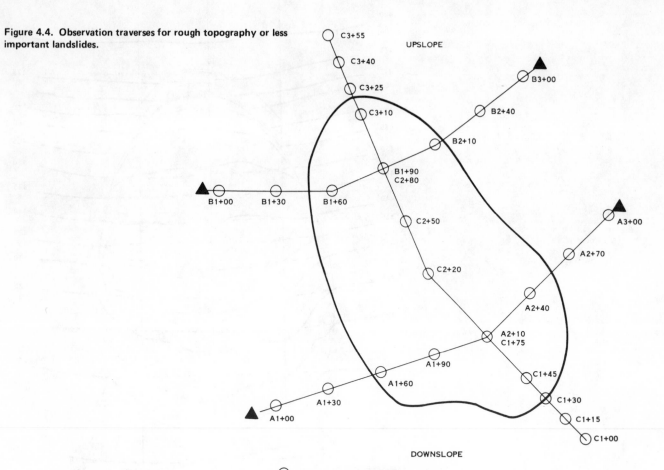

Figure 4.4. Observation traverses for rough topography or less important landslides.

○ POINT FOR MEASUREMENT STATIONS (m)

Note: 1 m = 3.3 ft.

passes the observed slide area plus all of the uphill and downhill cracks and seeps associated with the slide. Typically, the detailed map extends beyond the landslide uphill and downhill for a distance of half the length of the slide or to significantly flatter slopes. Horizontally, the detailed topography should extend at least half of the width of the slide area beyond the limits of the slide. Contour intervals in such detailed topography should be as close as 0.5 m (1 to 2 ft). The horizontal scale is typically 1:1000 (1 cm = 10 m; 1 in = 83 ft) or larger.

Profiles

In addition to the topographic map, profiles of the slide area are prepared (Figures 4.5 and 4.6). The most useful of these follow the lines of steepest slope of the slide area. Where the movement definitely is not in the direction of the steepest slope, two sets of profiles are necessary: One set should be parallel to the direction of movement and the other parallel to the steepest ground surface slopes. In small landslides, three profiles may be sufficient; these should be at the center and quarter points of the slide width (or somewhat closer to the edge of the slide than the quarter points). For very large slides, the longitudinal profiles should be obtained at spacings of 30 to 60 m (100 to 200 ft). It is particularly im-

portant that the profiles be selected so as to depict the worst and less critical combinations of slope and movement within the landslide area. To have at least one additional profile in the stable ground 15 to 30 m (50 to 100 ft) beyond the limits of the slide area on each side of the slide is usually desirable so that the effect on the movement of ground surface slope alone can be determined.

Each longitudinal profile of the landslide is generally plotted separately. If there are significant movements, the successive sets of elevations and consecutive profiles can be shown on the same drawing to illustrate the changing site topography. The original topography should be estimated from old maps and shown for comparison, where possible. However, to reference old maps precisely to the more detailed topography obtained for the landslide investigation is difficult. Differences between existing topography and preslide topography may represent survey mismatches as well as actual changes in the ground surface. Adjustments of the preslide profile from old maps to the profile from new surveys can be made by comparing old and new topography beyond the limits of observed slide movements.

Displacement Vectors and Trajectories

The survey grids and other critical points are entered on the

Figure 4.5. Landslide contours and profile location.

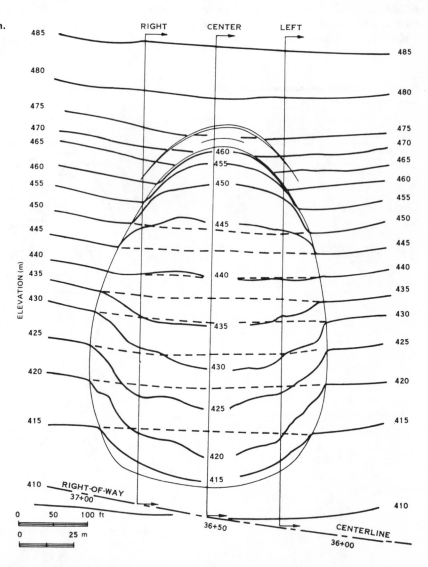

90

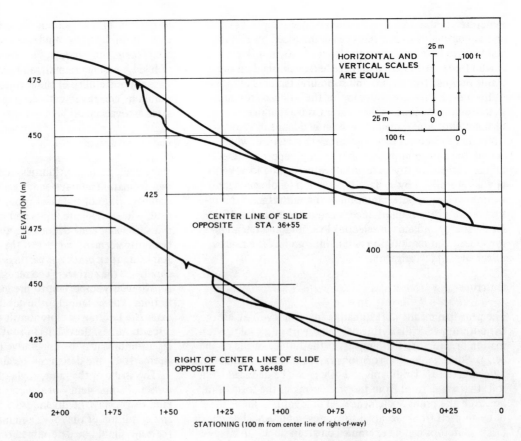

Figure 4.6. Landslide ground surface profiles.

HORIZONTAL AND VERTICAL SCALES ARE EQUAL

CENTER LINE OF SLIDE
OPPOSITE   STA. 36+55

RIGHT OF CENTER LINE OF SLIDE
OPPOSITE   STA. 36+88

STATIONING (100 m from center line of right-of-way)

more detailed topographic map of the landslide area. Both the topography and the depicted grid points should be referenced to the same data. From the consecutive readings on the survey grids and traverses, the horizontal and vertical displacements of the ground surface can be determined. If the movements are large, the subsequent positions of the reference points can be plotted on the topographic map. However, if the movements are small, the successive positions of the stakes may be plotted separately to a larger scale depicting vectors of movement. The vector map depicts only the slide outlines, reference points, and vectorial movements (Figure 4.7). Although the initial positions of the points are shown in their proper scale relations, the vectors of movement are plotted to a larger scale; this difference in scale should be noted. Elevations at successive dates can be entered beside the grid points.

Because the topography changes significantly with the continuing movement, the dates of the surveys should be noted on the maps. Furthermore, if there is a significant period of time between the dates of the surveys that establish the topography and the surveys that establish the movement grid, the elevations of the points on the grids will not necessarily correspond to those on the topographic map.

SUBSURFACE EXPLORATION

Geologic Reconnaissance

A thorough knowledge of the geology of the area is necessary to identify landslide-prone zones as well as to analyze and correct existing slides. The importance of recognizing

Figure 4.7. Movement vectors since beginning of measurement.

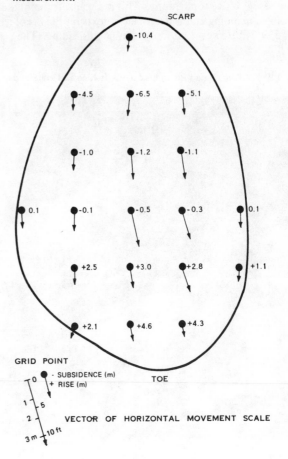

SCARP

GRID POINT

• - SUBSIDENCE (m)
+ RISE (m)

TOE

VECTOR OF HORIZONTAL MOVEMENT SCALE

91

structure, groundwater, and weathering in the prevention and evaluation of sliding has been emphasized by Philbrick and Cleaves (*4.9*), Záruba and Mencl (*4.18*), and Deere and Patton (*4.4*). Although data have been published on the major geologic formations and structural features of most of the world, they are seldom in sufficient detail for landslide studies. As long as their extent is comparable to that of the failure zone, small changes in bedding and geologically minor fractures or irregularities in structure, which are seldom shown in published maps and reports, influence failure as significantly as regional features. The local geology is determined by surface observation (geologic reconnaissance) and by interpretation of the subsurface exploration data. The reconnaissance generates an estimate of the local geology and aids in selecting locations and techniques for subsequent exploration and in interpreting the results of the overall investigation.

Outcrops

The principal means for estimating the geology of an area is the outcrop, which is defined as "that part of a geologic formation or structure that appears at the surface of the earth" (*4.1*). Some geologic formations can be identified indirectly by characteristic landforms and others can be deduced from soils that are derived from the weathering of the formation; however, the primary element used in identification is the outcrop itself (rock, residual soil, or even old landslide debris). Outcrops can be examined on steep slopes, in river channels, in highway and railroad cuts, and in quarries and borrow pits. The relation between geologic formations is determined by correlating structure, lithology, and unit thicknesses with site topography. This is a major technique for geologic mapping and is described in texts on field geology (*4.3, 4.7*). However, outcrops can be misleading. The

formations may have been displaced or distorted by ancient landslides or by rapid weathering of less-resistant strata. Some formations harden on exposure, especially in regions of highly seasonal rainfall, and give a false impression of the strength or stability of the formation. The outcrop data should be correlated with deep samples before final conclusions are reached.

Faults and Joints

Many geologic discontinuities, such as unconformities, cavities, formation contacts, and facies changes, can affect slope stability. Two of the most prominent, and perhaps most readily identified, are faults and joints. A fault is defined as a surface or zone of rock fracture along which there has been displacement parallel to the surface of the fracture. Joints are rock cracks or partings without displacement parallel to the surface. The recognition of faults and joints is particularly important to predesign and site or route selection. The potential for instability is greater in areas with extensive faulting or close jointing than in areas without discontinuities. Revisions in route or locations of engineering structures may be imperative if the potential is serious; alternatively, the design can be altered to fit the situation.

The strike of the fault, angle of dip of the fault plane, type and competency of the associated rock, extent of the fault (especially the thickness of the sheared or gouge zone), and condition of the rocks and materials on either side of the zone should be determined for both site selection and evaluation of an existing slide. If available, geologic maps may show the location of major faults. If detailed maps are not available, an experienced interpreter can often identify questionable areas from their topographic reflections on maps, aerial photographs, and LANDSAT imagery. These areas are then located in the field to determine their

Figure 4.8. Bedding-plane slide in hard metasiltstone and metasandstone in Tennessee.

92

possible significance with regard to site or slide.

Joint systems are seldom indicated on geologic maps, but they are often more troublesome than faults as far as slope stability is concerned. The strike, dip, and spacings of the various joint sets are determined by field mapping of outcrops and a supplementary examination of the rock cores. Statistical depictions of orientation, dip, and spacing or frequency (average number of joints per meter) can be correlated with the slope geometry, proposed cutting and filling, or observed slide features. Close joints in areas involving several rock types may be especially troublesome. In Figure 4.8, for example, sandstone overlies shale. The differences in density, permeability, strength, and rigidity of these materials in the near-vertical slope have caused rock falls. Had the position of these rock types been reversed or had the joints been less frequent, the problem would likely have been less severe. Had the geology of this cut been known in advance, a design could have been formulated to fit joint patterns and thus reduce the maintenance now required. Deere and Patton (*4.4*) have described slopes in interbedded shale and sandstone as one of the most common slide problems.

Inclined Bedding

Bedding or foliation inclined downward toward the slope face is a second major structural factor in instability. The dip and strike can be measured in outcrops; however, in areas of rock folding, there are frequently large local undulations that are far more unfavorable to stability than can be inferred from adjacent outcrops. As with faults, regional directions and angles of dip of the bedding sometimes can be discerned from geologic maps. However, as with faults and joints, the local structure that controls stability must be verified or revised by field study in test pits or by examination of rock cores (see later section on boring and sampling techniques and Table 4.1).

Generally, the preferred route selections should be in areas where the bedding dips into the wall of the excavation. Since this is not always possible, designs can be developed to compensate for the unfavorable bedding. Figure 4.9 shows a typical bedding-plane slide in hard rock (metasiltstone and metasandstone) in which the direction of dip is toward the roadway,

Relict Structure

Residual soils, as a result of the processes by which they were formed, sometimes exhibit the structural characteristics of the geologic formations from which they were derived. This relict structure often influences slope stability in soils in a manner similar to that in which bedding joints and faults influence rock stability. In such cases, the parameters that reflect soil strength, cohesion and angle of friction of the intact soil, may prove misleading. Strength tests should evaluate the surfaces of the relict structure as well as the intact soil; analyses must reflect the orientation of weaknesses defined by the relict structure.

Landforms

The study of landforms is an integral part of the study of

Figure 4.9. Slide in sandstone overlying shale in Tennessee.

outcrops and geologic structures. As pointed out in Chapter 3, this study can be done to a great extent with maps and aerial photographs, but to understand all the relations involved requires that the inferences from remote sensing be verified in the field. If the investigation is concerned with sliding that has already occurred, the investigator first reconstructs the conditions at the site prior to sliding. This can be done with the help of maps and aerial photographs. If sliding has not occurred and the investigator is concerned with slide prevention in route or site selection and design, similar landforms and comparable construction in the area should be studied. Failures and successes in areas with similar formations and conditions will help to indicate the behavior of future routes or sites.

Seeps, Springs, and Poorly Drained Areas

Drainage should be noted and mapped in as much detail as possible at the outset of the investigation. Much of this can be done with the aid of large-scale topographic maps and aerial photographs. Localized seeps, poor drainage, and wet-ground indicators, such as cattails and willows, cannot always be seen in photographs; these can be found through careful surface reconnaissance and their locations added to the detailed topographic maps.

Evidence of Past Instability

Many clues can often alert the investigator to past landslides and future risks. Some of these are hummocky ground, bulges, depressions, cracks, bowed and deformed trees, slumps, and changes in vegetation. The large features can

93

be determined from large-scale maps and aerial photographs; however, the evidence often is either hidden by vegetation or is so subtle and apparently inconsequential that it can only be determined by direct observation. Even then, only one intimately familiar with the soil, geologic materials, and conditions in that particular area can recognize the potential hazards.

Boring, Sampling, and Logging

Boring, sampling, and correlating of the data to develop the three-dimensional subsurface geometry are the most vital parts of the field investigation. The soils and rocks responsible for stability are hidden beneath the ground surface. Although their structure and physical properties can be inferred from the topography and outcrops, quantitative data must be obtained by more direct methods. The right kind of equipment and the capacity to change equipment are essential to a boring and sampling program. Hence, it is necessary for the investigative unit to have access to the appropriate types of drills and tools, tractors, bulldozers, backhoes, and other types of auxiliary machines to gain access to rugged terrain and to make test trenches and pits as needed.

Boring and Sampling Program

The subsurface investigation is only as good as the boring program; the success of the program is measured by the number and quality of samples collected and the reliability of the boring records produced. This phase of the investigation is both an art and a science: It is an art in that it requires a skilled craftsman to advance a boring through complex soil and rock formations aggravated by the distortions that are usually associated with landslides and to consistently collect samples that are representative. It is a science in that those materials and conditions must be properly identified and correlated if the correct interpretations are to be made of their significance. The objectives of the program are

1. Identify the weaker formations that are likely to be involved in movement;
2. Identify the stronger formations that offer significant resistance or that might limit the extent of the zone of movement or provide support for retaining structures;
3. Locate aquifers, define groundwater levels and pressures, and determine water chemistry; and
4. Obtain quantitative data on the physical properties of the formations for use in analyses of stability.

The layout and spacings of the borings depend on the area, configuration, and estimated depth of the slide or, where a suspected or potential slide situation exists, on the geometry of the ground surface. For a suspect area in which sliding has not developed, the boring layout is a grid that includes representative positions up and down the slope and along the length of the slope, as shown in Figure 4.10. Where a slide has occurred, the borings focus on the critical areas of the zone of movement, as well as on adjacent areas that have not yet failed, as shown in Figure 4.11. Philbrick and Cleaves (4.9) suggest that a profile of borings be developed along the centerline of the slide; the first boring should be placed between the midpoint and scarp or head of the slide. This profile should coincide with the

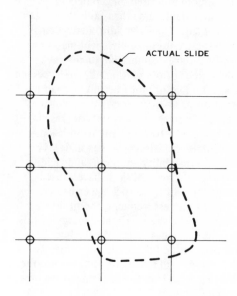

Figure 4.10. Grid of borings in suspect area before sliding commences.

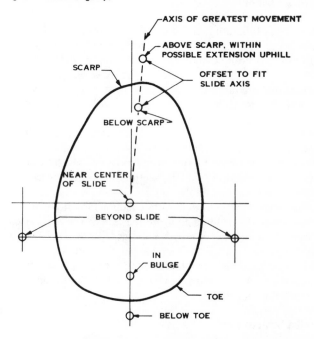

Figure 4.11. Location of borings in slide area (borings are on grid or offset slightly).

topographic profile of maximum slope (see earlier section on ground surveys). The next most important area to be explored, according to Philbrick and Cleaves, is the foot or bulge zone of the slide area. This is the area that usually continues to undergo the most change once sliding has occurred. Such changes are due primarily to the shear surface or zone being progressively softened because of fracture, rainfall infiltration, or blocked groundwater drainage. Moreover, this area is critical in the design of restraining structures that can help minimize continuing movement. Other borings should be distributed throughout the slide. The area outside the slide perimeter should also be drilled and sampled to provide a reference and to enable prediction of the amounts of material that might be involved in future movements.

Boring and Sampling Techniques

The boring and sampling techniques that can be employed are innumerable, and selection depends on the information required, the nature of the soil and rock, the topography and ground trafficability, and the amount of money available. ASTM standards, standards of other associations, books on drilling and sampling, and textbooks on geotechnical engineering (for example, *4.14, 4.16*) describe methods of boring and sampling and their applications and should be consulted for details. Those techniques particularly adapted to landslide studies are summarized in Table 4.1.

The type of sample obtained depends on the information needed. Oriented cores are required to determine the strike of inclined bedding or fractures. Samples from test borings are adequate for identification of formations, but samples from thin-wall tubes or rotary coring are required for good shear and consolidation tests. Accurate and detailed records of the conditions encountered are essential. If samples are lost, a description of the drilling peculiarities may imply the soil and rock condition. When samples are obtained, their value must be assessed from a detailed examination by a geologist or geotechnical engineer, guided by the driller's records. Because the act of sampling changes soils and rocks, the factors involved in those changes must be known to estimate the extent of alteration. Finally, the samples must be carefully preserved, packed, and handled so they will not be unduly changed by exposure and shipping.

Drilling Equipment

One of the problems common to most landslide investigations is accessibility. The same instability that generates movement makes accessibility difficult. For this reason, the investigative unit should be equipped with some type of all-terrain vehicle. Too often, materials and conditions within a particular slide may be underestimated or overlooked simply because of inadequate borings at critical, but inconvenient, inaccessible points. The drilling industry, in collaboration with manufacturers of specialized vehicles, has been able to develop subsurface exploration equipment that is highly versatile and mobile for work in rugged, steep terrain.

Borehole Logging

Because of the extreme variations in the character of the soil and rock and perhaps because of inadequacies on the part of those who perform the boring and sampling and test pit work, there may be gaps in the underground profiles established from any of the borings or geophysical studies. In some instances, the techniques of borehole logging borrowed from the petroleum exploration industry are fruitful. All are based on lowering a sensing device into an open borehole and measuring the soil and rock characteristics at closely spaced intervals of depth. A number of different qualities can be observed, including self-potential of the ground, electrical resistivity within a relatively short vertical distance, nuclear radiation, sound wave or impulse response, density based on nuclear absorption, and water content based on hydrogen ion reaction.

The product of this logging is a graph of each property plotted as a function of depth. The results may have significance in identifying the ion concentration and density of the strata. These aid in identifying soils and rock directly from the boring data. They are particularly useful where there are gaps in the data, such as core loss in broken rock. An even more valuable use of this information lies in establishing the correlation of strata between one boring and another. For example, it is difficult to compare samples obtained from two different holes to determine whether soils or rocks that have similar classification characteristics represent the same stratum. However, by comparing the continuous borehole logs, one can match the patterns of the different properties; similar patterns suggest similar stratification. Thus, while borehole logging may have limited engineering significance by itself in one hole, it is a significant tool for boring interpretation and correlation when used in adjacent holes.

Test Pits and Trenches

One of the best ways to sample and determine the structure and other physical properties of soils and weathered or weak rock foundations is by excavating pits or trenches. Shallow test pits can be excavated with hand tools. Mechanical equipment, such as backhoes, clamshells, draglines, or tractors equipped with front-end loaders, are required for deep pits or long trenches. The sides of the excavation should be sampled, logged, and photographed in detail to provide a three-dimensional picture of the materials and structure. Pits and trenches also provide excellent sites for in situ testing. When the observations and tests are complete, the excavation should be filled or, in some cases, incorporated in the remedial design by serving as a drainage outlet.

Geophysical Studies

Geophysical exploration uses the changes in certain force systems in the earth to define possible boundaries between different materials as well as to estimate some of the engineering properties. The forces include elastic shock waves, gravity, and electric current, none of which is directly concerned in landslide behavior. Thus, the methods are sometimes termed indirect exploration; the data of interest are inferred from the properties measured. Geophysical methods do not replace borings and sampling or test pits and trenches. Rather, they supplement these procedures and greatly reduce the time and cost and the environmental problems that often result from large-scale drilling operations.

Resistivity

Resistivity measurements are made by passing electric current through the ground and measuring the resistance of the various formations to current flow. The current flow is largely electrolytic in that it is dependent on moisture and dissolved salts within the soils and rocks. The Wenner method, which is the simplest and most commonly used, uses four electrodes, spaced equally in a straight line at the ground surface. Current is passed into the ground through the outer two electrodes, and the difference in potential generated by the resistance to current flow is measured between the inner two electrodes. The greater the spacings

Table 4.1. Boring, core drilling, sampling, and other exploratory techniques (4.2, 4.8, 4.11, 4.14).

Method and Reference	Procedure	Type of Sample	Applications	Limitations
Auger boring, ASTM D 1452	Dry hole drilled with hand or power auger; samples preferably recovered from auger flutes	Auger cuttings; disturbed, ground up, partially dried from drill heat in hard materials	In soil and soft rock; to identify geologic units and water content above water table	Soil and rock stratification destroyed; sample mixed with water below water table
Test boring, ASTM D 1586 (Figure 4.15)	Hole drilled with auger or rotary drill; at intervals samples taken 36-mm ID and 50-mm OD driven 0.45 m in three 150-mm increments by 64-kg hammer falling 0.76 m; hydrostatic balance of fluid maintained below water level	Intact but partially disturbed (number of hammer blows for second plus third increment of driving is standard penetration resistance or N)	To identify soil or soft rock; to determine water content; in classification tests and crude shear test of sample (N-value a crude index to density of cohesionless soil and undrained shear strength of cohesive soil)	Gaps between samples, 30 to 120 cm; sample too distorted for accurate shear and consolidation tests; sample limited by gravel; N-value subject to variations depending on free fall of hammer
Test boring of large samples	50 to 75-mm ID and 63 to 89-mm OD samplers driven by hammers up to 160 kg	Intact but partially disturbed (number of hammer blows for second plus third increment of driving is penetration resistance)	In gravelly soils	Sample limited by larger gravel
Test boring through hollow-stem auger	Hole advanced by hollow-stem auger; soil sampled below auger as in test boring above	Intact but partially disturbed (number of hammer blows for second plus third increment of driving is standard penetration resistance or N); N-value may be distorted by auger and should be compared with ASTM D 1586	In gravelly soils (not well adapted to harder soils or soft rock)	Sample limited by larger gravel; maintaining hydrostatic balance in hole below water table more difficult
Rotary coring of soil or soft rock	Outer tube with teeth rotated; soil protected and held by stationary inner tube; cuttings flushed upward by drill fluid (Denison—fixed cutter on outer tube; Pitcher—spring-loaded on outer tube; Acker air-mud core barrel—larger clearances for viscous drilling fluid)	Relatively undisturbed sample, 50 to 200-mm wide and 0.3 to 1.5 m long in liner tube	In firm to stiff cohesive soils and soft but coherent rock	Sample may twist in soft clays; sampling loose sand below water table difficult; success in gravel seldom occurs
Rotary coring of swelling clay or soft rock	Similar to rotary coring of rock; swelling core retained by third inner plastic liner	Soil cylinder 28.5 to 53.2 mm wide and 600 to 1500 mm long encased in plastic tube	In soils and soft rocks that swell or disintegrate rapidly in air (protected by plastic tube)	Sample smaller; equipment more complex
Rotary coring of rock, ASTM D 2113	Outer tube with diamond bit on lower end rotated to cut annular hole in rock; core protected by stationary inner tube; cuttings flushed upward by drill fluid	Rock cylinder 22 to 100 mm wide and as long as 6 m depending on rock soundness	To obtain continuous core in sound rock (percentage of core recovered for distance drilled depends on fractures, rock variability, equipment, and skill of driller)	Core lost in fracture or variable rock; blockage prevents drilling in badly fractured rock; dip of bedding and joints evident but not strike
Rotary coring of rock, oriented core	Similar to rotary coring of rock above; continuous grooves scribed on rock core with compass direction	Rock cylinder, typically 54 mm wide and 1.5 m long with compass orientation	To determine strike of joints and bedding	Method may not be effective in fractured rock
Rotary coring of rock, wire line	Outer tube with diamond bit on lower end rotated to cut annular hole in rock; core protected by stationary inner tube; cuttings flushed upward by drill fluid; core and stationary inner tube retrieved from outer core barrel by lifting device or "overshot" suspended on thin cable (wire line) through special large diameter drill rods and outer core barrel	Rock cylinder 36.5 to 85 mm wide and 1.5 to 4.6 m long	To recover core better in fractured rock, which has less tendency for caving during core removal; to obtain much faster cycle of core recovery and resumption of drilling in deep holes	Same as ASTM D 2113 but to lesser degree

Table 4.1. Continued.

Method and Reference	Procedure	Type of Sample	Applications	Limitations
Rotary coring of rock, integral sampling method (*4.10*)	22-mm hole drilled for length of proposed core; steel rod grouted into hole; core drilled around grouted rod with 100 to 150-mm rock coring drill (same as for ASTM D 2113)	Continuous core reinforced by grouted steel rod	To obtain continuous core in badly fractured, soft, or weathered rock in which recovery is low by ASTM D 2113	Grout may not adhere in some badly weathered rock; fractures sometimes cause drift of diamond bit and cutting rod
Thin-wall tube, ASTM D 1587	75 to 1250-mm thin-wall tube forced into soil with static force (or driven in soft rock); retention of sample helped by drilling mud	Relatively undisturbed sample, length 10 to 20 diameters	In soft to firm clays, short (5-diameter) samples of stiff cohesive soil, soft rock and, with aid of drilling mud, in firm to dense sands	Cutting edge wrinkled by gravel; samples lost in loose sand or very soft clay below water table; more disturbance occurs if driven with hammer
Thin-wall tube, fixed piston	75 to 1250-mm thin-wall tube, which has internal piston controlled by rod and keeps loose cuttings from tube, remains stationary while outer thin-wall tube forced ahead into soil; sample in tube is held in tube by aid of piston (Osterberg-type activates piston hydraulically; Hong-type by a ratchet)	Relatively undisturbed sample, length 10 to 20 diameters	To minimize disturbance of very soft clays (drilling mud aids in holding samples in loose sand below water table)	Method is slow and cumbersome
Swedish foil	Sample surrounded by thin strips of stainless steel, stored above cutter, to prevent contact of soil with tube as it is forced into soil	Continuous samples 50 mm wide and as long as 12 m	In soft, sensitive clays	Samples sometimes damaged by coarse sand and fine gravel
Dynamic sounding (*4.6, 4.12,* Figure 4.15)	Enlarged disposable point on end of rod driven by weight falling fixed distance in increments of 100 to 300 mm	None	To identify significant differences in soil strength or density	Misleading in gravels or loose saturated fine cohesionless soils
Static penetration (*4.6, 4.12,* Figure 4.15)	Enlarged cone, 36-mm diameter and 60° angle forced into soil; force measured at regular intervals	None	To identify significant differences in soil strength or density; to identify soil by resistance of friction sleeve	Stopped by gravel or hard seams
Borehole camera	Inside of core hole viewed by circular photograph or scan	Visual representation	To examine stratification, fractures, and cavities in hole walls	Best above water table or when hole can be stabilized by clear water
Pits and trenches	Pit or trench excavated to expose soils and rocks	Chunks cut from walls of trench; size not limited	To determine structure of complex formations; to obtain samples of thin critical seams such as failure surface	Moving excavation equipment to site, stabilizing excavation walls, and controlling groundwater may be difficult
Rotary or cable tool well drill	Toothed cutter rotated or chisel bit pounded and churned	Ground	To penetrate boulders, coarse gravel; to identify hardness from drilling rates	Identifying soils or rocks difficult
Percussion drilling (jack hammer or air track)	Impact drill used; cuttings removed by compressed air	Rock dust	To locate rock, soft seams, or cavities in sound rock	Drill becomes plugged by wet soil

are between the electrodes, the greater the depth of influence measured relative to the center of the electrode line or spread will be. Dense rock with few voids and little moisture, such as most granites, will have high resistance, but saturated clay will have low resistance. Sometimes the failure surface of a landslide can be detected as a zone of low resistance due to the concentration of moisture. The major advantages of resistivity lie in the portability of the instrument and the fact that large areas can be covered at a relatively small cost. The major disadvantage is that data interpretation is difficult and largely conjectural where strata are not horizontal or uniform in thickness and where contrasts in the resistivities of the materials are not sharp.

Seismic Refraction

The seismic refraction method is based on the measurement of the time required for a shock compression wave to pass from one point to another through the earth. The shock waves are generated by hammer impact or by detonating an explosive at or just below the ground surface. Some of the waves are deflected or refracted by the more rigid, deeper formations and return to the surface where their times of arrival are recorded. In most seismic work involving landslides, a multichannel seismograph system is used. It includes a number of detectors or geophones that are placed at the surface at varying distances from the shock source, amplifiers that enhance the signals, and a recording oscillograph that produces a time-based record of the signals received from all the detectors simultaneously.

When the shock wave from the explosion reaches each geophone, it appears on the recording as a pronounced change in the trace and is termed the first arrival. The time of first arrival at each geophone is used to compute the depth to successively more rigid strata. Seismic measurements are like resistivity measurements in that the environment is not disturbed, the equipment is portable, and rather large areas can be covered at relatively small cost. But, interpretation of seismic measurements, also like that of resistivity measurements, is conjectural where the geology is complex and where velocities of the various materials are not in sharp contrast. The technique is limited to strata that are successively more rigid with depth; it cannot differentiate softer strata below rigid ones.

Gravity

Precise measurements of the earth's gravity field can detect areas of low density. Colluvium or old landslide debris is usually less dense than the virgin materials. Large bodies of loosened rock can be identified where the density contrast is great; boundaries cannot be defined except by borehole logging.

Correlation Representation

The most complete survey or study of an actual or potential landslide is of little value unless it can be depicted in forms that will aid in analyzing and correcting the conditions. These include maps, profiles, cross sections, and three-dimensional representations. Each of these is discussed below.

Geologic Maps

As indicated in the earlier section on geologic reconnaissance, geologic maps provide information on physical features, lithology, and geologic structure. However, existing maps seldom have sufficient detail for the evaluation of an actual or anticipated problem. A geologic map has as its base a topographic map and depicts formations that immediately underlie the ground surface. The projection of each formation to the ground surface is developed from the outcrop mapping and boring data. In problem areas, detailed engineering geologic maps should be prepared on a large scale—the same as that for topographic maps, such as 1:1000 to 1:6000 (1 cm = 10 m to 1 cm = 60 m; 1 in = 83 ft to 1 in = 500 ft). The engineering geology map should also depict features such as rock outcrops, depths to bedrock, strike and dip of beds, faults, and joints, and locations of seeps and springs. The area covered in the detailed geologic map will depend on terrain and topography, alignment and grade of roadway, anticipated soil and geologic conditions, and extent of the slide, if one has already occurred.

Profiles

Centerline profiles of roadway location (made for design) often include data on soil and rock formations as interpreted from geologic maps and borings at representative cross sections. In landslide-prone areas, the number of these routine borings and map interpretations should be increased. The spacings along the centerlines of routine surveys are usually 15 to 150 m (50 to 500 ft). In landslide-prone areas, spacings should be on the order of 8 to 30 m (25 to 100 ft). The profile should show each of the materials encountered in each boring and appropriate additional information, such as moisture content, penetration resistance, strength, and rock core recovery.

Cross Sections

Cross sections, such as the one shown in Figure 4.12, are beneficial in depicting detailed subsurface conditions for landslide studies. These should show soil and rock in and below the slide, dip of the strata, groundwater, moisture contents, and sliding surface or zone of rupture. In all cases, the sections should extend from well into the stable ground above the scarp or crown of the slide to some distance beyond the toe. Geologic cross sections should be plotted by using an undistorted scale (horizontal scale equal to vertical scale); this is essential for a quantitative evaluation of the subsurface geometry that influences the slide.

Three-Dimensional Diagrams and Models

Since landslides are three-dimensional phenomena, the materials and conditions should be viewed in this perspective. To do so, diagrams of the slide from various viewpoints can be prepared in which all available surface information as well as the logs from at least three borings not in a straight line are used. Because of the problem of irregular boundaries and variations in material thicknesses, developing an accurate diagram is difficult. One form of three-dimensional

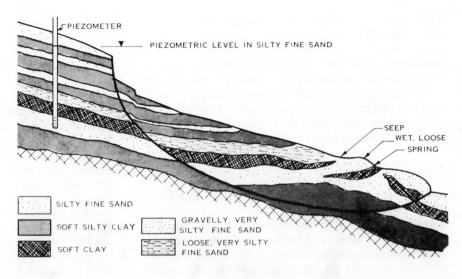

Figure 4.12. Cross section through center of slide.

PIEZOMETER

▼ PIEZOMETRIC LEVEL IN SILTY FINE SAND

SEEP
WET. LOOSE
SPRING

SILTY FINE SAND

SOFT SILTY CLAY

SOFT CLAY

GRAVELLY, VERY SILTY FINE SAND

LOOSE, VERY SILTY FINE SAND

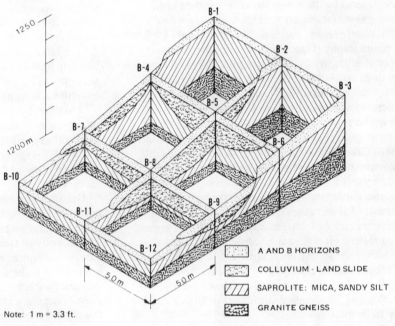

Figure 4.13. Three-dimensional representation of stratification extrapolated from boring.

A AND B HORIZONS

COLLUVIUM - LAND SLIDE

SAPROLITE: MICA, SANDY SILT

GRANITE GNEISS

Note: 1 m = 3.3 ft.

approach in developing a model involves the use of peg-board and string. The board represents some elevation well below the sliding surface. Holes are drilled in the board to plan scale. Pegs are inserted that represent the borings; the top of the peg is at ground level. The pegs are colored to depict the changes in materials and structure in the boring. The scale in all cases should be the same in all three directions. Strings with different colors or thin plastic sheets can then be laced between the pegs to depict the different boundaries, strata, ground surface, and shear surfaces. Such models can be especially beneficial in the analysis of complex slides or in exhibits for litigation. A fence diagram is a second form of three-dimensional representation (Figure 4.13). The borings are displayed as vertical lines on an isometric plot of location. An exaggerated vertical scale can be helpful in visualizing three-dimensional relations, but it distorts the perspective of the slopes.

Slide Enlargement

Landslides usually continue to increase in size after the ini-

tial failure, and the amount of increase can be depicted by using a series of transparent overlays on the original base map or by simply color coding successive new scarps and fractures on the original map. Often there is a considerable time lapse (months to years) between the failure and the correction. In the meantime, the slide sometimes increases in size so that the proposed corrective measures are no longer valid. Thus, it is imperative that the slide continue to be monitored and the displays revised until repair is complete. Some observations should be continued after repairs are made so that the effectiveness of the measures selected can be evaluated and the risk of future movement detected.

SURFACE WATER AND GROUNDWATER

Importance of Water

Next to gravity, water is the most important factor in slope instability. Therefore, identification of the source, movement, amount of water, and water pressure is as important as identification of the soil and rock strata. Yet, those who conduct the field investigations often pay insufficient attention to water for two reasons. First, the investigative techniques, particularly wet drilling, often obscure groundwater. Second, water conditions change, depending on the weather at the time of investigation and on the cumulative effects of rainfall, snowmelt, surface runoff, infiltration, evaporation, and transpiration throughout the year and during long-term climatic cycles.

Surface Water

In many landslides and potential landslides, the surface runoff is a major factor contributing to groundwater and the soil and rock pore-water pressures that lead to reduced soil strength and movement. Moreover, the diversion or blocking of intermittent and continuous surface streams by the landslide movement can add surface water (which originally flowed elsewhere) to the groundwater in the slide area. The permanent surface-water streams can be identified from aerial photography, particularly with the aid of color infrared photography (Chapter 3). Intermittent streams can be identified as significant water sources only during runoff periods in wet weather. However, the suspicion of intermittent surface water for streams can be raised by topographic details, such as gullying and surface washes.

In some arid regions where the surface-water channels are above the general groundwater table, infiltration from stream beds directly into the ground may be considerable and be aggravated when the stream bed has been loosened, fractured, distorted, or dammed by landslide activity. When water courses appear uphill or adjacent to a landslide, the establishment of flow gauges both upstream and downstream from suspected zones of surface-water loss into the ground is usually prudent. If a stream is very small, a sheet metal V-notch weir and a stilling basin upstream (to establish the appropriate approach velocity) are sufficient. The head is measured by a ruler, staff gauge, or continuous recorder. In larger streams, water-measuring flumes or large weirs may have to be established.

In addition, visual observations of surface water, including sheet runoff, during periods of high rainfall are desirable. An engineer or geologist who can adequately describe and record the flow should make the observations, even though the task is seldom convenient or pleasant. Evidence of sheet runoff sometimes can be found by mud lines and debris that become lodged in tall grass and shrubs. Experience shows that the contributions of sheet runoff and intermittent streams to groundwater and pore pressures influencing landslides are often overlooked in investigations. Securing evidence of surface water and groundwater is a major objective of reconnaissance.

Hints of excessive soil moisture frequently can be found in the character of the vegetation or wildlife in the vicinity of the suspect or sliding area. For example, cattails do not ordinarily grow in low dry areas, and bullfrogs require water for their reproduction and life cycle. The locations of permanent and intermittent surface flows are recorded on topographic maps of the slide or incorporated on map overlays. Attention should be paid to the relation of the changes in these water courses to the continuing changes in the slide and slide area topography.

Groundwater

Groundwater can be defined broadly as all water below the ground surface. Generally, however, the term is restricted to the water that is not restrained in the soil by capillary tension, partially immobilized in the stress field surrounding clay minerals, or linked to the soil or rock minerals. Although soil capillary moisture, absorbed water, and water of hydration may not be considered true groundwater, they are a part of the total ground moisture system. During wet weather, infiltration reduces capillary tension, allowing the groundwater to rise and encompass part of the capillary zone; during dry weather, transpiration depletes capillary moisture, increases capillary tension, and decreases the level of free groundwater. Capillary tension increases the effective stresses among soil particles and temporarily increases soil strength. Loss of capillary tension either by saturation or by drying causes loss of soil strength. An increase in absorbed moisture is a major factor in the decrease in strength of cohesive soils and some weakly cemented rocks. An increase in the water of hydration of minerals such as anhydrite is accompanied by expansion, which can destroy the bonding between soil or rock particles and decrease their strengths. Thus, in most cases, an increase in soil or rock moisture is accompanied by a decrease in strength.

A sudden moisture increase in a dry soil can produce a pore-pressure increase in trapped pore air accompanied by local soil expansion and strength decrease. The slaking or sudden disintegration of hard dry clay or clay-bonded rock is caused both by an increase in absorbed water and by pore air pressure.

Groundwater has been more narrowly defined as that part of the soil-rock-water system that is free to move from point to point under the influence of gravity. The surface of that body of free water, which is at atmospheric pressure, is the groundwater table. In simple terms, the groundwater table is the elevation of zero (atmospheric) water pressure; water is in tension above the water table in the zone of capillary saturation and in the unsaturated capillary fringe, and is under pressure below that table. At any level, the water pressure is equal to the unit mass of water multiplied by the distance (z) below the water table. Above the water table, z is negative and the computed pressure is negative.

Aquifers

In engineering usage, an aquifer is a soil or rock stratum that is significantly more pervious than the adjoining strata. An aquifer can also be an opening in the soil or rock formation, such as an animal burrow, a shrinkage crack, voids left by rotting of vegetation, joints, fracture zones, and other discontinuities that provide localized ability to transmit water. As the term implies, an aquifer contains water, the source of which may be infiltration from precipitation, infiltration from streams, leaking water pipes, or even the upward discharge from a deeper artesian aquifer. Potential aquifers can be identified from detailed records of the soil and rock boring that should describe all of the more pervious strata or fracture systems that are capable of transmitting water. The absolute permeability of the stratum does not determine whether it is an aquifer. Instead, its relative permeability compared to the strata above and particularly the strata below is more significant. For example, a stratum of silty fine sand could be an aquifer if it were confined between clay strata, but would be an aquiclude compared to a stratum of coarse clean sand. The identification of some discontinuities, such as vertical cracks and joints that are potential aquifers, is more difficult because borings have little chance statistically of encountering them unless the borings are inclined. Test pits and test trenches described in the earlier section on subsurface exploration are far more useful in determining the presence and spacing of such localized discontinuities. A thorough understanding of the geology of the individual formation also provides clues to potential discontinuities that might act as aquifers.

An aquiclude is a stratum or discontinuity that is sufficiently less pervious than the adjoining strata that it is a barrier to groundwater. For example, silt washing into a crack in the ground can produce a clastic dike or aquiclude that will block the flow in a sandy seam. The very movement of a landslide can shift pervious strata to align with impervious strata, generating localized aquicludes. These aquicludes change with the continuing movement of the landslide.

Once potential aquifers have been identified, one must determine whether they transmit water or are subject to water pressure. Long-term observations, particularly during periods of wet weather and high general groundwater levels, are necessary. Moreover, because of changing topography and changing interrelations of the strata with the movement of a landslide, an aquifer today can be dry tomorrow. Therefore, potential aquifers are evaluated by the variations in water pressure as measured by piezometers.

Piezometric Level

The piezometric level at a point is the elevation to which water eventually will rise in a small tube sealed into the appropriate aquifer. The water pressure (u) at a point is equal to the piezometric level (h_{piez}) minus the elevation of the point (h_{point}) multiplied by the unit weight of water:

$$u = \gamma_w (h_{piez} - h_{point}) \qquad [4.1]$$

The water pressure within a soil or rock stratum or crack, as reflected by the piezometric level, is a major factor in shear strength and the most significant single factor in landslide activity. A number of piezometric levels can be defined at any location. Although the two types of water tables (normal and perched) are well recognized by geologists and geotechnical engineers, the multiplicity of piezometric levels present in hillside areas, particularly in areas prone to landsliding, may surprise even experienced groundwater hydrologists.

A normal water table is the level to which the surface water infiltrates in the ground or the level at which the water pressure is atmospheric. Below this level, the groundwater is more or less continuous and pressure increases hydrostatically. A perched water table is one that is sustained above an underlying independent body of groundwater table by an aquiclude. Normal aquifers are sometimes converted to perched aquifers by the rotational

movement associated with landslides. The water table changes with rainfall, groundwater flow from outside the area, and movement of the landslide. Thus, the changing nature of a perched water table is one of its most important characteristics.

An artesian aquifer is one in which the level of atmospheric water pressure is higher than the upper surface of the aquifer, but the water is confined by an overlying aquiclude. If the water pressure level in a confined aquifer coincides with that of the aquifer above, a normal water table is present at that instant. However, drainage of the overlying aquifer may not necessarily affect an underlying confined aquifer; thus, a confined aquifer could be normal on one day and artesian on the next, without a change in pressure. Moreover, it might even become perched with more drastic changes in water distribution below the aquifer.

Groundwater Observations

Piezometers

Essentially, a piezometer is a small-diameter well in which the water level or water pressure in an aquifer can be observed. Piezometers have many forms. Simple borings with slotted or perforated casings become observation wells that reflect normal or perched water levels. Wells penetrating into a confined aquifer can measure artesian pressure if the casing is sealed into the aquicludes above. The design and the installation of an adequate system of piezometers require a thorough understanding of the location and permeability of the aquifer and of the surrounding aquicludes. If the volume of water within a piezometer tube or well is large with respect to the flow through an aquifer, the piezometer will be slow to respond to pressure changes. Thus, in aquifers of low permeability or flow, piezometers that require small water volume changes in order to respond to pressure changes are essential. In pervious aquifers, simple holes supported with perforated, screened plastic pipe surrounded by a filter of sand are adequate. Details of typical piezometers and an electric water-level detector are shown in Figure 4.14.

Monitoring Changes

The changing nature of groundwater is well recognized by geologists and geotechnical engineers. Changes in rainfall infiltration and changes produced by groundwater usage are common. However, the changes produced by a landslide are less well recognized. For example, the cracks associated with the landslide may create a more pervious zone of soil or rock that drains well-established aquifers. The opposite may also be true: Aquifers may be blocked by the landslide movements, and thus a normal water table becomes artesian. Groundwater levels must be observed throughout the period of slide investigation. During periods of dry weather, in which there is little movement of the landslide, observations at intervals of a week or two may be adequate. When the slide is moving rapidly, and particularly during and following periods of snowmelt and rainfall, daily, hourly, or continuous readings by a recorder may be desirable to correlate episodes of ground movement with groundwater changes.

Permeability

A knowledge of the ability of a formation to transmit water (permeability) is essential in the planning of drainage systems to correct landslide activity. However, the effects of minor variations in soil texture and particularly the effects of cracks and fissures cannot be easily determined. The permeability of the soil probably varies more from point to point than does any other soil property. Furthermore, the order of magnitude of the variation of permeability in common soils, ranging from gravels to clays, is greater than the variation of the other properties relevant to landslide analyses.

The simplest form of in-place permeability tests is a boring that is cased through the various soil strata down into the aquiclude above the aquifer whose permeability is to be measured. The hole is then drilled into the aquifer. The soil permeability is measured by adding water to the hole to a predetermined level and noting the rate at which the water level drops (4.17). The test should be repeated several times if it is suspected that the aquifer is not saturated; otherwise, the inflow during the first few trials may merely represent the filling of empty soil voids. Numerous modifications of this simple approach have been proposed to take into account the penetration of the hole into the aquifer, the hole diameter, and other geometric characteristics. However, experience shows that highly refined analyses of such a simple test are seldom justified.

A better field permeability test involves pumping water into the hole and measuring the rate of flow once equilibrium has been established. The water pressure is measured at several different locations and distances within the same aquifer at points surrounding the hole. Typical distances from the inflow hole are 5 and 15 m (15 and 50 ft); the piezometric holes are aligned in at least four directions. A more thorough investigation involves three sets of piezometers at distances such as 5, 10, and 20 m (15, 33, and 65 ft) and at least four different directions from the inflow hole (4.14). If the piezometric level in the aquifer is sufficiently high, a pump-out test can be used instead of a pump-in test. The arrangement of observation wells or piezometers is similar. The test hole must be sufficiently large so that the pump can be placed inside it; otherwise, the water can be lifted only about 8 m (25 ft). Pump-out tests are sometimes more reliable than pump-in tests because any soil fines that accumulate in the well are flushed out by the flowing water. In pump-in tests, those fines can accumulate in the soil or rock pores and give a false indication of low permeability. In pump-in tests, care must be taken not to contaminate an aquifer that is used for drinking water. For example, river water for pump-in tests should be chlorinated.

If the groundwater is definitely discharged through seeps and springs whose flow can be collected and measured, the permeability of an aquifer from those discharges and the gradient found within the aquifer may possibly be estimated by a grid of piezometers. Typically, the flow is downhill through an aquifer toward the toe of the slope. The piezometers, therefore, should be in lines parallel to the direction of maximum slope in order to measure the hydraulic gradient. Of course, a single piezometer might be introduced into the aquifer on the assumption that the water

pressure at the point of exit is approximately atmospheric. However, the actual water pressure just within the aquifer at the point of exit is usually greater than atmospheric, and such estimates are likely to be high. When several such determinations are made, they can yield the order of magnitude of the permeability of the stratum under its natural conditions of flow.

Springs and Seeps

The intersection of an aquifer with the ground surface produces either a concentrated flow in the form of a spring or a diffused flow in the form of a seep. Both of these represent the exit or discharge of the aquifer and may be regarded as safety valves for the release of groundwater pressure. So long as springs or seeps flow freely, an unusual buildup of groundwater pressure in the aquifer that supports the spring or seep will not be likely. If the rate of discharge of the spring or seep is known and if the permeability of the formation can be estimated, an estimate is even possible of the pore-water pressure within the aquifer at varying distances within the hillside. Unfortunately, engineers and geologists who evaluate landslides often regard a spring or seep as a causative factor (it can be, if the effluent of a spring above a slide infiltrates downhill into the slide zone), but generally the total effect is more beneficial than detrimental. For example, a sudden stoppage of a spring or seep may be the precursor of landslide activity. Increased flow in a seep or spring frequently indicates that an aquifer is draining, the piezometric pressures are reducing, and the stability is increasing. Springs and seeps suggest which potential aquifers contain water and could be involved in the buildup of pore-water pressure. Thus, a spring or seep once located should be identified with the particular soil stratum or rock formation that produces the flow.

A significant discharge of a spring or seep should be collected and the quantity monitored by a V-notch weir or similar device. The discharge from small springs can be collected by use of 5 to 10-cm (2 to 4-in) plastic pipes embedded in gravel-filled collecting wells. The flow can be piped to a buried oil drum, which serves as a catch basin for silt; the edge of the oil drum can be cut to form a V-notch weir, which serves as a measuring device. It is sometimes helpful to install a continuous water-level recorder to indicate rapid changes in the spring or seep discharge. As previously pointed out, infrared photography, as well as thermal sensing, can show where springs and seeps are located, even though they may be partially obscured by vegetation or colluvium cover.

Correlation

The location of aquifers and springs or seeps should be shown on both the topographic maps of the landslide area and the various landslide cross sections that are plotted from topographic data. Since water coming either from runoff or from infiltration or groundwater originating elsewhere is a major factor in most landslides, the evaluation of the groundwater aquifers and the changes in the piezometric level are a vital part of the investigation. Unfortunately, experience shows that these groundwater changes seldom

Figure 4.14. Groundwater observation devices.

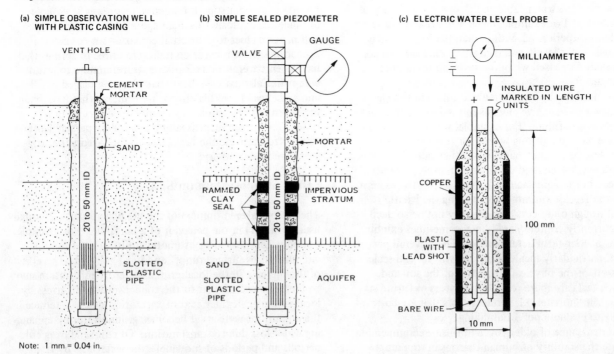

(a) SIMPLE OBSERVATION WELL WITH PLASTIC CASING

VENT HOLE
CEMENT MORTAR
SAND
20 to 50 mm ID
SLOTTED PLASTIC PIPE

(b) SIMPLE SEALED PIEZOMETER

VALVE
GAUGE
MORTAR
RAMMED CLAY SEAL
IMPERVIOUS STRATUM
20 to 50 mm ID
SAND
SLOTTED PLASTIC PIPE
AQUIFER

(c) ELECTRIC WATER LEVEL PROBE

MILLIAMMETER
INSULATED WIRE MARKED IN LENGTH UNITS
COPPER
100 mm
PLASTIC WITH LEAD SHOT
BARE WIRE
10 mm

Note: 1 mm = 0.04 in.

are evaluated in sufficient detail to obtain a complete picture of the factors leading to landslide activity.

ENVIRONMENTAL FACTORS

Both natural and human changes in the environment have a profound effect on landslides. The history of the area, both during human occupation and during recent geologic times, determines the conditions leading to land movement; moreover, as mentioned previously, the historical behavior provides clues to potential instability. Therefore, the total environment must be investigated to provide both the historical background and a key to future changes.

Weather

The climate of the area, as expressed in the various components of weather, is the ultimate dynamic factor influencing most landslides. The data ordinarily available for landslide investigations from weather stations within 100 to 200 km (60 to 120 miles) include rainfall (15 min, hourly, and daily), temperature (daily maximum and minimum and daily and monthly mean), evaporation (daily and monthly), wind (maximum, hourly, daily, and monthly), snowfall (daily and monthly), relative humidity (daily and monthly), and barometric pressure (maximum and minimum daily). Unfortunately, the weather station may be too distant for the data to be fully representative of the site. The effects of these factors can seldom be evaluated analytically because the relations are too complex. Empirical correlations of one or more of the weather factors (particularly rainfall, snow, and melting temperatures) with episodes of movement or movement rates can point out those environmental influences that must be controlled to minimize movements.

Human Changes Before Construction

Many areas of the world have been altered by human activities, such as terracing for agriculture, diversion of streams, mining, leveling for housing and industrial construction, and cuts and fills for highways, airfields, and railroads. Another human activity is alteration of the groundwater table. The piezometric level is lowered significantly by wells. In areas adjacent to large cities that depend on groundwater or in areas in which wells are used for irrigation, the level has been depressed more than 100 m (300 ft); the affected area may extend for many kilometers. This lowering of the piezometric level within the soil and rock imposes an increased effective stress, which in some instances contributes to the stability of the slope. Paradoxically, lowering the groundwater table can also trigger sliding because of the difference in the total and buoyed weight of the soil. Draining land for agriculture may lower the water table, but this is seldom of significance in landslides because such drainage is generally in bottomlands, which are not subject to sliding.

A rise in the water table can ultimately have an adverse effect on slope stability. Such a rise may occur around small towns that begin to use surface water instead of groundwater to meet increased water demands. Irrigation in arid regions has significantly elevated the groundwater table and thereby significantly decreased the strength of soils that are cemented by water-soluble agents such as calcium carbonate or dry clay. For example, in certain areas of India, hilly land, which for years had been stable, suddenly became landslide-prone after irrigation began. In one area of the Himalayas, a mountain road was destroyed each year by landslides for a distance of about 10 km (6 miles) because of irrigation of the terraced mountainsides above. The irrigation water penetrated through the closely jointed rock and produced flow slides. Similarly, water impoundments can raise the water table and change the stability of nearby hillsides. For example, a highway fill across a reservoir suddenly experienced sliding when the soil was saturated at high water levels, and then stress and localized porewater pressure increased when the reservoir was rapidly drawn down.

Changes in slope by excavation and filling in the vicinity of the study area may give clues as to the long-term effects of excavation and filling. For example, some rocks, such as shale, are hard and strong when first excavated. However, exposure to air and changing weather causes the shale to break down, soften, and slide. Such failures have been experienced in the Appalachian Mountains, on the eastern flanks of the Rocky Mountains, and along the Pacific Coast. Although older cuts and fills may not be so deep as those currently contemplated, they sometimes exhibit symptoms of long-term deterioration. Their overall performance, particularly local sliding, constitutes full-scale, in-place tests of the physical properties of the soil and rock. Such tests are more reliable indicators of future stability than the short-term laboratory tests that are customarily made to evaluate potential sliding.

A reconnaissance of older cuts and fills is recommended to evaluate the stability of human changes in topography. Calculations based on information regarding unstable areas will provide valuable data on the ultimate strength of the materials and on the degree and rate of change that can occur because of weathering. This reconnaissance of existing cuts and fills has been sorely neglected in most landslide studies, yet the data obtained are probably the most reliable of all in evaluating the strength of the materials. To make a valid correlation between the performances of old cuts or fills and those proposed requires that the geologic similarity of the two areas be established. Moreover, the climate at the area of old construction must be correlated with that at the new area if the older area is not in the immediate vicinity of the site under construction.

Changes Brought by Construction

New construction in areas in which landslide activity is suspected or has commenced should be monitored to determine whether changes predicted by studies made before construction actually occur, to evaluate the methods used in such studies, and to predict future landslide activity in an area where movement is just commencing. The major purpose of monitoring, however, is to determine what changes the construction actually produces. Because changes in the design of slope and drainage features are often made in the field, the completed project is usually different from that shown in the plans and specifications. Moreover, fills may not be compacted as specified or intact materials in virgin cuts may be loosened by uncontrolled blasting. Sometimes, the person responsible for changes in the plans and specifications will take no notice of them because of ignorance of their importance or will hide them because of fear of criticism or recrimination. Thus, the investigator has difficulty in finding out exactly what changes might have been made to the provisions of the plans and specifications and is often placed in the position of a detective ferreting out information.

A detailed study of the daily logs of inspectors, daily reports of project superintendents, records of blast-hole advance and explosives used, and journals of the contractor will provide clues to the actual construction conditions. Sometimes these records contain references to springs that have been covered by embankments, references to unsuitable materials that have been inadvertently placed in fills, and evidence of overblasting or movements that have been forgotten in the scramble to complete construction on time. The written records supplemented by personal interviews often provide the best clues to the factors that triggered landslide activity.

Effect of Ecosystem on Sliding

The biological environment of the site plays a part (although usually minor) in the behavior of a landslide. For example, a good vegetation cover promotes infiltration of rainfall and minimizes surface runoff and local gullying. The effect of the increase in the local groundwater by infiltration may be less serious than that of the concentrations of stress by local gullying. A thick mat of vegetation will also reduce the amount of water that becomes groundwater by enhancing water loss due to transpiration. Of course, during cold periods and periods of snowmelt, the water loss by transpiration is negligible.

The reinforcing effect of a strong root mat is significant in the scarp area of landslides. In some marginally stable slopes, the root mat can be the difference between sliding and creep. Moreover, a well-developed root system significantly reduces retrogression of the scarp up the hill. In highway excavation, the root mat may be affected by factors other than the slope. For example, the intercepting ditch, which is sometimes excavated above a deep highway cut on a hillside, may cut through the root mat and become the focal point for future sliding. Haul roads above cuts similarly damage the roots and promote local sliding, which leads to retrogression.

Animal burrows may also play some part in small slides. For example, the interlacing burrows of rodents can weaken the soil and provide channels for concentration of surface water and its infiltration into the ground. However, there is little documented evidence of such effects.

Overgrazing of hillsides reduces the vegetation cover and frequently promotes more rapid infiltration and localized sliding, which then triggers more profound movements, particularly in semiarid regions where vegetation does not recover rapidly from the overgrazing and where the rainfall may be extremely intense during short periods.

Effect of Sliding on Ecosystem

Landslides may also change the ecosystem of an area. Groundwater flows are altered by landslide movement; locations of springs and seeps change, and these changes are reflected in differences in vegetation. For example, wet-area vegetation, such as cattails, often develops within a few months in the depressed areas commonly associated with landslides. Similarly, the cracks and the scarp areas above a landslide provide local drainage of the topsoil and an associated loss in the ability of this material to sustain growth of vegetation during periods of dry weather. Thus, scarps that are hidden by vegetation are revealed by a reduction in the vigor of that vegetation. This can be sensed by color infrared photography during dry weather. On the other hand, in areas of bare rock, open joints in which moisture is trapped and roots penetrate provide the only zones that can sustain life. In one particular major rock slide in a steep mountainous area, the joints were clearly delineated by lines of small shrubs, but the remaining rocks were bare. Those joints became the scarps of multiple rock slides. The continuing enlargement of uphill joints above the rock slide could be seen in the gaps between the root mats and the walls of the cracks. Thus, a study of the vegetation of an area may provide supplementary clues to the rate of earth movement and its focal points, such as seeps, springs, cracks, and fissures in the ground.

FIELD TESTING

To evaluate the potential stability of a slope that has not failed or to assess the effectiveness of corrective measures for a slope that has failed requires that the physical properties of the materials involved be measured. Of course, measurements are most conveniently done in laboratory tests of undisturbed samples secured from the site, but obtaining representative samples is difficult for a number of reasons.

1. Discontinuous samples with relatively small diameters can miss thin critical strata (such as the slickensided surface of movement of an ancient slide) that control sliding.
2. The laboratory tests can integrate neither the effects of the discontinuities, such as cracks, within the soil nor the effects of localized hard spots, such as gravel, in a clay matrix.
3. Distortion, disturbance, and moisture and stress changes are always associated with taking a sample out of the ground, handling it, transporting it, and preparing it for laboratory testing.

To eliminate these difficulties, various in-place tests have been devised. An important difference between laboratory and in-place tests is their relation to the initial in situ state of stress. A laboratory test must often reproduce this state of stress whereas a field test inevitably begins at this state of stress. In either case, the in situ state of stress must be evaluated, which is often difficult and expensive. For some soil deposits, such as a deep, normally consolidated clay with a horizontal surface, the horizontal stress will be a fraction of the vertical stress, depending on the coefficient of earth pressure at rest. Unfortunately, such a simple situation is seldom valid for landslide studies because the ground surface is not level and other significant stresses may be imposed by desiccation, artesian water pressures, tectonic forces, residual forces from strains produced by past earth movements, and changes produced in any of the above by erosion or construction.

The fact that any in-place test must be conducted with reference to the existing in situ state of stress is an important limitation in itself. Although the stress field is altered somewhat by boring or introducing some testing device into the soil, the existing stress field cannot be changed significantly. Thus, in-place testing usually cannot simulate the large changes in stress that accompany environmental

changes, cutting, filling, and landslides. Although laboratory tests can simulate an almost unlimited range of stress changes, the sample tested has already been subjected to a significant stress cycle (unloading during sampling and reloading during testing) that is not necessarily similar to the stress changes involved in a landslide. These inherent stress limitations of both in-place and laboratory tests must be understood by those who use soil or rock test data in evaluating slides.

Borehole Tests

Certain tests have been devised to be performed in the same bore holes that are drilled for identifying the soil strata and for securing the small-diameter samples. Although borehole tests suffer from the limited volume of material tested, they do allow the soil to be tested without the disturbance produced by removing the sample from the ground, taking it to the laboratory, and preparing it for testing. However, some disturbance and sloughing are caused by stress relief in the bore-hole walls.

Dynamic Penetration Test

Driving a device into the ground by impact measures the resistance of the soil to rapid or dynamic displacement. Thus, it indirectly measures shear strength under the same form of impact loading produced in the test. If the dynamic and static shear strengths are similar, the test can be an indicator of the static shear strength, which may also be empirically related to other soil properties, such as the relative density of sands or the compression index of clays. Typical correlations are summarized by Sowers and Sowers (*4.14*).

The standard penetration test (ASTM D 1586) is an adjunct to split-tube sampling. A split-tube sampler 37 mm (1.4 in) ID, 50 mm (2 in) OD, and 660 mm (26.0 in) long (Figure 4.15) is driven 450 mm (18 in) in three 150-mm (6-in) increments into undisturbed soil at the bottom of a borehole by blows of a 63.5-kg (140-lb) hammer falling 760 mm (30 in). The sum of the blows for the second and third increments is the standard penetration resistance (N); it is expressed in blows per 300 mm (blows per 1 ft). Before the test, the sampler is seated 20 to 40 mm (8 to 16 in) in the hole bottom. Because of cuttings or other weakened material in the bottom of a bore hole, the first 100 to 150 mm (40 to 60 in) may not be meaningful. Therefore, the standard penetration test includes gaps in the penetration resistance record. The N-value encompasses both hard and soft seams in the 300-mm distance. Although some investigators have attempted to drive such sampling tubes as far as 1.8 m (5 ft), counting blows for each 150-mm increment, the accumulating skin friction and the buildup of soil resistance within the samples usually produce resistances that increase with each successive increment until the sample is withdrawn and the borehole cleaned out.

A more sensitive dynamic test involves driving a cone point that is 25 to 100 mm (2 to 4 in) in diameter and has a point angle of 60° into the soil by means of a weight [typically 50 to 100 kg (110 to 220 lb) falling 0.5 to 1 m (1.5 to 3 ft)]; the drive rod is 25 to 30 mm (1 to 1.2 in) in

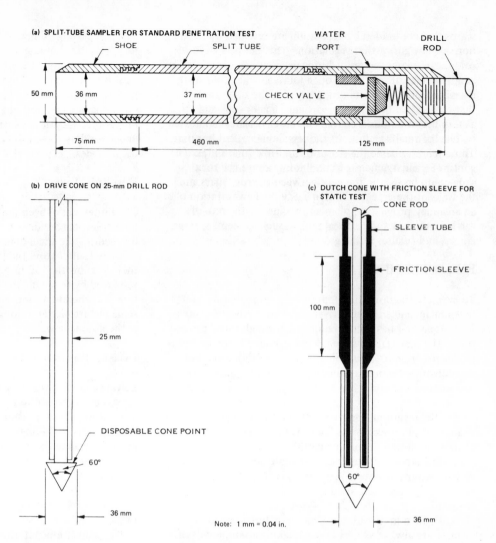

Figure 4.15. Penetrometers.

(a) SPLIT-TUBE SAMPLER FOR STANDARD PENETRATION TEST

SHOE SPLIT TUBE WATER PORT DRILL ROD

50 mm 36 mm 37 mm CHECK VALVE

75 mm 460 mm 125 mm

(b) DRIVE CONE ON 25-mm DRILL ROD

25 mm

DISPOSABLE CONE POINT

60°

36 mm

(c) DUTCH CONE WITH FRICTION SLEEVE FOR STATIC TEST

CONE ROD

SLEEVE TUBE

FRICTION SLEEVE

100 mm

60°

36 mm

Note: 1 mm = 0.04 in.

diameter. By such a procedure, the penetration resistance for each 100 to 300 mm (4 to 12 in) can be measured without the accumulation of soil friction along the drive rod. In many soils, a 50-mm (2-in) point driven with the 63.5-kg (140-lb) hammer falling 760 mm (30 in) gives penetration resistances for 0.3 m (1 ft) of driving that are equivalent to N. The drive cone on a 25-mm rod is also shown in Figure 4.15.

Dynamic penetration resistance is generally correlated empirically with soil properties measured by either laboratory tests or field tests of the same material. In this way, large numbers of low-cost penetration tests supplement the more limited information obtained by more expensive laboratory tests. Although many relations between resistance and soil properties such as angle of internal friction in sands and undrained shear strength (cohesion) of saturated clays have been published (*4.14, 4.16*), these should not be used indiscriminately. Instead, a new correlation should be established from the data obtained on the site in question, or the data should be used to verify the accuracy of the published relations (*4.5, 4.13*).

Static Penetration Test

Static penetrometers measure the resistance of the soil to displacement of a point by a static or slowly increasing load. Most tips are in the form of cones having point angles

ranging from 30° to 90° and diameters from 36 to 50 mm (1.4 to 2 in), as shown by Hvorslev (*4.6*). One form of cone widely used in the Netherlands and generally termed the Dutch cone employs a cylindrical sleeve that is 100 mm (4 in) long and 36 mm (1.4 in) in diameter and is above a 60°, 36-mm (1.4-in) conical point, as shown in Figure 4.15. The cone is forced ahead slowly by a steady pressure that is measured. Simultaneously or successively, depending on the design, the frictional resistance of the sleeve against the soil is measured. The cone directly provides information on the point bearing capacity of the soil. It can be interpreted in terms of the point bearing of piles. Although at one time the cone resistance of soil was believed to be identical to that of a pile, experience has shown that cone resistance may be double that of piles. The cone does provide detailed information on the relative strength of the soil at small intervals. For example, some cones have an electronic readout that generates a continuous graph or electromagnetic tape showing both cone resistance and sleeve resistance as a function of depth. The record is well suited to identifying weak zones, such as the shear surface of the slide, which may be less than a few centimeters thick. By way of contrast, conventional sampling might not find such a thin zone of weakness. The ratio of cone resistance to sleeve resistance is an indicator of the type of soil (*4.12*). The cone penetrometer is an extremely valuable supplement to the more direct boring and sampling techniques. It helps to identify changes in stratification and to pinpoint weak materials that should be investigated in more detail by direct methods.

Borehole Dilation Test

A number of field tests have been based on the resistance of a cylindrical borehole to dilation from applied internal pressure. The best known of these is the Menard pressuremeter (Figure 4.16). Most devices use a cylindrical rubber tube that closely fits the inside of the borehole. The tube is inflated with a fluid under pressure, and the expansion of the hole is measured by the volume of fluid that exceeds that required to fill the original hole. A plot of fluid volume (converted to hole diameter) as a function of pressure is used to compute the in-place deformation characteristics of the soil; the results may be interpreted to determine the in-place shear strength. In the Menard device, the end effects in the measuring cylinder are minimized by means of additional similar cylindrical rubber tubes above and below; these tubes are inflated to the same pressure as the test cylinder, thereby providing a two-dimensional stress configuration rather than a three-dimensional one, which has a more complicated elliptical zone of strain. Other similar devices omit the end tubes and depend on the theoretical interpretation of the elliptical zone of stress and strain. Still others employ mechanical sleeves and strain sensors to measure pressure and displacement.

Although it is claimed that such devices can provide the user with all of the necessary soil properties to evaluate shear and consolidation, the interpretation is largely empirical and certainly open to question in variable materials. Typically, the lengths of these devices limit the stress zone to a length of about 0.6 m (2 ft) and diameter of about 0.3 m (1 ft). The use of the test results without confirmation by other means would be unwise.

Borehole Shear Test

The borehole shear test measures the shear strength of the soil in an annular zone surrounding the boring. The device consists of an expandable plug with serrations on its outer surface to grip the soil walls of the hole when pressure of known magnitude is applied internally by a hydraulic system. The soil is then sheared by pulling the device upward through the hole. If several such tests (essentially undrained direct shear) are made on the same stratum at varying internal pressures, the Mohr failure envelope can be obtained. The test is limited because it shears the soil in a different direction than that involved in the landslide process. Therefore, if the soil has anisotropic properties (usually the axis of weakness is parallel to the greatest extent of the surface of shear movement and more or less perpendicular to the direction of shear in the borehole shear tests), the results may be misleading. There is usually some soil smear in the walls of the borehole; thus, the soil involved in the test may be partially disturbed. However, the disturbance from this cause is likely to be less than that resulting from rough handling of soil samples. The size of the device is such that it integrates the effect of soil irregularities over a cylindrical surface with a diameter of 76 mm (3 in) and a length of about 300 mm (12 in).

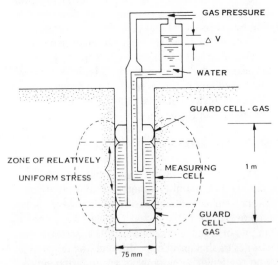

Figure 4.16. Menard pressuremeter for borehole dilation test (not to scale).

Note: Gas in guard cells; water in measuring cell.
1 m = 3.3 ft; 1 mm = 0.04 in.

Figure 4.17. Vane device.

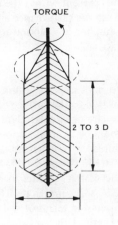

Vane Shear Test

Vertical blades at the end of a thin rod produce a vertical cylindrical surface of shear when rotated (Figure 4.17). The torque required to initiate continual rotation is a measure of the peak undrained strength of the soil, and the torque required to maintain rotation after several revolutions measures the residual or disturbed strength. To minimize end effects, the length of the vane should be at least twice its width. The blades should be sufficiently thin that there is a minimum soil disturbance due to displacement and sufficiently thick that they do not bend under load. In very soft soils, the vane and its torque rod are forced into the soil to each level to be tested. A reference test using the torque rod without the vane is required so that the torque necessary to overcome rod friction can be subtracted from the total torque measured when the soil is tested. In firm soils or at great depths, the test is made in undisturbed soil 300 to 760 mm (12 to 30 in) below the bottom of a bore hole; hence, the resistance of the torque rod in the hole generally is negligible. Numerous procedures and forms of equipment, ranging from simple torque wrenches to elaborate torque meters that apply a uniform angular strain rate, have been utilized (4.6). Our opinion is that the increased accuracy of the results does not justify the use of elaborate procedures and complex equipment. Caution should be exercised when one interprets the peak strength; in some cases, the strength measured in a vane test has been found to be as much as 30 percent greater than that measured by other methods. Moreover, safety factors computed from such strengths have been found to be unrealistically high. Vane data should be correlated with other shear data for use in analysis and design.

Large-Scale Pit Tests

As previously stated, one of the major limitations of laboratory tests is their inability to integrate the variations in the soil, particularly in zones with weak or hard spots. This can be overcome by large-scale, in-place tests performed in pits or trenches excavated to the questionable strata or zones of slickensiding. Although the range of stresses and particularly the range of groundwater pressures that can be evaluated by such tests are limited, the tests permit large volumes of soil to be evaluated under the conditions present within the total mass without the problems of sample disturbance and exposure inherent in small-scale sampling and laboratory testing.

Load Test

The oldest form of in-place test is the plate load test (4.14). A pit is excavated to the surface of the stratum in question, and a rigid square or circular plate is placed on the ground. Its width should be as great as possible, but no wider than about two-thirds the thickness of the stratum whose strength is to be evaluated. The plate is loaded incrementally so that at least 10 successively greater loads are applied before the plate shears the soil beneath it. The results of such a test can be interpreted in terms of soil bearing capacity to give the shear strength of the soil along a curvilinear surface (which may crudely approximate the failure surface of a landslide). However, there are many different interpretations of such tests, all yielding different values for the shear strength parameters. Therefore, the test has limited value in determining the strength of the soil involved in the instability of large earth masses.

Large-Scale Direct Shear Test

A large-scale direct shear test can be performed in a pit at the level of the suspected weak stratum; in the case of an existing landslide, the test may be performed on the actual failure surface of the soil. The pit is excavated to the level of the stratum or shear surface to be evaluated and should be large enough to allow engineers and technicians to work around the sides without disturbing the soil to be tested in the center. All the soil within the pit is excavated, except that to be tested, which is left in the form of a block or crude stump above the bottom of the pit. The size of the block is dictated by the engineer's or geologist's evaluation of the variations of soil strength. It should be large enough to be representative of the stratum as a whole and not just its weaker or stronger segments. A double box is placed around the block in question. If there is a definite plane

of weakness, the sides of the box should be perpendicular to that plane, and the plane should lie between the top and bottom halves of the box. Such a test setup is shown in Figure 4.18. Good contact is secured between the soil and the box either by careful trimming of the soil or by pouring plaster to fill the space between the box and the soil. Because of the difficulties in trimming gravelly soils or soft rock, the plaster filling is recommended.

A normal load is placed on the block to be tested by means of a plate, the dimensions of which are slightly smaller than that of the box, fitting just inside the upper half of the box. The load is applied by jacking against a piece of heavy machinery above the pit or against a heavy steel beam anchored to the ground by earth anchors. The anchors must be sufficiently far from the test zone so that the stresses around them do not influence the test. The bottom half of the box is anchored securely in place by packed soil, concrete, or plaster in the bottom of the test pit. The top half is then jacked sideways by a calibrated system so that the amount of lateral movement and the load causing the movement can be measured. The direction of jacking should be parallel to that which is suspected in a potential landslide zone or parallel to the movement that has occurred in an actual landslide (as indicated by slickensides on the shear surface). The same surface can be tested at several different normal loads if the test for each vertical load increment is stopped soon after peak strength or significant movement develops. At this point, a larger normal load is applied and the test is resumed. Such a direct shear test determines the average shear stress required to produce failure on a predetermined failure plane. If the peak strength of the soil must be determined, separate tests should be performed on fresh sample blocks for each normal load applied. This will require a large test trench or more than one test pit.

The results of a large-scale shear test simulate the shear strength of the soil along an actual failure surface. The results integrate the effects of both hard and soft zones if the test sample is sufficiently large. Unfortunately, it is difficult to include the effects of changing water pressure. However, our experience shows that meaningful shear test data have been obtained from such tests, particularly if done during the wet season. Correlating the results of an in-place test with those of smaller laboratory tests on similar soils makes it possible to extend the data obtained from the in-place tests to include the effects of changing water pressure introduced under controlled laboratory conditions. Thus, a more reliable combination of data for evaluating the strength of the soil mass involved in the landslide is provided by a combination of judiciously selected large-scale field tests and laboratory tests than by laboratory tests alone.

Standard equipment for making such tests is ordinarily not available. Instead, the equipment is fabricated to fit the size of the sample needed for the particular situation, the space available within the test pits or trenches, and the geometry of the shear surface. It can be improvised out of steel angles, channels, and plates at a reasonable cost. The loading is provided by calibrated hydraulic jacks or jacks with load cells; and the movements are measured by micrometer dial gauges. Although such measurements may be characterized as crude, their lack of precision is more

Figure 4.18. Direct shear test of strength along failure surface. (a) PLAN VIEW

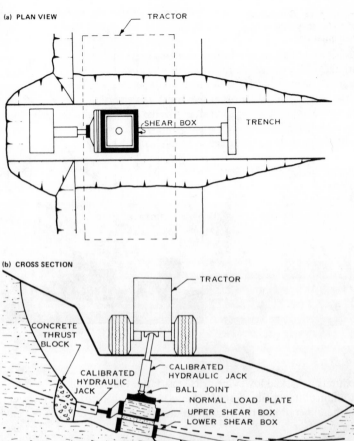

(b) CROSS SECTION

Note: Trench shored for safety and support of tractor.

than compensated for in their realistic representation of field conditions.

Borehole Dynamics

The effect of a landslide on a borehole, particularly in squeezing or shearing, can be a valuable indicator of rates of movement and locations of shear surfaces. Although equipment for measuring such changes is available, much meaningful information also can be obtained from drillers' observations in those borings that are not instrumented. Among the factors to be observed are loss of drilling fluid, gain of water or fluid, gas, squeezing of hole, damage to drilling tools, loss of samples, blocking of cores, drill rod drop, broken core, slickensided samples, and unusually weak zones.

Geophysical Tests

The geophysical studies previously described can be used not only as a means to define the stratification but also as a direct measure of certain physical properties of the soils. Although these properties may not necessarily be those needed to evaluate stability, an empirical correlation among the properties of interest can sometimes be established in the stability analysis of the landslide.

The seismic refraction technique yields the compression wave velocity of the soil or rock. If the density is known, the dynamic modulus of elasticity in compression can be

calculated. Since the density of various soils and rocks does not vary greatly from point to point, calculations based on estimates may be sufficiently reliable for estimates of the rigidity of the mass. The seismic compression wave velocity is particularly valuable as an indication of discontinuities in rock, because it is reduced by cracks or microfissures.

The refraction technique does not allow identification of a weak fractured zone beneath a sounder, higher velocity stratum. However, this identification can be done by cross-hole seismic measurements, utilizing borings from 3 to 12 m (10 to 40 ft) apart. The seismic impulse is generated at the depth of the stratum measured by an explosion, and the time required for the shock wave to travel through the material is measured in the second hole by a geophone suspended at the level of the stratum in question. Although engineering properties relevant to landslide evaluation, such as shear strength, cannot be determined directly by such techniques, the suspect stratum can be identified and a crude empirical correlation can be developed between seismic velocity and engineering properties. Below the water table the compression wave velocity in materials of low rigidity is obscured by the compression wave velocity in water.

The apparent resistivity of a soil can be related empirically to soil type and soil moisture. An aquifer with highly ionized water exhibits a far lower apparent resistivity than a dry stratum in which the minerals do not ionize readily. For example, the shear surface within a landslide frequently

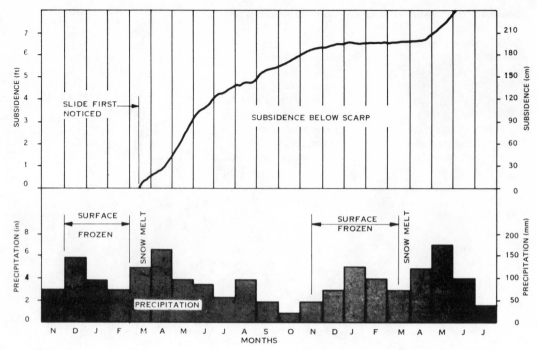

Figure 4.19. Correlation of slide movement with precipitation.

CORRELATION OF DATA

During an investigation, considerable data will be accumulated. The variations of these data, which may occur randomly or in some definite pattern, should be studied in three dimensions because the mass of soil or rock involved in a landslide is three dimensional. Some of the ways in which this can be accomplished are discussed briefly below.

Areal Variations

Some of the data obtained will vary with geographic position. These data can be depicted best in the form of maps. If the variations in the data are systematic, they can be depicted as contours, such as lines of equal strength or lines of equal groundwater pressure. These variations are frequently plotted on overlays to the topographic map of the site; several overlays may be desirable.

Cross Sections

Depicting the data in the form of cross sections is particularly useful because the stability of the mass is usually evaluated by use of cross sections (Figure 4.12). Cross sections parallel to the direction of the maximum slope, to the

exhibits a marked lower electrical resistance than the surrounding intact soils, probably because of the disturbance of the minerals and the accumulation of water along the shear surface.

At best, however, geophysical tests are only indirect supplements to the more direct means of evaluating the qualities of the soil and rock in place. The tests are simple, can be performed in a short period of time, and are relatively inexpensive. However, interpretation of the results must be done with extreme caution; too often interpreters jump to conclusions and obtain misleading results.

maximum water pressure gradient, or to the observed trajectories of movement are the beginning point for a stability analysis of an earth mass. The various strata encountered and the groundwater levels or pressures are depicted on these cross sections. In addition, overlays should show the engineering properties relevant to the stability analyses. The surfaces of failure, as deduced from ground surface observations, sophisticated instrumentation, borings, test trenches, and pits, define a body of soil on which various force systems act. The external forces depicted on an overlay to the cross sections provide the beginnings of the analyses described in Chapter 7.

Time-Based Observations and Correlations

The significance of the different factors involved in a landslide can frequently be found from empirical correlations between movement and observed forces or environmental factors. For example, a time-based graph of both landslide displacement and rainfall or accumulated rainfall and snowmelt may show a visual relation (Figure 4.19). Similar time-based relations can be observed between construction operations and movement. In some cases, plots of the observed phenomena as functions of the logarithm of time are instructive. For example, stochastic processes, such as the readjustment of coarse particles through creep and secondary soil compression, often can be approximated by a straight line on a plot of movement as a function of the logarithm of time. Thus, the time graph or the logarithm of the time graph becomes a diagnostic tool in analyzing the causes of movement.

CONCLUSIONS

The field investigation identifies potential problem areas and defines the features involved in an existing landslide.

More intensive investigation uncovers the soil and environmental factors that produced the movement. The mathematical analysis of a landslide is based on the field investigation and obviously can be no more accurate than the data obtained from the field work; likewise, the corrective measures based on the analyses can be no more effective than the quality of the data used in the analyses. Unfortunately, our experience has been that field investigations are often inadequate and that geologists and engineers sometimes jump to conclusions before the data have been obtained and evaluated; thus, they merely perpetuate their past mistakes. A technically sound solution can be derived only from technically sound data.

REFERENCES

4.1   American Geological Institute. Glossary of Geology. AGI, Washington, D.C., 1972, 805 pp.

4.2   American Society for Testing and Materials. Annual Book of ASTM Standards: Part 19—Soil and Rock; Building Stones; Peats. ASTM, Philadelphia, 1977, 494 pp.

4.3   Compton, R. R. Manual of Field Geology. Wiley, New York, 1962, 378 pp.

4.4   Deere, D. U., and Patton, F. D. Slope Stability in Residual Soils. Proc., 4th Pan-American Conference on Soil Mechanics and Foundation Engineering, San Juan, American Society of Civil Engineers, New York, Vol. 1, 1971, pp. 87-170.

4.5   Fletcher, G. F. A. Standard Penetration Test: Its Uses and Abuses. Proc., Journal of Soil Mechanics and Foundations Division, American Society of Civil Engineers, New York, Vol. 91, No. SM4, 1965, pp. 67-75.

4.6   Hvorslev, M. J. Subsurface Exploration and Sampling of Soils for Civil Engineering Purposes. U.S. Army Engineer Waterways Experiment Station, Vicksburg, Miss., 1949, 521 pp.

4.7   Lahee, F. H. Field Geology. McGraw-Hill, New York, 6th Ed., 1961, 926 pp.

4.8   Lambe, T. W. Soil Testing for Engineers. Wiley, New York, 1951, 165 pp.

4.9   Philbrick, S. S., and Cleaves, A. B. Field and Laboratory Investigations. In Landslides and Engineering Practice (Eckel, E. B., ed.), Highway Research Board, Special Rept. 29, 1958, pp. 93-111.

4.10   Rocha, M. A Method of Integral Sampling of Rock Masses. Rock Mechanics, Vol. 3, No. 1, 1971, pp. 1-12.

4.11   Sanglerat, G. The Penetrometer and Soil Exploration. Elsevier, New York, 1972, 464 pp.

4.12   Schmertmann, J. H. Static Cone Penetrometers for Soil Exploration. Civil Engineering, Vol. 37, No. 6, 1967, pp. 71-73.

4.13   Schultze, E., and Melzer, K. J. The Determination of the Density and the Modulus of Compressibility of Non-Cohesive Soils by Soundings. Proc., 6th International Conference on Soil Mechanics and Foundation Engineering, Montreal, Vol. 1, 1965, pp. 354-358.

4.14   Sowers, G. B., and Sowers, G. F. Introductory Soil Mechanics and Foundations. Macmillan, New York, 3rd Ed., 1970, 556 pp.

4.15   Sowers, G. F. Remote Sensing: A New Tool in Site Investigation. Law Engineering Testing Co., Marietta, Ga., Bulletin G-5, 1973.

4.16   Terzaghi, K., and Peck, R. B. Soil Mechanics in Engineering Practice. Wiley, New York, 2nd Ed., 1967, 729 pp.

4.17   U.S. Bureau of Reclamation. Design of Small Dams. U.S. Bureau of Reclamation, Denver, 1973, 816 pp.

4.18   Záruba, Q., and Mencl, V. Landslides and Their Control. Elsevier, New York, and Academia, Prague, 1969, 205 pp.

PHOTOGRAPH CREDITS

Figure 4.8   Courtesy of Tennessee Department of Transportation
Figure 4.9   Courtesy of Tennessee Department of Transportation

Chapter 5

Field Instrumentation

Stanley D. Wilson and P. Erik Mikkelsen

The development of field instrumentation has had a significant impact on geotechnical engineering in the past 15 years. Of particular importance is the notable contribution of instrumentation to landslide-oriented problems (*5.50*). The usefulness of field instrumentation for the identification of landslide movements, for the monitoring of slides that have been remedially treated, and for other geotechnically oriented measurements has been described in the literature (*5.1, 5.8, 5.18, 5.36, 5.45, 5.48, 5.50, 5.56*). Although the earlier Highway Research Board Special Report on landslides (*5.20*) mentioned the use of simple instrumentation (which is by no means obsolete and is described in this chapter), its publication generally preceded the introduction of modern instruments. The economic impact of slope failures created a demand for better instrumentation and led to its general acceptance and use. This chapter describes the types of instrumentation currently available, their limitations, and their specific applications to slope instability problems. In addition, traditional approaches to the determination of slide movements and groundwater levels are reviewed.

The need to gather certain quantitative data to analyze slope-stability problems and to design remedial measures is discussed in other chapters. Topographic mapping, geologic mapping, subsurface soil and groundwater investigations, and laboratory strength testing are normally performed to aid in determining the cause of the slide, the mode of failure, and the physical and engineering characteristics of the soil and rock involved. Such investigations are necessarily performed on already developed landslides, the characteristics of which may be observed and noted. However, small movements of a soil mass prior to or even at incipient failure are usually not visually evident; so the value of information that can be obtained at the ground surface is limited. However, instrumentation can provide valuable information on incipient, as well as fully developed, landslides. In this respect, instrumentation is not intended to replace field observations (*5.38*) and investigative procedures discussed in other chapters. Instead, it augments other data by providing supplementary information and by warning of impending major movements. Typical situations for which various instruments have been used are

1. To determine the depth and shape of the sliding surface in a developed slide so that calculations can be made to determine the available soil strength parameters at failure and so that remedial treatments can be designed;
2. To determine absolute lateral and vertical movements within a sliding mass;
3. To determine the rate of sliding (accelerating or decelerating movements) and thus warn of impending dangers;
4. To monitor the activity of marginally stable natural slopes or cut slopes and the effects of construction activity or precipitation on them;
5. To monitor groundwater levels or pore pressures normally associated with landslide activity to enable effective stress analyses to be performed;
6. To provide remote digital readout or a remote alarm system that would warn of possible dangers; and
7. To monitor and evaluate the effectiveness of various control measures.

In the last situation, savings are often realized in remedial treatment by a planned and monitored sequence of construction. For example, drainage might be initially installed and its effect monitored to determine whether a planned buttress is actually necessary.

INSTRUMENTATION PLANNING

Adequate planning is required before a specific landslide is instrumented. The steps are (a) determine what types of measurements are required, (b) select the specific types of instruments best suited to make the required measurements,

(c) plan the location, number, and depth of instrumentation, and (d) develop the recording techniques.

Initially the planning process requires the development of ideas on the causes of the landslide and the probable limits of the depth and outer boundaries of the movements. Of course, to have the answers before the system is planned would be helpful. Reconnaissance of the area, study of the geology, review of rainfall records, and observation of topographic features, especially recent topographic changes, will often provide clues. Unfortunately, no two slides are alike in all details, and experience alone without the application of judgment may lead to erroneous concepts.

An instrumentation system in which the instruments do not extend below the zone of movement, were installed at the wrong locations, or are unsuitable is a waste of time and money. Loss of time may mean that corrective treatment is started too late to save the project.

Types of Measurements Required

Landslides, by definition, involve movement, and the magnitude, rate, and distribution of this movement are generally the most important measurements required. Equally important in many slide problems, however, are measurements of pore-water pressures within the slide area, particularly in layered systems in which excess hydrostatic pressures may exist between layers.

If the depth of sliding is readily apparent from visual observations, surface measurements may be sufficient for obtaining the rate of movement. The surface measurements should extend beyond the uppermost limit of visual movement so that possible extension in advance of cracking can be monitored. Vertical and horizontal measurements of movement of the ground surface at various locations within the slide area should be obtained. Vertical offsets, widening of cracks, and toe heave should be monitored. The direction of movement can often be inferred from the pattern of cracking, particularly by the matching of the irregular edges of the cracks. If the depth and thickness of the zone of movement are not apparent, inclinometers or similar devices that can detect the movement with depth must be used. Pore pressures at or near the sliding surface must be measured to enable an effective stress analysis to be performed and to assess the adequacy of drainage measures. Rapid-response piezometers are advantageous, particularly in impervious soils.

Selection of Instrument Types

Many types and models of instruments are available for measuring the changing conditions in a landslide; they vary in degree of sophistication, particularly in regard to readout capabilities. Instruments have been developed to measure vertical and horizontal earth movements, pore-water pressures, in situ stresses and strains, dynamic responses, and many other parameters. However, in most landslide problems, the measurement of horizontal movement of the foundation soil or rock and the measurement of pore-water pressures are of primary concern. Instruments commonly used for these purposes are described in this chapter.

If the movement is known to be along a well-defined shear plane, such as a bedding plane or fracture, simple probe pipes will suffice to determine the depth. If the movements are large and rapid, accuracy is not an essential requirement and even relatively crude inclinometers may suffice. When the rate of movement is small and the depth and distribution are not known, more precise instrumentation is required. Carefully installed precision inclinometers are best in such instances, although there may be cases in which extensometers or strain meters can be used to advantage.

The Casagrande type of piezometer is the most useful general purpose pore-pressure measuring device, but may give too slow an initial response in fine-grained soils; therefore, pneumatic or electric types may be preferable. High-air-entry, low-flow piezometers should be used in clays or clay-shales in which permeabilities are low or suctions may be present because of unloading (5.53).

The types of instruments, layout, and monitoring schedules are usually determined by the specific needs of a project. Several basics, however, should be thoroughly evaluated for any system. Instruments should be reliable, rugged, and capable of functioning for long periods of time without repair or replacement. They must also be capable of responding rapidly and precisely to changes so that a true picture of events can be maintained at all times. High sensitivity is usually a prerequisite when performance is monitored during construction, since it is often the rate of change rather than the absolute value that provides the key to proper interpretation. The location of instruments requires a thorough understanding of the geologic and subsurface conditions if meaningful data are to be obtained. This is particularly true of pore-pressure recording devices that are intended to measure pressures in specific zones of weakness or potential instability. Since most measurements are relative, a stable base or datum must be provided so that absolute movements can be determined.

In this chapter, the primary emphasis is on measurement techniques and instruments for landslides; less emphasis is placed on conventional surface surveying techniques. Conventional surveying methods are fundamental to displacement measurements at the surface (5.23); this has been discussed more fully in Chapter 4. Modern deformation instruments for landslides have augmented the methods of surface measurements, but more important they have expanded capabilities in terms of accuracy, simplicity, and ease or convenience of operation. In addition, an ever-expanding range of instruments makes possible the measurements of subsurface deformations (5.13, 5.17, 5.21, 5.25, 5.44). Table 5.1 gives summaries of various survey methods, instruments for measuring subsurface deformations, and piezometers.

SURFACE SURVEYING

Conventional Surveying

In an active slide area, surface movements are normally monitored to determine the extent of slide activity and the rate of movement (5.21, 5.32). Optical instrument surveys and tape measurements are used to determine lateral and vertical movements. Bench marks and transit stations, located on stable ground, provide the basis for which subsequent movements of hubs can be determined optically and

Table 5.1. Surveying methods, horizontal movement devices, extensometers, and piezometers (5.13).

Method or Measurement Instrument	Type	Range	Accuracy	Advantages	Limitations and Precautions	Reliability
Method	Chaining Ordinary, third order Precise, first order	Variable	±1/5000 to 1/10 000 of distance ±1/20 000 to 1/200 000 of distance	Is simple and inexpensive; has direct observation	Requires clear, relatively flat surface between points and stable reference monuments Corrections for temperature and slope should be applied and a standard chain tension used	Excellent
	Electronic distance measurement (EDM)	20 to 3000 m	±1/50 000 to 1/300 000 of distance	Is precise, long range, fast, usable over rough terrain	Accuracy is influenced by atmospheric conditons; accuracy at short ranges (<30 to 90 m) is limited for most instruments	Good
	Optical leveling Ordinary, second and third order		±3 to 6 mm	Is simple, fast, particularly with self-leveling instruments	Has limited precision; requires good bench mark nearby	Excellent
	Precise, first order (parallel plate micrometer attachment, special rod)		±0.6 to 1.2 mm	Is more precise	Requires good bench mark and reference points and careful adherence to standard procedures	Excellent
	Offsets from a baseline Theodolite and scale	0 to 1.5 m	±0.6 to 1.5 mm	Is simple; has direct observation	Requires baseline unaffected by movements and good monuments; accuracy can be improved by using a target with a vernier and by repeating the sight from the opposite end of the baseline	Excellent
	Laser and photocell detector	0 to 1.5 m	±1.5 mm	Is faster than transit	Is seriously affected by atmospheric conditions	Good
	Triangulation		±0.6 to 12 mm	Is usable when direct measurements are not possible; is good for tying into points outside of construction area	Requires precise measurement of base distance and angles; requires good reference monuments	Good
	Photogrammetric		±1/5000 to 1/50 000 of distance	Can record hundreds of potential movements at one time for determination of overall displacement pattern	Weather conditions can limit use	Good
Horizontal movement instrument	Long base strain meter with electrical readout	150 mm	0.3 mm	Is precise; can be used to check horizontal movement at top of other devices, such as inclinometers	Has risk of electrical failure; has limited application for tunnels	Good
	Fixed multipoint borehole deflectometers		±20 arc s[a]	Is available in portable version and double pivot version to measure movement along two axes	Is complex; does not measure continuous profile	
	Fixed multipoint borehole inclinometers		±0.03 mm in 3 m	Is precise; can be removed for repairs or reuse; uses standard inclinometer casing	Is complex; does not measure continuous profile Accuracy is lost when removed and replaced	
	Portable borehole inclinometers Wheatstone bridge pendulum	±12°, optional to ±25°	±20 mm in 30 m	Has long experience record; is not sensitive to temperature	Requires lengthy calculations; reads one axis at a time; has no provisions for automatic readout	Very good

114

Table 5.1. Continued.

Method or Measurement Instrument	Type	Range	Accuracy	Advantages	Limitations and Precautions	Reliability
	Accelerometer	±30°, optional to ±90°	±5 mm in 30 m	Reads two axes at a time; has automatic readout and recording provisions	Requires lengthy calculations without automatic readout; requires manual check of data for errors with automatic readout	Good
	Vibrating wire	±15° or 20°	±10 mm in 30 m	Is available in single or double axis models	Requires lengthy calculations. Errors due to zero drift are possible	Good
	Bonded resistance strain gauge	±20°	±10 mm in 30 m	Has adjustable range on some models; uses ordinary square tubing for casing on one model	Requires lengthy calculations; reads one axis at a time. Errors due to zero drift, temperature, or electrical connections are possible	Fair
Extensometer	Tape	1 to 30 m	±0.03 to 0.3 mm	Is simple, precise, portable; is good for measuring tunnel diameter changes	Has accuracy limited by tension adjustment; requires temperature correction	Excellent
	Portable rod	1 to 8 m	±0.03 to 0.3 mm	Is simple, precise, portable	Has limited span; has accuracy limited by sag; invar tubes can be used to minimize temperature corrections	Excellent
	Weight-tensioned wire	Variable	±5 to 20 mm	Is simple	Has creep in wire that leads to errors; is vulnerable to damage in tunnel	Fair
	University of Illinois rod	150 mm	±0.03 to 0.13 mm	Is simple, precise, and large range (some models can be reset and extended through linings placed later); can be quickly installed; is more blast and damage resistant; has easily adjusted anchors	Is not adaptable to remote reading; has two anchor units only	Good
	Interfels rod		Variable	Is simple, precise; can have multiple anchors; can accept remote readout transducers	Has projecting head that is vulnerable to damage	Good
	Variable-tensioned wire	15 to 90 mm[b]	±0.05 to 0.13 mm	Has multiple anchors, up to 6 or 8 (some models are designed for remote readings with transducers using bonded resistance or vibrating-wire strain gauges)	Has variable tension that requires varying calibration factors, wire friction and hysteresis that can seriously affect accuracy, risk of electrical failure, and projecting head that is vulnerable to damage	Fair; short term
	Constant-tensioned wire	50 mm[b]	±0.05 to 0.13 mm	Has multiple anchors; is designed for remote reading using potentiometers; has constant calibration factor	Has wire friction and hysteresis that can seriously affect accuracy, risk of electrical failure, and projecting head; is complex mechanically	Fair; short term
Piezometer	Open system		±3 mm head	Is simple, inexpensive, and adequate for most earth problems	Central observation system cannot be used	Excellent
	Well point			Is simple, inexpensive, and universally available; can be driven in place	Has large time lag in low porosity materials (k < 1 μm/s) and metallic elements that may corrode; cannot measure negative pore pressures	
	Casagrande			Is simple, inexpensive; has no metallic elements, long service life, and provisions for offset riser pipe and flushing tip	Cannot measure negative pore pressures; requires borehole and carefully placed bentonite or grout seal	

Table 5.1. Continued.

Method or Measurement Instrument	Type	Range	Accuracy	Advantages	Limitations and Precautions	Reliability
	Geonor			Can be pushed or driven into soft ground; can be placed in borehole with filter zone to reduce time lag	Cannot measure negative pore pressures	
	Cambridge			Has simple drivable piezometer with inexpensive tip and shield to protect tip during driving	Cannot measure negative pore pressures; has metallic elements that may corrode	
	Closed system		_[c]	Allows central observation system to be used; can measure negative pore pressures; is usable in low permeability soils	Is more difficult to install than open system piezometers; requires frequent and careful deairing	Excellent
	USBR			Is simple, inexpensive, and readily available; has long experience record		
	Bishop			Is simple, and designed for less frequent deairing than USBR type; can be pushed into soil from bottom of borehole		
	Diaphragm		±1% of full scale[c]	Has small time lag; is usable in low permeability situations; allows central observation system to be used; can measure negative pore pressures	Is costly and difficult to install and operate	Good to fair
	Pneumatic			Is not subject to freezing; uses smaller, less expensive tubing	Cannot measure negative pore pressures; has leakage and moisture in lines	
	Hydraulic			Is simple and easy to seal against leakage	Cannot measure negative pore pressures; requires constant volume pumps or flow control valves	
	Electrical resistance strain gauge		±1% of range[d]	Can often be locally fabricated from commercially available parts; is adaptable to automatic data recording; can measure negative pore pressures	Has often limited service life; is susceptible to wiring damage	
	Vibrating-wire strain gauge			Is more reliable than resistance strain gauge type; is adaptable to automatic data recording		

Note: 1 m = 3.3 ft; 1 mm = 0.39 in.

[a] Total displacement accuracy depends on pivot spacing.

[b] Can be reset.

[c] Depends on pressure gauge used.

[d] Depends on transducer.

by tape measurement. As shown in Figure 5.1, transit lines can be established so that the vertical and horizontal displacements at the center and toe of the slide can be observed. Lateral motions can be detected by transit and tape measurements from each hub. When a tension crack has opened above the top of a slide, simple daily measurements across the crack can be made between two hubs driven into the ground. In many cases, the outer limit of the ground movements is not known, and establishing instrument setups on stable ground may be a problem.

Various techniques and accuracy achieved in optical leveling, offset measurements from transit lines, chaining distances, and triangulation have been discussed extensively in the literature (5.8, 5.23), particularly for dams, embankments, and buildings. Although conventional surveys, par-

ticularly higher order surveys, can define the area of movement, more accurate measurements may be required in many cases. Terzaghi (5.47) stated that "if a landslide comes as a surprise to the eyewitness, it would be more accurate to say that the observers failed to detect the phenomena which preceded the slide." The implication is that the smallest movements possible should be measured at the earliest possible time. Terzaghi (5.47) also describes the movements that precede a landslide.

The detection of small surface movements when cracking is not apparent requires a trained observer. If the ground surface is covered with rocks, or if there is a rock embankment, horizontal stretching will result in local instability of individual rocks such that walking over the slope gives one a sense of insecurity. Overturned rocks can be detected by a change in coloring or surface weathering. Trees inclined at the base but changing to vertical trunks a meter or so above the ground may indicate old slide movements. Inclined but straight trunks indicate recent movements. Cracks covered over with leaves or surface duff can be detected by walking over the area and noting the firmness of support. Frequently animals will avoid grazing in a potential landslide area because of uncertain support or hidden fissures. Small openings on the downhill side of structures, or next to tree trunks, may indicate creep. Overtaut or excessively sagging utility lines or misalignment of fence posts or utility poles are excellent indicators of ground movements. Such movements, when accurately monitored, serve as an important tool in assessing the potential hazard to structures, nearby residents, and the public.

Other Types of Surface Surveying

There are three rapidly developing surveying techniques today, and these will undoubtedly find increasing use in field measurements. Some are already in extensive use (e.g., electronic distance measurement and lasers); others are in limited use or are in the experimental stage (e.g., terrestrial photogrammetry). Electronic distance measurement (EDM) has changed surveying practices more than anything else in the last 100 years (5.15). EDM devices have proved particularly suited for use over rugged terrain; they perform more accurately and much faster than ordinary surveying techniques and require fewer personnel. Lightweight EDM instruments can be used efficiently under ideal conditions for distances as short as 20 m (66 ft) and as long as 3 km (2 miles); errors are as small as ±0.0032 m (±0.010 ft) (5.28, 5.43). Larger instruments using light waves or microwaves can be used at much longer distances. The accuracy of EDM is influenced by weather and atmospheric conditions; comparative readings with three different instruments are described by Penman and Charles (5.40).

EDM can be used to monitor large slides with large movements and provide a rapid way to survey many points on the mass from a single, readily accessible location. An example of such an installation involves an ancient landslide in Washington along the Columbia River, where the boundaries of the active slide are more than 0.6 km (1 mile) in width and 4.8 km (3 miles) in length. Yearly movements vary from only 1 m to 6 to 9 m (3 ft to 20 to 30 ft) and depend on the time of year and the rainfall. A permanent station, readily accessible all year, has been set up on the opposite side of the river, and monthly distance readings are taken to 14 points on the slide area and to 2 points located outside the slide zone. The distances involved vary from about 1500 to 6000 m (5000 to 20 000 ft). Figure 5.2a shows the movements (changes in distance) recorded by the EDM (electrotape) during a 1-year period for 2 selected points at the Columbia River slide, based on monthly readings. At the end of the year, the points were resurveyed by triangulation. The discrepancy is about 10 cm (4 in), which, although larger than anticipated, is quite satisfactory considering the total movements. Figure 5.2b shows recorded changes for 2 points believed to be on stable ground. The variation of monthly readings is seen to be ±0.06 m (±0.2 ft); variation is no greater for a 4765-m (15 630-ft) length than for a 1844-m (6050-ft) length.

Laser instruments are already widely used for setting alignments, and they are well suited for setting a reference line for offset measurements to surface monuments. Laser beams are also used with some EDM instruments. It should be possible to measure offsets with an error no greater than 0.003 to 0.006 m (0.01 to 0.02 ft) (5.23).

Terrestrial photogrammetry has been used in some cases on dam projects (5.35) and in mines, but no landslide measurements using this method have been found in the literature. Phototheodolites are used to take successive stereophotographs from a fixed station along a fixed camera axis; movements are identified in a stereocomparator, and accuracies of 0.006 to 0.009 m (0.02 to 0.03 ft) have been

Figure 5.1. Movement measurements in a typical slide area.

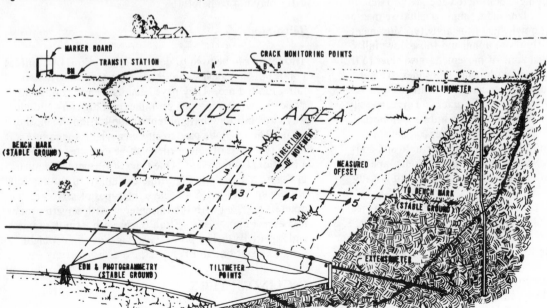

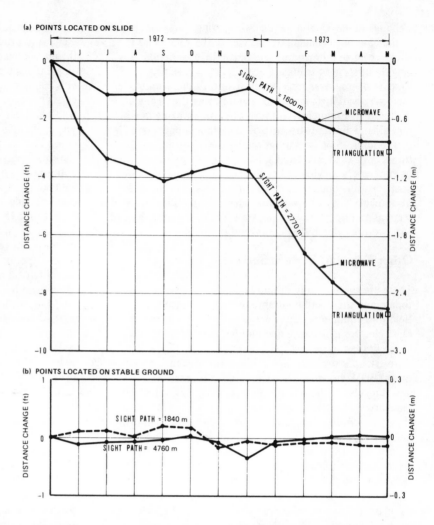

Figure 5.2. Microwave measurements of slide movements along Columbia River, Washington.

reported. Moore (*5.35*) reports average errors of about one order of magnitude greater than those from precise ground surveying.

Crack Gauges

When geologic mapping is considered adequate to describe the area affected by a landslide, simple qualitative measurements can provide knowledge of the activity. Movements on cracks, particularly those uphill and those downhill from the well-defined zone of movement, give clues to the increasing size associated with many landslides. Therefore, it is desirable to monitor the change in width, as well as the change in elevation, across the cracks. This can easily be done by direct measurement from hubs set on both sides of the cracks. Crude, simple gauges can be constructed in the field to provide a more accurate and continuing indication of crack movement. Periodic tape measurements can detect movements (and rates) not visually apparent.

In the areas of shear along the sides of the sliding area, a similar crack indicator system is used, except that three-point stations are often installed. However, because of the greater width of the shear zone, such simple devices may not be satisfactory. In jointed rock, the change in joint spacing can be determined by scribing marks or bonding washers on the rock surface on both sides of the joint. These indicator marks should be joined by a straight line

so that both shear and opening or closing of the joint can be monitored. Another simple device is a hardwood wedge lightly forced into the joint crack with a mark placed on it at the level of the rock. If the joint opens, then the wedge will fall to a deeper level in the rock. However, if the joint is closing or subject to shear movement, such a wedge indicator may not be helpful.

Tiltmeters

Tiltmeters can be used to detect tilt (rotation) of a surface point, but such devices are relatively new and have had fairly limited use. They have been used mostly to monitor slope movements in open pit mines and highway and railway cuts, but they have potential application in landslide areas. They may be used in any area where the failure mode of a mass of soil or rock can be expected to contain a rotational component. One type of tiltmeter is shown in Figure 5.3, and sample tiltmeter data from a mine slope are shown in Figure 5.4. The same types of servo accelerometers are used with the tiltmeter as with sensitive inclinometers (a fairly rugged transducer with a similar range and sensitivity). The prime advantages are light weight, simple operation, and compactness at a relatively low cost.

INCLINOMETERS

The development of inclinometers has been the most im-

118

portant contribution to the analysis and detection of landslide movements in the past 2 decades. Although it has been used most extensively to monitor landslides, the inclinometer has gained widespread use as a monitoring instrument for dams, bulkheads, and other earth-retaining structures and in various areas of research (*5.1, 5.13, 5.17, 5.34, 5.44, 5.55, 5.59*). Since the introduction in the early 1950s of the pendulum-actuated inclinometer operating in grooved plastic casing (*5.45, 5.50*), the same basic concepts have been applied to instruments manufactured by at least 12 U.S. and foreign firms. Although the principle of operation has remained unchanged (Figure 5.5), inclinometers have been improved in accuracy and ease of operation and have undergone numerous modifications to adapt them to individual projects.

The inclinometer measures the change in inclination (or tilt) of a casing in a borehole (Figure 5.5) and thus allows the distribution of lateral movements to be determined as a function of depth below the ground surface and as a function of time (Figure 5.6). The application of the inclinometer to landslides is readily apparent, namely, to define the slip surface or zones of movement relative to the stable zones. Inclinometers have undergone rapid development to improve reliability, provide accuracy, reduce weight and bulk of instruments, lessen data acquisition and reduction time, and improve versatility of operations under adverse conditions. Automatic data-recording devices, power cable reels, and other features are now available.

An inclinometer system has four main components.

1. A guide casing is permanently installed in a near vertical borehole in the ground. The casing may be made of plastic, steel, or aluminum. Circular sections generally have longitudinal slots or grooves for orientation of the sensor unit, but square sections are used with some types.

2. A portable sensor unit is commonly mounted in a

Figure 5.3. Portable tiltmeter.

Figure 5.4. Tilt at ground surface due to advance of longwall mining face.

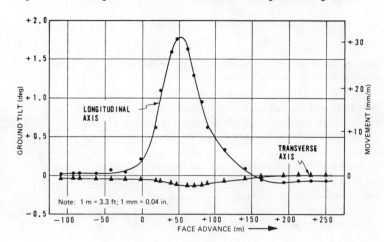

Figure 5.5. Principle of inclinometer operation.

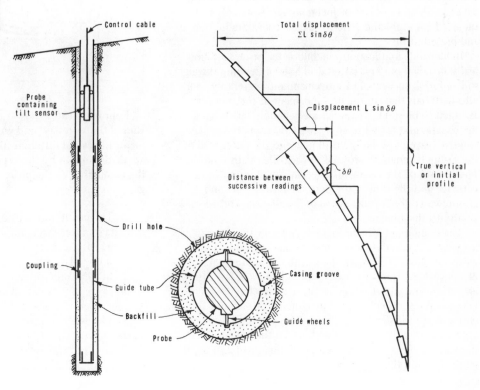

119

Figure 5.6. Measurement of slide movements with inclinometer.

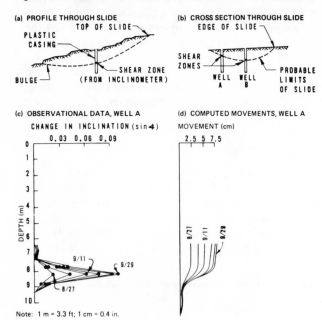

(a) PROFILE THROUGH SLIDE

PLASTIC CASING
TOP OF SLIDE
SHEAR ZONE (FROM INCLINOMETER)
BULGE

(b) CROSS SECTION THROUGH SLIDE

EDGE OF SLIDE
SHEAR ZONES
WELL A   WELL B
PROBABLE LIMITS OF SLIDE

(c) OBSERVATIONAL DATA, WELL A

CHANGE IN INCLINATION (sin ∢)

(d) COMPUTED MOVEMENTS, WELL A

MOVEMENT (cm)

Note: 1 m = 3.3 ft; 1 cm = 0.4 in.

Figure 5.7. Inclinometer.

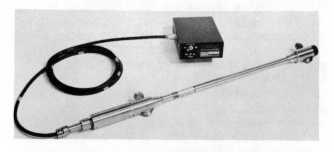

Figure 5.8. Pendulum-activated Wheatstone bridge circuit for typical inclinometer.

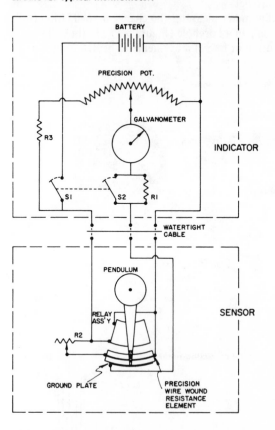

carriage designed for operation in the guide casing (often referred to as the probe or torpedo).

3. A control cable raises and lowers the sensor unit in the casing and transmits electrical signals to the surface. For accurate depth control of the sensor unit, the cable is usually graduated or lowered on a separate surveyor's steel tape.

4. A portable control and readout unit at the surface supplies power, receives electric signals, and displays readings in dial or digital format.

Components other than these four are accessories. The need for accessories increases with the volume of output data (depth and number of inclinometers installed) and the need for rapid and efficient data retrieval, reduction, and presentation.

In landslide applications, inclinometer casings are normally installed in exploration drill holes extending through soil and rock suspected of movement and preferably well into materials that, in the best judgment at the time, are assumed to be stable. Samples are generally taken from the borings and tested to confirm such assumptions. The annular space between the drill hole and the outer wall of the casing is generally grouted or backfilled with sand. The success of the casing installation depends on the experience and the skill of the personnel; soil, rock, and groundwater conditions; depth of installation; and accessibility to the area.

The inclinometer instrument sensor unit (Figure 5.7) is lowered and raised on an accurately marked cable, its wheels or guides following the oriented, longitudinal slots of the casing. The response to slope changes in the casing are monitored and recorded at the surface. Readings are taken at fixed, usually equal, increments throughout the entire depth. Instruments differ primarily in the type of sensor used, in the accuracy with which the sensor detects the inclination, and in the method of alignment and depth control within the borehole.

Four basic categories of inclinometers are in use in the United States today, and within each category are instruments of several different manufacturers. These categories are distinguished by the type of sensor selected to measure the inclination of the probe relative to the direction of gravity.

1. The pendulum-activated Wheatstone bridge circuit consists of a free-swinging, magnetically damped pendulum moving across a resistance coil; measurement for each sensor is in one plane (*5.24, 5.33, 5.54*).

2. In the strain-gauge type, bonded or vibrating-wire strain gauges are mounted around a stiff pendulum; measurement for each sensor is in one plane (*5.24*).

3. The accelerometer type is a closed-loop, servo-accelerometer circuit; measurement for each sensor is in one plane, and the use of two sensors per unit (biaxial) is common (*5.34*).

4. The photographic type photographically records the projection of a pendulum with two degrees of freedom on a target oriented by a magnetic compass; no orientation is needed within the casing or borehole. This type is not often used because its accuracy is generally much lower than that of previous types and data reduction is much more time consuming. Measurements with this instrument are described by Hanna (5.25).

With the exception of the photographic type, most inclinometers measure the inclination of the casing in two mutually perpendicular near-vertical planes. Thus, horizontal components of movement, both transverse and parallel to any assumed direction of sliding, can be computed from the inclinometer measurements. Deviations from this ideal occur because of limitations and compromises in manufacturing and because of installation circumstances. Instruments of different manufacture cannot generally be used interchangeably. In the interest of accuracy, the interchangeable use of instruments (probes) of the same manufacture also should be avoided.

The function of the inclinometer is to detect the change in inclination of the casing from its original near-vertical position. Readings taken at regular preestablished depths inside the casing allow the change in slope at various points to be determined; integration of the slope changes between any two points yields the relative deflection between those points. Repeating such measurements periodically provides data on the location, magnitude, direction, and rate of casing movement. The integration is normally performed from the bottom of the boring, since the bottom is assumed to be fixed in position and inclination (Figure 5.5).

All too often, confusing or unexplainable results appear because of instrumentation problems. These problems are common to all inclinometers; therefore, they are discussed in some detail so that, even without a detailed understanding of the inner workings of the instruments, one will be able to recognize these anomalies, the limits of accuracy, and the meaning of the results. In simple terms, the person closest to the job should be able to answer the basic questions: Is the slide active? How fast is it moving? How deep is it?

Inclinometer Sensors

Before a description is given of some of the sophisticated types of inclinometers, one device deserves mention. The borehole probe pipe, also called the slip indicator or poor boy, is one of the crudest of the subsurface measuring devices. It typically consists of a 25-mm (1-in) diameter semirigid plastic tube in a borehole. Metal rods of increasing length are lowered inside the tubing in turn, and the rod length that is just unable to pass a given depth gives a measure of the curvature of the tubing in the vicinity of that point. Toms and Bartlett (5.49) describe the use of this technique for the location of slip surfaces in unstable slopes. This type of measurement can easily be performed in the riser pipe of an observation well or open piezometer. If there are several shear planes, or if the shear zone is thick, a section of rod hung on a thin wire initially can be left at the bottom of the pipe; subsequently, the rod is pulled up to detect the lower limit of movement.

As originally conceived, the first modern inclinometers used pendulum-activated resistors to convert inclinations into electrical measurements; instruments of this type have had long and successful experience records (5.45). A typical inclinometer of this type consists of a pendulum, the tip of which makes contact with a resistance coil, subdividing the coil into two resistances that form one-half of a Wheatstone bridge (Figure 5.8). The other half of the bridge, contained in the portable control box, includes a precision potentiometer. The potentiometer readings are proportional to the inclination of the torpedo. The system has a sensitivity of about 3 min of arc and generally demonstrates a precision of ±13 to 25 mm (±0.50 to 1.0 in) over a 33-m (100-ft) casing (1:1000). The instrument is adaptable to most measurement programs; however, the data acquisition time is at least twice that of the more recent accelerometer types. Cornforth (5.14) and Green (5.24) describe the performance of this type of inclinometer in more detail.

The most sensitive transducer commercially available is the servo accelerometer. A servo accelerometer is composed of a pendulous "proof mass," that is free to swing within a magnetic field (Figure 5.9). The proof mass is provided with a coil or "torquer," which allows a linear force to be applied to the proof mass in response to current passed through the coil. A special sensing unit, called a pick-off, detects movement of the proof mass from a vertical position. A signal is then generated and converted by a restorer circuit, or servo, into a current through the coil, which balances the proof mass in its original position. In this manner, the current developed by the servo becomes an exact measure of the inertial force and thus of the transducer's inclination. The proof mass consists of either a pendulum with jewel-bearing support or a flexure unit that operates on the cantilever principle. Jewel-bearing accelerometers are fragile and subject to frictional interference at their bearings. Flexure accelerometers generally offer the same precision and have greater durability. Accelerometer systems have a sensitivity of approximately 18 s of arc in ranges from 30° to 90°, and they generally demonstrate a precision of ±1.3 to 2.5 mm (±0.05 to 0.10 in) over a 33-m (100-ft) casing (1:10 000). One excellent example of the precision of such measurements is shown in Figure 5.10. These data were taken with a servo-accelerometer sensor by the U.K. Transport and Road Research Laboratory during the advancement of a tunnel (5.6).

Temperature has only a negligible effect on readings taken with pendulum or accelerometer transducers. However, a marked variation of reading with temperature has been demonstrated for transducers using bonded resistance strain gauges (5.24), and errors resulting from temperature changes, as well as variations in zero drift, have been reported for inclinometers using vibrating-wire transducers (5.17). Even so, Burland and Moore (5.11) used a strain-gauge-based inclinometer to obtain some precise results for a diaphragm wall.

A recently developed inclinometer that is also used in a square casing (45-mm, 1.8-in, internal dimension) is described by Phillips and James (5.41). The unit has a cantilevered pendulum device with a four-arm resistance gauge bridge. Over a 5° range, the repeatability under ideal conditions is reported to be ±5 s of arc. However, the relatively small range may render it impractical for use in a

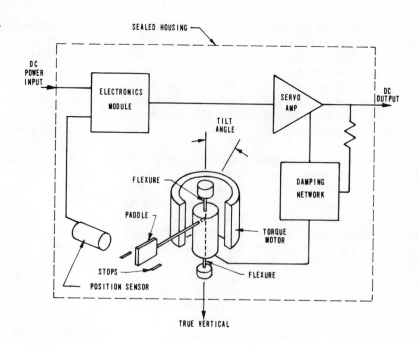

Figure 5.9. Servo accelerometer for inclinometer measurements (5.13).

Figure 5.10. Lateral ground movement at borehole C1 of bentonite tunneling experiment (5.6).

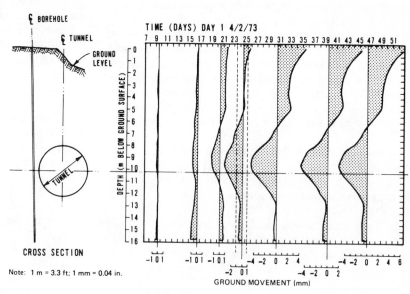

Note: 1 m = 3.3 ft; 1 mm = 0.04 in.

landslide area where boreholes often have significant inclinations.

Casing Installation

Inclinometer casings generally may be installed for the full depth normally encountered in landslides. Some known installations are close to 300 m (1000 ft) deep (5.7). Since the measurement is referenced with respect to the bottom of the casing, the casing bottom must be extended about 6 m (20 ft) or more into soil that will not undergo lateral displacement. If any doubt arises as to the stability of the casing bottom, movement of the casing top should be checked by precise surveying methods. The accuracy of the observations is generally limited not by the sensitivity of the inclinometer, but by the requirement that successive readings be made with the same orientation of the instrument at the same point in the casing. Because the casing must provide a reliable orientation and a continuous track-

ing for the sensor unit, proper casing installation is of paramount importance.

It is important for the observation accuracy that the original installation be as close to vertical as practicable. The error in inclinometer surveys is proportional to the product of the casing inclination and angular changes in the sensor alignment. For inclined casings, angular changes in sensor alignment on the order of 1° to 2° may produce errors of several centimeters per 10 m (several inches per 100 ft) of casing, as shown in Figure 5.11. Sensor-alignment change occurs from time to time because of one or several factors, such as wheel play in the groove, wear of the sensor carriage (particularly wheel assemblies), internal change in the sensor itself, and change in the alignment between sensor and carriage.

The casing for the inclinometer is usually installed in 1.5 and 3.0-m (5 and 10-ft) lengths, which are joined together with couplings and either riveted or cemented or both to ensure a firm connection (Figure 5.12). Each coupling represents a possible source of leakage for grout or mud, which can seep into the casing and be deposited along the internal tracking system. For this reason, couplings should be sealed with tape or glue. The bottom section of the casing is closed with an aluminum, plastic, or wooden plug, which also should be sealed. If the drill hole is filled with water or drilling mud, the inclinometer casing must be filled with water to overcome buoyancy. Sometimes, extra weight (sand bags, drill stem) may be needed.

The annular space between the boring wall and the inclinometer casing may be backfilled with sand, pea gravel, or grout. The selection of backfill depends mostly on soil, rock, and groundwater conditions (i.e., whether the borehole is dry, wet, caving, stable without mud). The type of drilling technique (e.g., rotary, hollow-stem, cased holes) is also important. Poorly backfilled casings can introduce a scatter in initial measurements, and, if maximum precision is required, great care should be taken in the selection of the proper backfilling material. Grouting is generally preferred, but may not always be possible, particularly in pervious geologic materials, such as talus. Grouting may be facilitated by use of a 20 to 25-mm (¾ to 1-in) diameter plastic tube firmly attached to the casing bottom through which grout is pumped from the ground surface until the entire hole is filled. In a small clearance, drill-hole grout can be tremied through drill rods via a one-way valve at the inside bottom of the casing.

The as-manufactured casing may have some spiral to the grooves. During installation, the casing can become even more twisted or spiraled so that, at some depth, the casing grooves may not have the same orientation as at the ground surface. Spiraling as great as 18° in a 24-m (80-ft) plastic casing has been reported (5.24); 1°/3-m (10-ft) section is not uncommon. Because significant errors in the assumed direction of movement may result, deep inclinometer casings should be checked after installation by using spiral indicators available from the inclinometer manufacturer. For a particular installation, groove spiraling is generally a systematic error occurring for each section in the same amount and the same direction. In any event, spiraling, whether it is due to manufacture or to torque of the casing during installation, can be measured and is thus not really an "error" in inclinometer measurement. Spiraling is important only when the true direction of the movement is determined and at that time can and should be measured and the results adjusted accordingly.

Small irregularities in the tracking surface of the casing can lead to errors in observation, especially if careful repetition of the depth to each previous reading is not exercised. If plastic casing has been stored in the sun before installation, each section may be locally warped (opposite grooves not parallel) and large measurement errors can occur with minor variations at the depths at which the readings are taken (5.23).

Aluminum casings should be used with some caution because severe corrosion may occur if the casing is exposed

Figure 5.11. Measurement error as a result of casing inclination and sensor rotation.

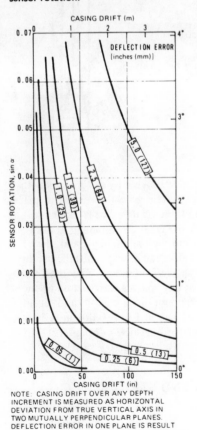

NOTE: CASING DRIFT OVER ANY DEPTH INCREMENT IS MEASURED AS HORIZONTAL DEVIATION FROM TRUE VERTICAL AXIS IN TWO MUTUALLY PERPENDICULAR PLANES. DEFLECTION ERROR IN ONE PLANE IS RESULT OF DRIFT IN OTHER PERPENDICULAR PLANE.

Figure 5.12. Details of inclinometer casing.

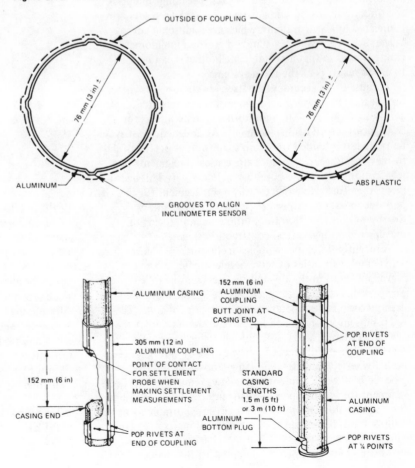

123

to alkaline soil, corrosive groundwater, or grout. Epoxy coatings help to minimize this problem. Burland and Moore (*5.11*) recommend cleaning the casing with a stiff brush before taking readings. Frequent flushing with water is also helpful. The top of the casing is generally capped with a tight-fitting plug to prevent intrusion of debris. In addition, the inclinometer casing should be protected and padlocked at the ground surface. When selecting protective casing or monument covers for use around the top, one should keep in mind the casing-mounted pulley or similar device.

In-Place Inclinometers

An inclinometer that remains in place in the borehole to continually monitor displacements normal to the borehole axis at discrete points along the borehole is called an in-place inclinometer. The sensors are permanently positioned at intervals along the borehole axis and may be more closely spaced (to increase resolution of the displacement profile) in zones of expected movement. Total movements are determined by summing the relative movements measured at each sensor along the borehole. Since fixed units consist of a number of sensors, they are more expensive and complex than portable inclinometers.

The major advantage of the fixed borehole inclinometer over the portable unit is the elimination of problems of tracking inaccuracies and repeatable positioning. If the fixed unit is removed for repairs, the overall accuracy will be reduced to that of a portable unit or less. Also, accelerometers may have long-term drift. With portable units, these and other effects are canceled by taking readings in opposite directions. Fixed units can be monitored remotely or connected to alarm systems, which are additional advantages. In some systems, a standard grooved inclinometer casing is used so that a portable inclinometer can be used to check readings of the fixed system.

Because of its accuracy, the fixed borehole inclinometer may be used to measure small movements in rock. Its adaptability to various remote-monitoring systems allows for a continuous record of displacement. As designed, most systems are not intended to obtain a continuous profile of deformation. Rather, they measure critical movements of a few sections within the borehole inclinometer. This instrument, therefore, is not necessarily a replacement for the portable units, but rather has its own purpose and is worthy of consideration for continuous monitoring of movements during critical construction stages.

A recently developed in-place inclinometer (Figure 5.13) employs a series of servo-accelerometer sensors; it is available with both one- and two-axis sensors (*5.13*). The sealed accelerometer packages are spaced along standard grooved inclinometer casing by a series of rods. The rods and sensors are linked by universal joints so that they can deflect freely as the soil and the casing move. The sensors are aligned and secured in the casing by spring-loaded fins or wheels, which fit the casing grooves. Use of the grooved casing and guide wheels allows removal of the instrument for maintenance, adjustments, and salvage. Since the casing is standard inclinometer casing, portable inclinometers can be used in the casing when the fixed instrument is removed.

Readings are obtained by determining the change in tilt

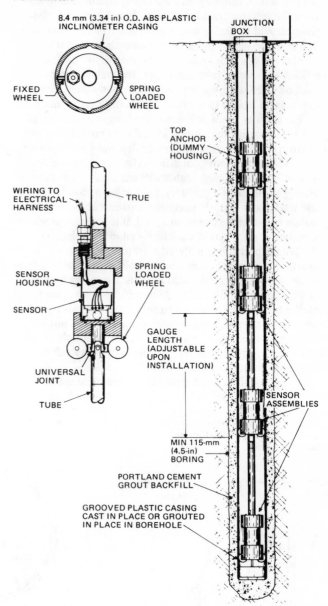

Figure 5.13. Installation and detail of multiposition in-place inclinometer.

of the accelerometer and multiplying by the gauge length or spacing between sensors. This gives the relative displacement of each sensor, and these relative displacements can be summed to determine the total displacement at each sensor. Since the grooves in the casing may spiral, the spiral should be checked after the casing is installed so that the exact direction of displacement measurement is known. Any number of sensors at any spacing can be used in one borehole. Maximum deflection range is ±30°, and sensitivity is reported to be ±0.01 mm in 1 m (±0.001 in in 10 ft), but this is not usually attained because of long-term drift. Monitor and alarm consoles and telemetry systems are available.

EXTENSOMETERS AND STRAIN METERS

Extensometers and strain meters measure the increase or

decrease in the length of a wire or rod connecting two points that are anchored to the soil in the borehole and whose distance apart is approximately known. One commercially available device is shown in Figure 5.14. When gauge lengths are on the order of a meter or less, these devices are often referred to as strain meters rather than as extensometers. When they are used as extensometers, the accuracy and repeatability depend on the type of sensing element and its range of travel and also on the type of connecting wire or rod and the methods used to control the tension. Dead weights are best for maintaining constant tension in wires; if these cannot be used, constant tension springs are acceptable, although there may be some hysteresis. Relatively inexpensive wire-wound or conducting plastic linear potentiometers are often used as sensors, and relatively simple, battery-operated Wheatstone bridge circuits are used for manual readout. Sensitivity is on the order of 0.1 percent of the range of travel, but repeatability and accuracy may be no better than 0.5 mm (0.02 in), depending on the type of anchor and connecting member; this is usually sufficient, however.

When the devices are used as strain meters, the repeatability and accuracy are essentially the same as the sensitivity. Thus, for a grouted-in-place assembly 3 m (10 ft) long with an invar rod and a range of 25 mm (1 in), unit strains as low as 0.000 01 may be detected with relatively inexpensive instrumentation. Horizontal stretching of embankments has been observed by installing anchors at various positions at a given elevation and attaching horizontal wires to dead weights on the downstream face; this was done at Oroville Dam (5.55). Care is required to ensure that the wires (or rods) do not get pinched off if localized vertical shear movements occur. Other cases of application are described by Dutro and Dickinson (5.19) and Heinz (5.26).

PORE-PRESSURE AND GROUNDWATER MEASUREMENTS

Observation Well

The most common water-level recording technique, despite more sophisticated methods, is the observation of the water level in an uncased borehole or observation well. A particular disadvantage of this system is that a perched water table or artesian pressure can occur in specific strata that may be interconnected by the borehole so that the recorded water level may be of little significance.

Piezometers

Pore pressures and groundwater levels in a slide area can be measured by a variety of commercially available piezometers (5.25). The selection of the best type for a particular installation involves several considerations.

1. Piezometers are usually installed in a difficult environment and may become inoperative because of pinching off of the tubes, blockage by air bubbles, electrical short circuits, and malfunctions of buried moving parts. Therefore, reliability and durability are often of greater importance than sensitivity and precision. For example, in some cases it does not matter that the actual head may be in error by 0.3 m (1 ft) or so as a result of time lag, provided the piezometer is functioning properly. If a malfunction occurs, the fact that the apparent head can be recorded to a millimeter is of little importance.

2. The basic problem with a piezometer is that the energy required to operate it prevents the instrument from recording a pore-pressure change immediately. For a given pressure change, this energy is proportional to the volume of pore water that must flow into the instrument. A piezometer with an open standpipe, for example, requires a much greater volume of water than one with a stiff diaphragm and therefore requires a correspondingly greater time to respond to a change of pore pressure. Hvorslev (5.27) discusses time-lag effects for open-standpipe piezometers; Brooker and Lindberg (5.10) evaluated these effects for closed types; and Penman (5.39) and Vaughan (5.52) studied the response times of various types of piezometers. In slides in which piezometers are placed in boreholes, it is usually possible to provide a large collecting volume of porous material (sand) around the tip and thus to reduce the time lag in open standpipes.

3. Partially saturated soils (fills, soils above the water table, and soils in which gas is generated from organic matter) pose particular problems because the gas in the voids exists at higher pressures than the water as a result of surface tension (capillarity). Thus, one must know whether the piezometer is reading the pore-air pressure or the pore-water pressure. To prevent the entry of air into the piezometer tip and thus to ensure that pore-water pressure

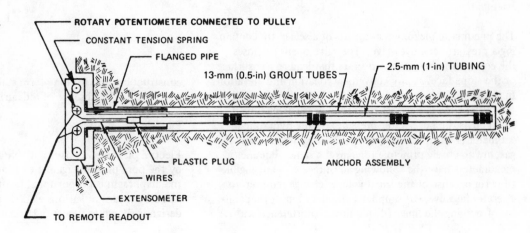

Figure 5.14. Extensometer.

ROTARY POTENTIOMETER CONNECTED TO PULLEY
CONSTANT TENSION SPRING
FLANGED PIPE
13-mm (0.5-in) GROUT TUBES
2.5-mm (1-in) TUBING
PLASTIC PLUG
ANCHOR ASSEMBLY
WIRES
EXTENSOMETER
TO REMOTE READOUT

125

is being measured, fine ceramic tips, with high air-entry values, can be used (5.2). Even with a high air-entry value filter, air will slowly enter the piezometer cavity by diffusion and, if reliable long-term readings are required, deairing facilities must be provided (5.52).

4. Additional problems resulting from diffusion of either air or water vapor through the walls of connecting tubing are described by Bishop, Kennard, and Penman (5.2). Furthermore, either settlement or horizontal movement may pinch off the connecting tubing or, in the case of open standpipes, may prevent the lowering of a probe to detect the water level.

The features and uses of various types of piezometers are discussed below.

Open Standpipe

Open-standpipe piezometers vary mainly in diameter of standpipe and type and volume of collecting chamber. The simplest type (Figure 5.15) is merely a cased or open observation well in which the elevation to which the water rises is measured directly by means of a small probe. In this case, the static head is the average head that exists over the depth of the inflow part of the well below the water table. This measured head may be higher or lower than the free water table and, in the case of moderately impervious soils, may be subject to a large time lag. Although the open standpipe is not satisfactory in impervious soils because of time lag or in partially saturated soils because the significance of the measured head may be difficult to evaluate, its simplicity, ruggedness, and overall reliability dictate its use in many installations.

Casagrande

The Casagrande type of piezometer (Figure 5.16) consists of a porous stone tip embedded in sand in a sealed-off portion of a boring and connected with a 1-cm (³⁄₈-in) diameter plastic riser tube (5.45). When properly installed, this type of piezometer has proved successful for many materials, particularly in the long term, for it is self-deairing and its nonmetallic construction is corrosion resistant. The reliability of unproven piezometers is usually evaluated on the basis of how well the results agree with those of adjacent Casagrande piezometers.

Pneumatic

The pneumatic piezometer consists of a sealed tip containing a pressure-sensitive valve. The valve opens or closes the connection between two tubes that lead to the surface, or the slope face, at any convenient location and elevation. In the piezometer shown in Figure 5.17, flow of air through the outlet tube is established as soon as the inlet-tube pressure equals the pore-water pressure. In the hydraulic piezometer (Figure 5.18), hydraulic fluid is used instead of gas, but the basic principle remains the same. Pneumatic piezometers have the following advantages: (a) negligible time lag because of the small volume change required to operate the valve, (b) simplicity of operation, (c) capability of purging the lines, (d) minimum interference with

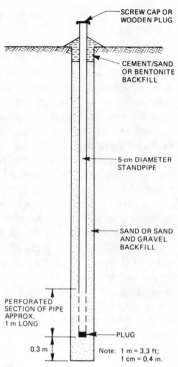

Figure 5.15. Open-standpipe piezometer.

SCREW CAP OR WOODEN PLUG

CEMENT/SAND OR BENTONITE BACKFILL

5-cm DIAMETER STANDPIPE

SAND OR SAND AND GRAVEL BACKFILL

PERFORATED SECTION OF PIPE APPROX. 1 m LONG

PLUG

0.3 m   Note: 1 m = 3.3 ft; 1 cm = 0.4 in.

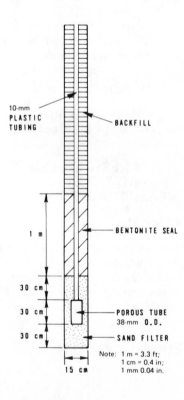

Figure 5.16. Casagrande borehole piezometer.

10-mm PLASTIC TUBING

BACKFILL

1 m

BENTONITE SEAL

30 cm

30 cm   POROUS TUBE 38-mm O.D.

30 cm   SAND FILTER

Note: 1 m = 3.3 ft; 1 cm = 0.4 in; 1 mm 0.04 in.

15 cm

construction, and (e) long-term stability. Their main disadvantage is the absence of a deairing facility.

Electric

Electric piezometers have a diaphragm that is deflected by the pore pressure against one face. The deflection of the diaphragm is proportional to the pressure and is measured by means of various electric transducers. A typical design is shown in Figure 5.19. Such devices have negligi-

Figure 5.17. Pneumatic piezometer.

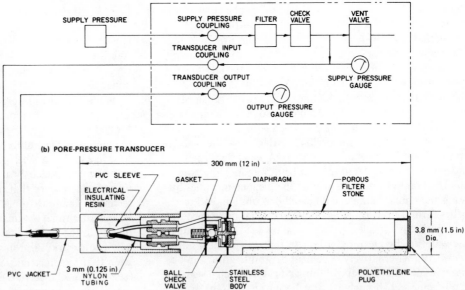

(a) PORTABLE PORE-PRESSURE INDICATOR

(b) PORE-PRESSURE TRANSDUCER

Figure 5.18. Hydraulic piezometer.

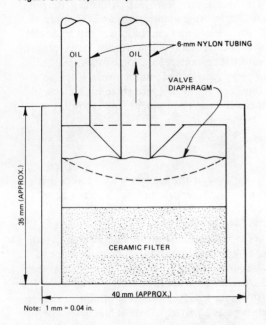

Note: 1 mm = 0.04 in.

Figure 5.19. Electric piezometer.

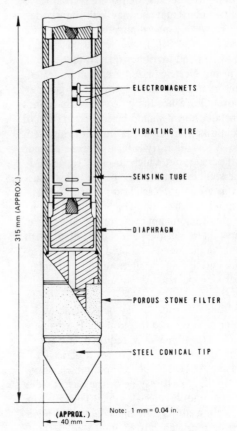

Note: 1 mm = 0.04 in.

ble time lag and are extremely sensitive. Piezometers of this type are described by Shannon, Wilson, and Meese (*5.45*), Cooling (*5.12*), Brooker and Lindberg (*5.10*), Penman (*5.39*), Bishop, Kennard, and Vaughan (*5.3*), and Hanna (*5.25*). Because they are affected by the environment and have poor long-term stability, electric piezometers are not generally recommended for installations in which reliable readings are required during an extended period of time. Generally, they cannot be deaired nor can the sensitive electric transducers be recalibrated in situ, although there are exceptions (*5.16*). For short-term observations at installations in which transmission is over limited distances, standard resistance strain-gauge pressure transducers are suitable. A new type of electric piezometer, using a pressure-sensitive crystal, has been developed, but its long-term stability has not yet been verified.

Piezometer Sealing

Piezometers in landslides can be installed in boreholes advanced into soil or rock. Based on the drill log of materials encountered and the estimated position of the water table and of the sliding surface, the depth of each piezometer is selected. A simple open-standpipe piezometer device consists of a tube placed in the drilled hole extending to the ground surface. The bottom of the tube should be slotted or made of a porous material or a sufficiently fine screen that it will restrict movement of soil into the piezometer but at the same time permit free access of water. Only that part of the tube that is actually within the stratum in question should be pervious. When the tube is centered in the drilled hole, a known volume of clean sand is placed around the piezometer; the purpose is to create a sand filter between the soil and the piezometer tip.

Construction of an impervious barrier above the piezometer tip and sand pocket is essential. A well-established procedure is to drop balls of soft bentonite into the hole and then tamp those balls around the piezometer tubes by using an annular hammer. Prepared bentonite in ball or pellet form is now available commercially; it has a specific gravity that is sufficiently large to allow it to sink through the water in the piezometer hole so that a hammer is not necessary. Alternatively, a cement-bentonite grout can be tremied into the hole above the sand filter. Such a seal can be pumped through a 1.3 to 1.9-cm (½ to ¾-in) diameter pipe adjacent to the piezometer tube. Installing only one, or at most two, piezometers in a borehole has generally been found preferable because bentonite seals are sometimes difficult to construct and may leak slightly despite precautions. However, special-purpose, multiple-point piezometers (with four tips) have been used successfully by Vaughan (5.51) and merit consideration.

Both pneumatic and electric piezometers can be sealed in a similar manner to an open-standpipe or a Casagrande piezometer in a borehole. Since the piezometer tip cannot be deaired after installation, it should be soaked in deaired water beforehand and kept in the water until it is lowered into the borehole. A low air-entry filter tip can be placed in a saturated sand pocket, but a high air-entry tip should be pushed into the soil beyond the base of the hole or surrounded by a porous grout, such as plaster of paris. This will ensure more rapid equalization, which may be of considerable value in obtaining reliable pore-pressure measurements soon after installation in low-permeability clay. Both electric and pneumatic piezometers should be checked for malfunction before and during installation, and particular care is needed in driving electric piezometers to prevent overpressure.

In soft and medium-stiff soils, it may be preferable to push the piezometer directly into the ground. Flush-coupled heavy steel water pipe will be required, and the piezometer tip must be robust enough to resist damage to the well point and should be designed to minimize disturbance around the tip (5.34). It may be of the Casagrande, pneumatic, or, less commonly, electric type. This drive-in piezometer is self-sealing and rapidly installed and, if pneumatic, has a rapid response. Recently, an improved method of sealing piezometers in boreholes was described by Vaughan (5.51). Instead of compacting bentonite balls just above the sand filter, the borehole is completely grouted with cement-bentonite grout. Even if the permeability of the grout is significantly higher than that of the surrounding soil, little error will result because of the relatively large grouted length. In many cases, a sand filter need not be included, and the piezometer tip can be grouted directly with little error.

SYSTEMS FOR MONITORING ROCK NOISE

Use of the rock-noise detection method for observing the stability of soil and rock materials is a relatively new concept, and most practical applications so far have been in mines (5.4, 5.5, 5.30, 5.60). More work has been done in basic research than in practical application of the technique, although sufficient laboratory and field tests have been performed to demonstrate the importance and usefulness of the method. The ability to detect the occurrence of distress in a rock mass before the development of measurable movement is obviously a significant technological advance. Experimental work that has led to the development of techniques and instrumentation for measuring microacoustic, microseismic, or subaudible rock noise has been carried on in the United States since the early 1940s (5.5). All of this work, regardless of terminology, has been related to measuring transient noise disturbances in earth (soil and rock) materials for the purpose of establishing the relative stability and, in some instances, locating zones of weakness in earth materials (5.29, 5.31, 5.42, 5.60). In the mid-1960s, Goodman, Blake, and others (5.4, 5.22, 5.31) adapted the techniques developed by the U.S. Bureau of Mines to civil engineering applications and the study of soil and rock slides. Particular emphasis was placed on the determination of relative stability, location of failure planes, determination of the epicentral location of specific disturbances, and improvement of field equipment. Rock-noise detection systems can be divided into three basic elements: sensors, signal-conditioning equipment, and recording and data-acquisition equipment.

Sensors

The frequency response of the sensors is a critical design parameter. Rock noise is reported to occur over a broad range of frequencies between about 50 and 10 000 Hz (5.4); however, disturbances at lower frequencies (20 to 40 Hz) have been observed at Downie slide, British Columbia, and are believed to be important. The occurrence of rock noise at frequencies above 10 kHz has been observed in the laboratory when rock specimens are placed under high compressive loads (5.30). Because the frequency range of noise phenomena at a particular site is unknown, the frequency response of the sensors must be sufficiently broad to ensure that the system will respond properly to any meaningful noises that occur. In addition, the sensors must be capable of (a) producing a high-level signal that is proportional to the amplitude and frequency of the exciting noise and is without spurious response and (b) transmitting the signal through long electric cables. For this reason, the output impedance of the sensor must be relatively low to prevent the cable from acting as an antenna that could pick up radio, television, and ignition interference.

Signal-Conditioning Equipment

Signal-conditioning equipment includes amplifiers and filters for the enhancement of the rock-noise signals and rejection of unwanted noise or interference. It also includes the alteration of the received signal to whatever form is required to operate data acquisition equipment, usually analogue or event pulse form. Rock-noise events are minute bursts of energy; therefore, high amplification is required by the signal-conditioning equipment to make the signals useful. Further, the amplifiers must be of a type that generates little internal noise and responds accurately to rock-noise signals over the entire frequency range specified for the sensors. Sharp cutoff filtering for both high and low frequency must be incorporated into the signal-conditioning equipment to reject unwanted noise that might interfere with the rock-noise signals.

Recording and Data-Acquisition Equipment

After rock-noise signals are received, amplified, and conditioned, they must then be recorded or analyzed or both. A rock-noise signal is generally a sharp burst of acoustic energy that attenuates rapidly. It typically sounds like a snap, click, or grind and, in analogue form, has the appearance of the record of a miniature earthquake of short duration. To be able to record the signals in analogue form is important in their identification as rock noise rather than as noise generated from other sources. Recording in analogue form is usually not necessary on a continuous basis, but can be done only periodically to ascertain that interfering signals are not being received.

Data recorded in analogue form, which is the form in which rock noise data are usually presented, are generally inconvenient to use for routine analysis of noise rate. A convenient method of determining noise rate is to electronically convert each rock noise burst to a pulse signal, which is counted mechanically or electronically on a unit-of-time basis. The number of pulses or events per unit of time can then be recorded to document the time-rate history of rock-noise activity during a long period of time. Although time-rate data may need to be accumulated continuously in active areas, they are generally sampled during selected intervals of time, such as several minutes or an hour, and sampled periodically rather than continuously. Ambient background rock-noise data from a specific area should be studied before a reasonable estimate of recording duration and intervals can be recommended. Time-rate data can be accumulated and recorded by a variety of methods. The selection of the method depends on the cost and the required convenience to the user.

AUTOMATIC WARNING AND ALARM SYSTEMS

Installing a slide-warning system may be desirable in some instances. Such systems vary from simple slide fences, sometimes used by railroads (5.9), to more complicated in-place inclinometers, extensometers, and piezometers. Their purpose is to provide automatic warning in the event of a sudden change that could be indicative of an impending earth movement. The mechanics of such systems are relatively simple to devise. An in-place inclinometer or extensometer can be used to actuate a red light in a central location when the movement exceeds a certain threshold, which, for example, may occur along a specific shear plane in soil or a bedding plane in rock. A similar alarm could

be actuated when the piezometric level exceeds a certain elevation.

One problem with such warning systems is to determine in advance the boundary between tolerable and intolerable change. Should the extensometer be set for 0.25, 2.5, or 25 mm (0.01, 0.1, or 1.0 in)? Should each instrument be monitored individually, or do only a few key instruments need such signals? Probably no computer or automatic warning system will ever replace engineering judgment in the evaluation of data from field instrumentation systems. With respect to advance warning, impending failure is normally signaled by a long-term change in rate rather than by a short-term localized change. The data must be reviewed periodically by a geotechnical engineer experienced both in field measurements and in the design requirements of that particular project.

DATA ACQUISITION AND EVALUATION

Data-Acquisition Methods

The data obtained from the instrumentation described consist of a sequence of numbers that are converted by means of a calibration chart or other relation to length, volume change, pressure, or other parameter. These numbers, plus other information, such as date, depth, or instrument number, must be recorded in the field in some form that can later be properly identified and retrieved for analysis. There are several ways in which this can be accomplished.

Manual Recording at Sensor Location

Each individual sensor or cluster of sensors can be connected to its own gauge or panel at a nearby location (Figure 5.20a). The field crew carries a portable control box to that specific location, connects it to the individual sensor, obtains and records the pertinent data, and forwards the data sheets to the field office or home office for reduction and analysis. This procedure is time consuming on large jobs and requires a substantial amount of labor, and the resulting data sheets are voluminous and subject to errors in reading, recording, and subsequent retrieval and analysis. However, the initial capital investment is low.

Automatic Recording at Sensor Locations

Significant reduction in the time required for data acquisition and subsequent savings in analysis of the data can be effected by use of automatic recording devices that can be transported to the sensor location or a nearby panel. For example, inclinometer surveys require that the sensing torpedo be transported to the hole and lowered manually to the desired depths. However, if the site is accessible by panel truck, the cable, reel, and associated readout equipment can be transported readily to the site and the data can be recorded on magnetic or punched tape in response to manual command (Figure 5.20b).

Manual Recording at Central Terminal

If the sensors are of a type whose output signal can be

Figure 5.20. Data acquisition systems.

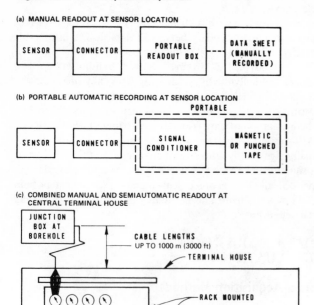

(a) MANUAL READOUT AT SENSOR LOCATION

(b) PORTABLE AUTOMATIC RECORDING AT SENSOR LOCATION

(c) COMBINED MANUAL AND SEMIAUTOMATIC READOUT AT CENTRAL TERMINAL HOUSE

transmitted for some distance over an electric cable or other direct connection or can be transmitted by wireless telemetry, all leads can often be brought into one or more central terminals. At those terminals, the signals are switched sequentially to the appropriate control boxes to produce a digital output. In the simplest form, the operator goes to the control terminal, activates the proper instruments, switches one at a time, and records the data manually. Additional savings in time and increased accuracy will result if the operator merely pushes a button to direct the output data to be recorded on punched tape (Figure 5.20c).

Completely Automatic Operation at Central Terminal

If extensive instrumentation is installed and frequent or continuous recording of the data is required, a completely automatic system may be necessary or desirable. This may vary from a comparatively simple continuous, slow-speed, strip-chart recorder to a system in which an event, such as an earthquake, triggers a mechanism that starts a recorder, which records the event for a predetermined interval of time. Several recent applications involve automatic systems in which accumulated totals or specific readings are systematically sampled at preset intervals and recorded on punched tape or some other form of record. One such arrangement systematically records at desired preset intervals the number of bursts of subaudible rock noise picked up remotely by sensors at varying depths in

several holes (5.26). Similar systems can readily be devised to record, for example, the readings of individual extensometers.

Inclinometer Observations

Since all inclinometer readings are referenced to an original set of measurements, extreme care must be taken to procure an initial set of reliable observations. Measurements of the original profile, to which all subsequent observations will be related, should be established by a double set of data. If any set of readings deviates from the previous or anticipated pattern, the inclinometer should be checked and the readings repeated. A successful record of measurements generally requires at least two trained technicians who are allowed the proper amount of time for setting up, recording observations, and undertaking maintenance of the equipment. Use of the same technicians and instrument for all measurements on a particular project is highly desirable.

General Recording Procedures

The wheels of the inclinometer torpedo provide points of measurement between which the inclination of the instrument is measured. If the reading interval is greater or less than the torpedo wheel base, correspondence between the position and orientation of the instrument will be only approximate in determining the total lateral displacement profile. Optimum accuracy is achieved only if the distance between each reading interval equals the distance between the upper and the lower wheels of the torpedo. Gould and Dunnicliff (5.23) report that readings at depth intervals as large as 1.5 m (5 ft) result in poor accuracy.

Inclinometer measurements generally are recorded as the algebraic sums or differences of 180° readings. For each depth of measurement, the reading in one vertical plane is taken and then repeated with the instrument turned through 180°. Computing the differences of these readings minimizes errors contributed by irregularities in the casing and instrument calibration. One excellent check on the reliability of each measurement is to compute the algebraic sum of 180° readings. The sum of 180° readings should be approximately constant for all measurements, except those readings made while one set of wheels is influenced by a casing joint. Differences between the sum of two readings and the observed constant sum usually indicate that an error has occurred or that opposite sides of the casing wall are not parallel. If 180° readings are summed during observation, errors resulting from mistaken transcription, faulty equipment, or improper technique can be eliminated. When the algebraic sum does not remain nearly constant, the sensor unit, readout, and casing should be rechecked before further use.

When an inclinometer is surveyed for the first time (initial set of data), a fixed reference for the torpedo should be selected so that each time a survey is repeated the torpedo will have the same orientation in the casing (i.e., each time the same set of grooves is used for alignment orientation). For the typical biaxial inclinometer (Figure 5.7), it is generally recommended that the A-component (or sensor) be oriented so that it will register the principal com-

130

ponent of the anticipated deformation as a positive change. For example, in an area suspected of landslide activity, the first set of readings is taken by orienting the fixed wheels of the torpedo in the casing groove closest to the downhill position. For a complete survey, the torpedo is reversed 180° after the first set and the readings are repeated. The algebraic difference of the two sets of readings is used to compute the profile of the casing or, more important, to determine any change compared to other surveys; using the algebraic difference is, in effect, the same as using the average of the two sets of readings.

Maintenance

Inclinometers are specialty items; consequently, the number of suppliers is limited. Repair or replacement of an inclinometer can be expensive and time-consuming and result in loss of important data. The best insurance against damage is careful use and systematic maintenance. The sensor unit should be checked frequently during operation and its wheel fixtures and bearings tightened and replaced as necessary. After each casing has been read, the guide wheels should be cleaned and oiled. Most important, the electric readout should be protected against water at all times. The introduction of a few drops of water into the readout circuitry can cause the galvanometer to drift in a pendulum inclinometer or can induce a drift of numbers on the digital voltage display in an accelerometer instrument. If readings are made too rapidly or if an automatic recorder is used, this drift may not be detected and erroneous measurements may result. When the data are analyzed at a later time, the unexpected results cannot be related to the cause.

Inclinometer Data Reduction

Hand reduction of inclinometer data is a tedious and time-consuming operation. A single movement profile for a 33-m (100-ft) casing can involve more than 200 separate computations. Because of the amount of effort involved in taking readings and, subsequently, in computing displacements, a successful measurement program depends primarily on organization and discipline. Field readings must be transposed to discernible measurement, preferably in the form of summary plots indicating successive movement profiles, as soon as possible after the field observation. This also provides perhaps the best check of instrument reliability. If a record of successive movement profiles is established, the consistency of new measurements can be referenced to previous readings and judged in light of anticipated soil behavior. Because of occasional error in the accumulation of field readings, the data reduction for a particular casing must be performed with a knowledge of both former movement and expected soil behavior. Excessively large 180° sums, which occur without precedence and are not explainable, should be discarded or the readings should be repeated.

Recording the data manually is adequate for many projects of small and intermediate size. A special form of data sheet is used and is turned over to operators who transcribe the data into computer language. Once this is done, the punched cards or tape can be handled by the computer

Figure 5.21. Sample field data sheet for transcribing inclinometer data to punched cards.

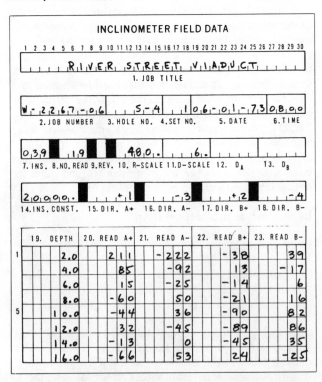

with any desired degree of sophistication. One must realize, however, that, in addition to the time delay (perhaps only a day or so), there are two potential sources of error: The first is in recording the data manually, and the second is in transcribing the data from digital form to punched card. Figure 5.21 shows a portion of a special field data sheet used for recording inclinometer data, which are to be subsequently transcribed onto punched cards.

Sources of error in recording and transcribing can be reduced by automatic printout directly onto punched tape or magnetic tape. Provided the recordings are without extraneous data, these tapes can then be returned to the office and fed directly into the computer, which plots the movement. An example of this application is shown in Figure 5.22 for an inclinometer survey. The printed output from the computer is obtained within a few hours after the data are taken. In the case of a central terminal unit with automatic printout, the end result is the same; the only difference is in the ease and speed of operation. The computer programs vary from those that simply reduce data and provide a digital printout of results to extremely sophisticated programs that reduce the data, compare the results with initial readings, plot the changes, and compare those changes with previous changes.

Many inclinometer systems, particularly servo-accelerometer types, can be adapted to automatic data recording for computerized data processing. Inclinometer manufacturers produce different systems that use punched paper tape, dual image tape, or magnetic tape for data recording. The inclinometer control boxes are modified to contain keys for recording the inclinometer well number and depth and an automatic switch to transfer the reading directly to the data tape. Automatic data recording and processing are mainly useful in reducing the time and labor involved in office computation of the data.

Automatic data recording contributes to the complexity of the measurement operation and introduces an additional source of error. With manual data recording, the technician can scan the data for face errors and make corrections or reread the casing on the spot. Most data-recording systems do not allow this advantage; the data must be scanned for errors after being printed out in the office and before computer processing. During instrumentation of the Washington, D.C., subway construction (5.13), an automatic data recording and processing system for inclinometer data from 34 installations was used briefly, but it produced no increase in efficiency because of the need for individual screening and interpretation to detect errors.

Evaluation and Interpretation

Inclinometers

One particular principle in movement monitoring with inclinometers should be emphasized: The instrument measures the change in slope over a certain depth interval during a period of time. An inclinometer will record this change in slope at any depth within the limitations of the cable on which the probe is lowered (i.e., weight, strength, and elongation characteristics of cable). Once the active zone has been detected from successive sets of data, the rate of deformation can be determined by plotting the change versus time. Usually, the slide zone is only a few feet thick; hence, the sum of the changes over a few consecutive intervals will often be representative of the magnitude and rate of movement of the entire slide.

Time plots of change at each reading depth are normally unwarranted and would be extremely time consuming and costly. As the data develop, however, such plots should be made for a selected number of intervals at which progressive change is evident. Despite its usefulness, this technique has seldom been used. Instead, the deformation, slope change, and casing profile most frequently are plotted versus depth. These plots, particularly those of change (difference) and cumulative change (deflection) versus depth, are important steps in detecting movement and visualizing what is occurring. As increasing numbers of data sets are plotted, the diagrams usually become increasingly cluttered by scatter of data and are difficult to interpret accurately. The most useful plots are those that show changes (differences) in inclination; the zones of movement can most readily be detected and time plots at each zone can be initiated.

Accuracy is usually discussed in terms of the repeatability of an integrated curve of deformation for a depth increment of 33 m (100 ft); this relates to the deflection versus depth plots. However, on deeper installations, this plot may be misleading to the interpreter. Although the instrument may be operating within its range of accuracy, over a period of time it may suggest tilting of several centimeters back and forth and perhaps a small kink may begin to develop somewhere in the curve. This situation is somewhat similar to an open-ended traverse. Primary concern should lie with the developing kink in the curve rather than with the overall tilt, which is probably related to instrument ac-

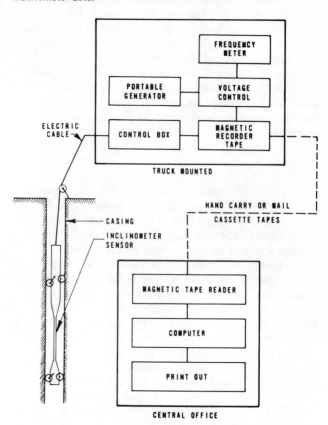

Figure 5.22. Schematic diagram of magnetic tape recording of inclinometer data.

curacy at the time of measurement. One must remember that the greatest asset of the inclinometer is its ability to measure change in inclination at a specific depth rather than to survey an exact profile of the borehole.

Extensometers

Up-to-date plots should be kept of changes in length from each anchor to the sensor and of computed changes between anchors. Particular attention should be given to the rate of change of length, for any increase in this rate may be an indication of impending failure. Extensometer readings are usually sensitive to temperature changes. If the connecting rods or wires are made of steel, any increase in temperature will result in an increase in its length and thus a reduction in the actual extensometer reading. Daily and seasonal temperature variations are likely to show similar variations in the gauge readings.

Wire extensometers, especially those with long distances between the anchor and the sensor, are particularly sensitive to changes in wire tension resulting from friction along the wire and hysteresis effects in the sensor or constant tension spring. For example, assume that a 16-gauge stainless wire 33 m (100 ft) long is subjected to a 67-N (15-lbf) pull. If the sensor requires 1.11 N (0.25 lbf) to actuate it, the change in length of the wire required to change the direction of movement of the sensor is 0.15 mm (0.006 in). An average change in temperature of -12°C (10°F) in the same wire will change its length by 3.0 mm (0.120 in). Not only does the length of the wire change with temperature, but

Figure 5.23. Geologic profile and inclinometer observations at failure plane of landslide on I-94, Minneapolis (5.57).

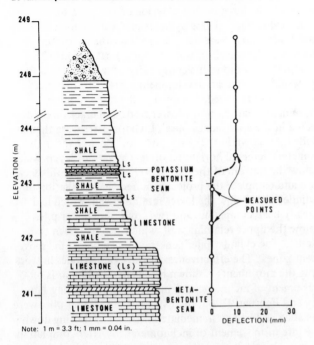

Note: 1 m = 3.3 ft; 1 mm = 0.04 in.

Figure 5.24. Section showing failure plane and movement distribution of landslide on I-94 (5.57).

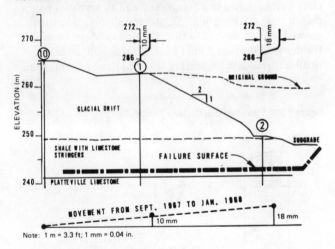

Note: 1 m = 3.3 ft; 1 mm = 0.04 in.

the ground itself will expand and contract as its temperature varies. At a hillside stability project in Montana, rock outcrops were found to expand and contract seasonally by as much as 1.5 mm (0.06 in). Electric lead wires also change their electric resistance with temperature and will cause erroneous readings unless such a change is properly taken into account.

Piezometers

Piezometric heads should be plotted on time graphs showing rainfall and other data that may influence the pore pressure. If drainage has been installed, the quantity of seepage should also be recorded and plotted. If possible, the response of each piezometer should be checked periodically to determine its recovery rate.

EXAMPLES

This section briefly describes several projects that were successfully instrumented to measure pore pressures and to detect the depth and rate of movement before corrective treatment was designed. The emphasis is on selection of instrumentation and analysis of data rather than on corrective treatment.

Minneapolis Freeway

In 1967, a landslide developed along a 335-m (1100-ft) section of I-94 on the east side of Minneapolis (5.56, 5.57). The movements were evidenced by upheaval of the subgrade and by cracking and slumping of the cut slope. Although the area had been investigated before construction, no instability problems had been anticipated. Therefore, determining the depth of movement was necessary before investigating the cause of the movements could begin. Fifteen borings were drilled to serve the dual purpose of providing subsurface soil information and holes for the installation of field instrumentation. Two sections in the vicinity of the slide were instrumented rather extensively. To measure the depth and rate of movement, 10 inclinometers were installed, and to evaluate groundwater conditions, 7 Casagrande piezometers were installed.

Several of the inclinometers detected movement within a few days, and within a month it was apparent that slope failure was occurring as a result of sliding along a near-horizontal plane approximately at elevation 243 m (798 ft). Figure 5.23 shows the geologic profile and the detailed movements recorded by one of the inclinometers, and Figure 5.24 shows the distribution of those movements along a cross section and the location of the failure plane. The piezometric data obtained during 2 months of observation showed that there were several perched water tables, but none with high uplift pressures. Water levels also were observed in the 10 inclinometer casings. However, this is not always a reliable means of groundwater observation because grout may tend to seal water either in or out of the aluminum well casing.

The data from the inclinometers demonstrated conclusively that slope failure was occurring along a near-horizontal plane only 3.3 to 4.5 m (10 to 15 ft) below the bottom of the subcut elevation, and, since the borings had failed to detect any unusual material at this depth, a decision was made to undertake a test-pit program to observe the materials along the failure surface. Eight test pits were dug, varying in depth from 3.3 to 5.5 m (10 to 18 ft). During test-pit inspection, the soil within the failure zone could be observed, sampled, and tested. The slide was found to be taking place on a thin seam of potassium bentonite, averaging less than 2.5 cm (1 in) in thickness; this material had been removed by the wash water during sampling and therefore had not been detected previously.

Stability computations were conducted on three representative cross sections where the limits of sliding were clearly defined by inclinometer data and by field observation of the heaving and cracking. In addition, the geology was more completely mapped at those locations. Since the movement developed along a horizontal plane surface, a sliding wedge analysis was used. Hydrostatic pressures,

133

based on available piezometric data, were assumed. After extensive analysis and review, the remedial scheme finally recommended and adopted was reinforced concrete buttresses cast in narrow slit trenches excavated normal to the roadway center line (5.57).

Potrero Tunnel Movements

The Potrero Tunnel of the Southern Pacific Railroad was constructed in 1906 and underlies a ridge that crosses the only rail entrance into San Francisco (5.46, 5.56). Excavation for a freeway above the tunnel in 1967 initiated movements that cracked the lining and caused the two walls of the tunnel to move closer together. However, visual observation and surface measurements did not indicate clearly (a) whether a massive landslide was involved and (b) what the relation was between freeway construction and tunnel movements. Those questions had to be answered in order to resolve legal responsibilities and to design and construct remedial measures. The following types of instruments and measuring systems were installed:

1. Inclinometers to measure horizontal ground movements,
2. Extensometers to measure ground extension adjacent to the tunnel,
3. Portable extensometers to measure closing of the tunnel sides and heave of the roof,
4. Bench marks inside the tunnel to measure changes in alignment and elevation of the track, and
5. Bench marks to measure surface movement.

The distribution and magnitude of the ground movements at a typical section are shown in Figure 5.25. Analysis of the data indicated that the ground downhill from the tunnel was stable and that the movements uphill were evenly distributed over a 9-m (30-ft) thick zone. The increased lateral earth pressure against the sides of the tunnel resulted from the freeway construction and was the cause of the tunnel distress. Corrective treatment consisted of placing two rows of heavily reinforced cast-in-place concrete piles alongside the tunnel with a connecting strut above the tunnel; these were installed from the subgrade elevation of the freeway. The same instruments that were used to determine the displacements were also read during and after construction to verify the effectiveness of the remedial measurements.

Seattle Freeway

In early 1960, a freeway constructed through downtown Seattle required the excavation of cuts into a gently sloping sidehill. A discovery was made at an early stage that even relatively shallow cuts, extending some distance along the hillside, could initiate serious ground movements. Extensive measurements, obtained primarily from inclinometers, demonstrated that the slides were progressive (5.56). After detailed study of the problem, the Washington State Highway Department developed the concept of cylinder-pile retaining walls, which consist of large-diameter concrete caissons cast in prebored, closely spaced holes placed uphill from the proposed cut before excavation. In each caisson,

a massive H-beam was inserted to provide the necessary structural strength and resistance to bending. Next, a curtain wall was hung on the exterior of the piles to provide a finished surface. These cylinder pile walls, which act as cantilever beams, were designed to limit the maximum deflection of the top of the wall to approximately 5 cm (2 in).

Figure 5.26 shows the complexity of modern freeway construction in urban development. After the material was partially removed, as shown in the upper portion, movement occurred in the lower slope and cracks were detected in the apartment house. Inclinometers were then installed, as shown in the lower portion. Most of the inclinometers detected horizontal offsets in various layers well below the bottom of the excavation. To stabilize this section and complete the project required the installation of cylinder piles below the lower retaining wall, which had already been completed, and two rows of cylinder piles below the upper retaining wall, which also had been completed. The cylinder piles penetrated well below the movement zones. The effectiveness of the cylinder piles in stopping the movements is shown by the detailed records of movement of inclinometer 5 in Figure 5.27. During January and February 1963, the two rows of cylinder piles were installed and had an immediate influence in slowing down the rate of movement of inclinometer 5. After completion of the pile installation, the movement stopped and the section has since been stable (5.37).

At the Tukwila interchange south of Seattle, inclinometers were installed in a completed wall to verify its performance after construction. Similar installations elsewhere showed that, in general, wall deflections seldom exceeded the design criteria of 5 cm (2 in). In this wall, however, small but gradually increasing horizontal movements were

Figure 5.25. Movements at Potrero Tunnel, San Francisco (5.56).

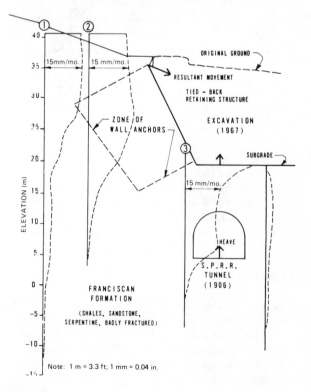

134

Figure 5.26. Seattle freeway problem (5.56).

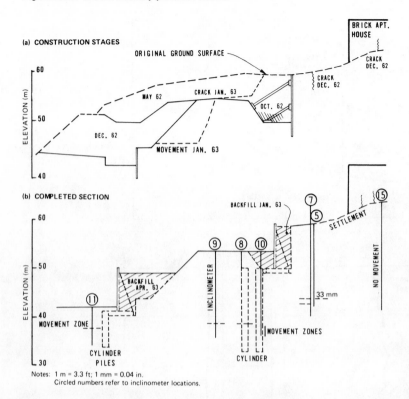

(a) CONSTRUCTION STAGES

(b) COMPLETED SECTION

Notes: 1 m = 3.3 ft; 1 mm = 0.04 in.
Circled numbers refer to inclinometer locations.

Figure 5.27. Movement of inclinometer 5, Seattle freeway (5.56).

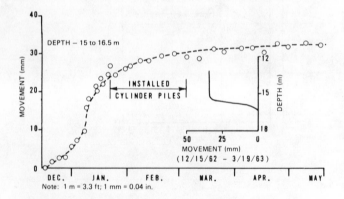

Note: 1 m = 3.3 ft; 1 mm = 0.04 in.

detected in two zones well below the tip of the cylinder piles. Fortunately, the movements were detected at an early stage, thereby permitting the construction of a deeper wall before a failure condition developed. In this instance, there was no visual indication of a potential problem and surface alignment surveys would have been difficult because the wall was on a long sweeping curve.

Fort Benton Slide

Along the Burlington Northern rail line from Great Falls to Havre in western Montana, the tracks are situated on a sidehill fill some 12 to 15 m (40 to 50 ft) above the flood-plain of the Missouri River. During a period of many years, the fill has experienced vertical and horizontal movement. These continuing movements have necessitated almost continuous maintenance, involving both realignment and res-

toration. In 1969, an investigation was undertaken to determine the probable cause of the continuing movements so that remedial treatment to improve stability could be initiated (5.58). To determine subsurface conditions beneath the site, several test borings were made and inclinometers, designated S-4 and S-5, were installed at the crest and toe of the slope respectively to determine the slide plane or zone of movement (Figure 5.28). Later in this study, for reasons described subsequently, a second inclinometer (S-6) was installed at the toe of the slope.

The borings for inclinometers S-4 and S-5 disclosed that the soils beneath the slope consisted of stiff to hard silts and clays of low to medium plasticity with intermittent zones or pockets of sand and gravel. These soils extended to a depth of about 16 m (52 ft) beneath the crest and 5 m (16 ft) beneath the toe. Below these depths, the soil encountered in both borings consisted of hard, mottled dark gray clay. It was first believed that the upper irregular zone of clay, silt, and sand was ancient slide debris or soil that had sloughed from the bluffs above and that the underlying hard clay was "original" ground. However, subsequent data obtained from inclinometer S-6 proved this was not the case. It was later determined that slide debris beneath the toe extended to a depth of about 11 m (37 ft) and was underlain by hard clay-shale that extended beyond the 26-m (85-ft) depth drilled. Although no additional drilling was accomplished at the top of the slope, movement recorded by inclinometer S-4 indicated that the surface of the hard clay-shale was probably at a depth of about 18 m (60 ft) beneath the tracks, about a meter below the bottom of the casing.

As stated earlier, movement of the sidehill fill had been occurring regularly for many years. Although no accurate

Figure 5.28. Section through Fort Benton slide, Montana (*5.58*).

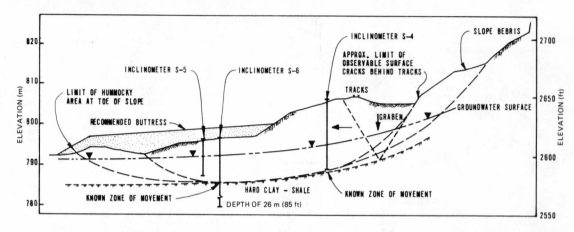

Figure 5.29. Movement of inclinometer S-5 at toe of slope, Fort Benton slide (*5.58*).

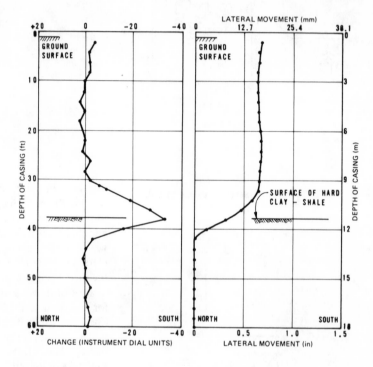

Figure 5.30. Movement of top of inclinometer S-4, Fort Benton slide (*5.58*).

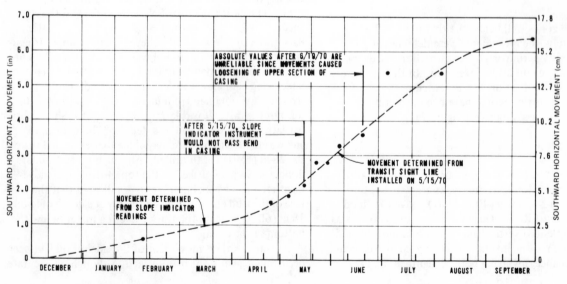

records of the magnitude and direction had been kept, the predominant track movement had been vertical and accumulated settlement during the years amounted to nearly a meter. In addition, the movement rate varied seasonally, being greatest in spring and summer and least during late fall and winter. By mid-February 1970 (installation was in November 1969), the inclinometer at the crest of the slope indicated about 1.5 cm (0.5 in) of lateral (southward) movement; however, no change was recorded at the toe of the slope during this 3-month period. Movement of inclinometer S-4 continued, accelerating somewhat in early May and decreasing in late July and August. Data from this same installation showed that movement was occurring at a depth of 18 m (59 ft) and, because the bend in the casing was relatively abrupt, was occurring within a relatively thin zone, probably only a few centimeters thick. Although reliable data were being obtained from the installation at the top of the embankment, no changes were recorded at the toe of the slope, and it became apparent that the zone of movement beneath the toe of the slope was passing below the 9-m (30-ft) depth of the casing. Hence, a second, deeper casing (S-6) was installed.

Soon after the additional installation was made, the data showed definite displacement at the toe of the slope; movement was occurring within a relatively narrow zone some 11 m (37 ft) below the ground surface. These data are shown in Figures 5.29 and 5.30. The zone of movement correlated with the surface of the hard clay-shale layer that was encountered at the same depth. Once the failure mechanism had been established from the instrumentation data, the design of corrective measures could be started. Buttressing was selected as the most practicable means of stabilizing the hillside (Figure 5.28). Had the field data revealed movement was occurring within a cohesive zone (e.g., through a thin seam of bentonite clay), some other approach to stabilization would have been necessary.

REFERENCES

5.1 American Society of Civil Engineers. Proc., Specialty Conference on Performance of Earth and Earth-Supported Structures, Purdue Univ., Lafayette, Ind., American Society of Civil Engineers, New York, Vol. 1, Part 1, 883 pp., Part 2, 1555 pp.; Vol. 2, 154 pp.; Vol. 3, 385 pp., 1972.

5.2 Bishop, A. W., Kennard, M. F., and Penman, A. D. M. Pore-Pressure Observations at Selset Dam. In Pore Pressure and Suction in Soils, Butterworths, London, 1961, pp. 91-102.

5.3 Bishop, A. W., Kennard, M. F., and Vaughan, P. R. Developments in the Measurement and Interpretation of Pore Pressure in Earth Dams. Proc., 8th International Congress on Large Dams, Edinburgh, Vol. 2, 1964, pp. 47-72.

5.4 Blake, W., and Leighton, F. Recent Developments and Applications of the Microseismic Method in Deep Mines. In Rock Mechanics: Theory and Practice, Proc., 11th Symposium on Rock Mechanics, American Institute of Mining, Metallurgy and Petroleum Engineers, New York, 1970, pp. 429-443.

5.5 Blake, W., Leighton, F., and Duvall, W. I. Techniques for Monitoring the Behavior of Rock Structures. U.S. Bureau of Mines, Bulletin 665, 1974, 65 pp.

5.6 Boden, J. B., and McCaul, D. Measurements of Ground Movements During a Bentonite Tunneling Experiment. U.K. Transport and Road Research Laboratory, Crowthorne, Berkshire, England, Rept. 653, 1974, 8 pp.

5.7 British Columbia Hydro and Power Authority. Inclinometer Data, Downie Slide. British Columbia Hydro and Power Authority, Vancouver, job file, 1976.

5.8 British Geotechnical Society. Field Instrumentation in Geotechnical Engineering. Wiley, New York, 1974, 720 pp.

5.9 Broms, B. B. Landslides. In Foundation Engineering Handbook (Winterkorn, H. F., and Fang, H. Y., eds.), Van Nostrand-Reinhold, New York, 1975, pp. 373-401.

5.10 Brooker, E. W., and Lindberg, D. A. Field Measurement of Pore Pressure in High Plasticity Soils. In Engineering Effects of Moisture Changes in Soils (Aitchison, G. D., ed.), Proc., International Research and Engineering Conference on Expansive Clay Soils, Texas A&M Univ., College Station, 1965, pp. 57-68.

5.11 Burland, J. B., and Moore, J. F. A. The Measurement of Ground Displacement Around Deep Excavations. In Field Instrumentation in Geotechnical Engineering (British Geotechnical Society), Wiley, New York, 1974, pp. 70-84.

5.12 Cooling, L. F. Second Rankine Lecture: Field Measurements in Soil Mechanics. Geotechnique, Vol. 12, No. 2, 1962, pp. 75-104.

5.13 Cording, E. J., Hendron, A. J., Jr., Hansmire, W. H., Mahar, J. W., MacPherson, H. H., Jones, R. A., and O'Rourk, T. D. Methods for Geotechnical Observations and Instrumentation in Tunneling. Department of Civil Engineering, Univ. of Illinois, Urbana-Champaign; National Science Foundation, Vol. 2, 1975, pp. 293-566.

5.14 Cornforth, D. H. Performance Characteristics of the Slope Indicator Series 200-B Inclinometer. In Field Instrumentation in Geotechnical Engineering (British Geotechnical Society), Wiley, New York, 1974, pp. 126-135.

5.15 Dallaire, E. E. Electronic Distance Measuring Revolution Well Under Way. Civil Engineering, Vol. 44, No. 10, 1974, pp. 66-71.

5.16 Dibiagio, E. Discussion: Session 2—Equipment. In Field Instrumentation in Geotechnical Engineering (British Geotechnical Society), Wiley, New York, 1974, pp. 565-566.

5.17 Dunnicliff, C. J. Equipment for Field Deformation Measurements. Proc., 4th Pan-American Conference on Soil Mechanics and Foundation Engineering, San Juan, American Society of Civil Engineers, New York, Vol. 2, 1971, pp. 319-332.

5.18 Durr, D. L. An Embankment Saved by Instrumentation. Transportation Research Board, Transportation Research Record 482, 1974, pp. 43-50.

5.19 Dutro, H. B., and Dickinson, R. O. Slope Instrumentation Using Multiple-Position Borehole Extensometers. Transportation Research Board, Transportation Research Record 482, 1974, pp. 9-17.

5.20 Eckel, E. B., ed. Landslides and Engineering Practice. Highway Research Board, Special Rept. 29, 1958, 232 pp.

5.21 Franklin, J. A., and Denton, P. E. The Monitoring of Rock Slopes. Quarterly Journal of Engineering Geology, Vol. 6, No. 3, 1973, pp. 259-286.

5.22 Goodman, R. E., and Blake, W. Rock Noise in Landslides and Slope Failures. Highway Research Board, Highway Research Record 119, 1966, pp. 50-60.

5.23 Gould, J. P., and Dunnicliff, C. J. Accuracy of Field Deformation Measurements. Proc., 4th Pan-American Conference on Soil Mechanics and Foundation Engineering, San Juan, American Society of Civil Engineers, New York, Vol. 1, 1971, pp. 313-366.

5.24 Green, G. E. Principles and Performance of Two Inclinometers for Measuring Horizontal Ground Movements. In Field Instrumentation in Geotechnical Engineering (British Geotechnical Society), Wiley, New York, 1974, pp. 166-179.

5.25 Hanna, T. H. Foundation Instrumentation. Trans Tech Publications, Clausthal, Germany, Series on Rock and Soil Mechanics, Vol. 1, No. 3, 1973, 372 pp.

5.26 Heinz, R. A. In Situ Soils Measuring Devices. Civil Engineering, Vol. 45, No. 10, 1975, pp. 62-65.

5.27 Hvorslev, M. J. Time Lag and Soil Permeability in Groundwater Observations. U.S. Army Engineer Waterways Experiment Station, Vicksburg, Miss., Bulletin 36, 1951, 50 pp.

5.28 Kern and Company, Ltd. The Mekometer, Kern ME3000, Electro-Optical Precision Distance Meter. Kern and Co., Ltd., Aarau, Switzerland, Technical Information 10, 1974, 4 pp.

5.29 McCauley, M. L. Microsonic Detection of Landslides. Transportation Research Board, Transportation Research Record 581, 1976, pp. 25-30.

5.30 Malone, A. W. Elastic Wave Measurement in Rock Engineering. Imperial College, Univ. of London, PhD thesis, 1968, 299 pp.

5.31 Mearns, R., and Hoover, T. Subaudible Rock Noise (SARN) as a Measure of Slope Stability. Transportation Laboratory, California Department of Transportation, Sacramento, Final Rept. CA-DOT-TL-2537-1-73-24, 1973, 16 pp.

5.32 Merriam, R. Portuguese Bend Landslide, Palos Verdes Hills, California. Journal of Geology, Vol. 68, No. 2, 1960, pp. 140-153.

5.33 Mexico Ministry of Public Works. Failures at the Tijuana-Ensenada Highway. Mexico Ministry of Public Works, Mexico City, 1975, 220 pp.

5.34 Mikkelsen, P. E., and Bestwick, L. K. Instrumentation and Performance: Urban Arterial Embankments on Soft Foundation Soil. Proc., 14th Symposium on Engineering Geology and Soils Engineering, Boise, Idaho, Idaho Department of Highways, Univ. of Idaho, and Idaho State Univ., 1976, pp. 1-18.

5.35 Moore, J. F. A. The Photogrammetric Measurement of Constructional Displacements of a Rockfill Dam. Photogrammetric Record, Vol. 7, No. 42, 1973, pp. 628-648.

5.36 Muñoz, A., Jr., and Gano, D. The Role of Field Instrumentation in Correction of the "Fountain Slide." Transportation Research Board, Transportation Research Record 482, 1974, pp. 1-8.

5.37 Palladino, D. J., and Peck, R. B. Slope Failures in an Overconsolidated Clay, Seattle, Washington. Geotechnique, Vol. 22, No. 4, 1972, pp. 563-595.

5.38 Peck, R. B. Observation and Instrumentation: Some Elementary Considerations. Highway Focus, Vol. 4, No. 2, 1972, pp. 1-5.

5.39 Penman, A. D. M. A Study of the Response Time of Various Types of Piezometer. In Pore Pressure and Suction in Soils, Butterworths, London, 1961, pp. 53-58.

5.40 Penman, A. D. M., and Charles, J. A. Measuring Movements of Embankment Dams. In Field Instrumentation in Geotechnical Engineering (British Geotechnical Society), Wiley, New York, 1974, pp. 341-358.

5.41 Phillips, S. H. E., and James, E. L. An Inclinometer for Measuring the Deformation of Buried Structures With Reference to Multitied Diaphragm Walls. In Field Instrumentation in Geotechnical Engineering (British Geotechnical Society), Wiley, New York, 1974, pp. 359-369.

5.42 Piteau, D. R., Mylrea, F. H., and Blown, J. G. The Downie Slide, Columbia River, British Columbia. In Natural Phenomena: Developments in Geotechnical Engineering—Rockslides and Avalanches (Voight, B., ed.), Elsevier, New York, Vol. 1, 1978.

5.43 St. John, C. M., and Thomas, T. L. The N.P.L. Mekometer and Its Application in Mine Surveying and Rock Mechanics. Trans., Institution of Mining and Metallurgy, London, Vol. 79, Sec. A, 1970, pp. A31-A36.

5.44 Schmidt, B., and Dunnicliff, J. C. Construction Monitoring of Soft Ground Rapid Transit Tunnels. Urban Mass Transportation Administration, U.S. Department of Transportation, Vols. 1 and 2, 1974; NTIS, Springfield, Va., PB 241 536.

5.45 Shannon, W. L., Wilson, S. D., and Meese, R. H. Field Problems: Field Measurements. In Foundation Engineering (Leonards, G. A., ed.), McGraw-Hill, New York, 1962, pp. 1025-1080.

5.46 Smith, T. W., and Forsyth, R. A. Potrero Hill Slide and Correction. Journal of Soil Mechanics and Foundations Division, American Society of Civil Engineers, New York, Vol. 97, No. SM3, 1971, pp. 541-564.

5.47 Terzaghi, K. Mechanism of Landslides. In Application of Geology to Engineering Practice (Paige, S., ed.), Geological Society of America, Berkey Vol., 1950, pp. 83-123.

5.48 Tice, J. A., and Sams, C. E. Experiences With Landslide Instrumentation in the Southeast. Transportation Research Board, Transportation Research Record 482, 1974, pp. 18-29.

5.49 Toms, A. H., and Bartlett, D. L. Applications of Soil Mechanics in the Design of Stabilizing Works for Embankments, Cuttings and Track Formations. Proc., Institution of Civil Engineers, London, Vol. 21, 1962, pp. 705-711.

5.50 Transportation Research Board. Landslide Instrumentation. TRB, Transportation Research Record 482, 1974, 51 pp.

5.51 Vaughan, P. R. A Note on Sealing Piezometers in Boreholes. Geotechnique, Vol. 19, No. 3, 1969, pp. 405-413.

5.52 Vaughan, P. R. The Measurement of Pore Pressures With Piezometers. In Field Instrumentation in Geotechnical Engineering (British Geotechnical Society), Wiley, New York, 1974, pp. 411-422.

5.53 Vaughan, P. R., and Walbancke, H. J. Pore Pressure Changes and the Delayed Failure of Cutting Slopes in Overconsolidated Clay. Geotechnique, Vol. 23, No. 4, 1973, pp. 531-539.

5.54 Wilson, S. D. The Use of Slope Measuring Devices to Determine Movements in Earth Masses. In Field Testing of Soils, American Society for Testing and Materials, Philadelphia, Special Technical Publ. 322, 1962, pp. 187-197.

5.55 Wilson, S. D. Investigation of Embankment Performance. Journal of Soil Mechanics and Foundations Division, American Society of Civil Engineers, New York, Vol. 93, No. SM4, 1967, pp. 135-156.

5.56 Wilson, S. D. Observational Data on Ground Movements Related to Slope Instability. Journal of Soil Mechanics and Foundations Division, American Society of Civil Engineers, New York, Vol. 96, No. SM5, 1970, pp. 1521-1544.

5.57 Wilson, S. D. Landslide Instrumentation for the Minneapolis Freeway. Transportation Research Board, Transportation Research Record 482, 1974, pp. 30-42.

5.58 Wilson, S. D., and Hilts, D. E. Application of Instrumentation to Highway Stability Problems. Proc., Joint ASCE-ASME National Transportation Engineering Meeting, Seattle, American Society of Civil Engineers, New York, and American Society of Mechanical Engineers, New York, 1971.

5.59 Wilson, S. D., and Squier, R. Earth and Rockfill Dams. 7th International Conference on Soil Mechanics and Foundation Engineering, Mexico City, State-of-the-Art Vol., 1969, pp. 137-223.

5.60 Wisecarver, D. W., Merrill, R. H., and Stateham, R. M. The Microseismic Technique Applied to Slope Stability. Trans., Society of Mining Engineers, American Institute of Mining, Metallurgical and Petroleum Engineers, New York, Vol. 244, 1969, pp. 378-385.

Chapter 6

Strength Properties and Their Measurement

Tien H. Wu and Dwight A. Sangrey

The methods of limiting equilibrium are frequently used to analyze the stability of a soil or a rock mass (Chapter 7). In such analyses, the shear strength of the material is assumed to be fully developed along the slip surface at failure. This chapter outlines the basic principles that govern the shear strength and the methods that may be used for its measurement.

GENERAL PRINCIPLES

Mohr-Coulomb Failure Criterion

The Mohr-Coulomb criterion is most widely used to define failure; it states that the shear strength (s) is

$$s = c + \sigma \tan \phi \qquad [6.1]$$

where

σ = normal stress on slip surface,
c = cohesion, and
ϕ = angle of internal friction.

In terms of principal stresses, the Mohr-Coulomb criterion becomes

$$\sigma_1 = \sigma_3 \tan^2 [(\pi/4) + (\phi/2)]$$
$$+ 2c \tan [(\pi/4) + (\phi/2)] \qquad [6.2]$$

where

σ_1 = major principal stress, and
σ_3 = minor principal stress.

Other failure criteria, particularly the modified Tresca and Von Mises, are sometimes used for soils, but their applica-tion to landslides has been limited (see later section on com-mon states of stress and stress change).

Effective Stress Versus Total Stress Analysis

Since the shear strength of soils and rocks is strongly influ-enced by the drainage conditions during loading, those con-ditions must be properly accounted for in the use of shear strength. A fundamental principle in soil engineering is the use of effective stress (σ'), which is defined as

$$\sigma' = \sigma - u \qquad [6.3]$$

where

σ = total stress, and
u = pore pressure.

The shear strength can be expressed consistently in terms of effective stress, or

$$s = c' + \sigma' \tan \phi' = c' + (\sigma - u) \tan \phi' \qquad [6.4]$$

where c' and ϕ' are the strength parameters for effective stress. The use of the effective stress parameters requires that the pore pressure be known so that σ' may be evaluated.

In general, pore pressure consists of the hydrostatic pore pressure related to groundwater level and the excess pore pressure due to applied loads. When soils are loaded under undrained or partially drained conditions, the tendency to change volume results in pore-pressure change. The excess pore pressure may be either positive or negative, depending on the type of soil and the stresses involved. Under the fully drained, long-term condition, the excess pore pressure is zero, and pore pressure due to groundwater flow can usu-ally be evaluated without serious difficulty. Hence, analysis with the effective stress description of shear strength (Equation 6.4) is most useful.

For partially drained and undrained conditions, the evaluation of excess pore pressure is often difficult. In some cases, a total stress description of shear strength may be used. One important case is the undrained loading of saturated soils. In this case, the undrained shear strength (s_u) can be used, where $s = s_u$. This is the shear strength description in the common $\phi = 0$ method of analysis. The shear strength usually changes as drainage occurs. If the change results in a higher strength, the short-term, undrained stability is critical and the stability can be expected to improve with time. On the other hand, if drainage produces a decrease in strength, the undrained shear strength can be used only for short-term or temporary situations.

Common States of Stress and Stress Change

The Mohr-Coulomb criterion does not indicate any effect of the intermediate principal stress (σ_2') on the shear strength. In practical problems, σ_2' may range from σ_3' to σ_1' depending on the geometry of the problem. The direction of the major principal stress also changes during loading. Many stability problems can be approximated by the plane-strain condition in which σ_2' lies near the midpoint between σ_3' and σ_1'. Experimental studies show that the relative value of σ_2' compared with σ_1' and σ_3' exerts some influence on the stress-strain characteristics and the shear strength.

Several common states of stress are shown in Figure 6.1. In the initial state, σ_z' is the effective overburden pressure, $\sigma_r' = K_0\,\sigma_Z'$ is the lateral pressure, and K_0 is the coefficient of earth pressure at rest. In the stress state beneath the center of a circular loaded area, the vertical stress ($\sigma_z' = \sigma_{z0}' + \Delta\sigma_z'$) is the major principal stress and the radial stress (σ_r') is the minor principal stress. In the stress state below the center of a circular excavation, the vertical stress is the minor principal stress and the radial stress is the major principal stress. For the circular load, the intermediate principal stress (σ_2') is equal to the minor principal stress (σ_3'); for the excavation, it is equal to the major principal stress (σ_1'). Slopes and retaining structures can be approximated by the plane-strain condition in which the intermediate principal strain (ϵ_2) is zero. Then, the intermediate principal stress (σ_2') is σ_y' and, in Figure 6.1d, lies between σ_1' and σ_3'.

Another important feature in many stability problems is the rotation of the principal axes during loading or excavation. It has been reported that this reduces the shear strength of some soft clays (6.50). The rotation of principal axes is shown in Figure 6.2. Before the excavation of the cut, the state of stress is represented by that shown in Figure 6.1a. After excavation, the major principal stress is in the horizontal direction at the toe (point a). Thus, the principal axes are rotated through an angle of 90°; at point b, a rotation of approximately 45° occurs. At point c, the original principal stress directions remain the same although the values of the stresses change.

Stress-Strain Characteristics

Two stress-deformation curves are shown in Figure 6.3; stress-strain curves are of similar form. In common practice,

Figure 6.1. Common states of stress in soil.

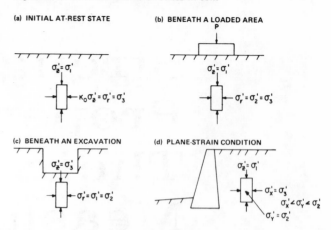

Figure 6.2. Rotation of principal stress axes in a slope.

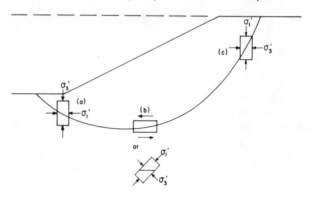

the strength of the soil is defined as the peak strength (points a and b) measured in the test. When this is used in a stability analysis, the tacit assumption is that the peak strength is attained simultaneously along the entire slip surface.

Many soils demonstrate strain-softening behavior, as illustrated by curve A. Any of several phenomena may explain the strength decrease, but it is important in design to account for this decrease. The lower limit to strength (point c) may be called the fully softened strength, remolded strength, or residual strength, depending on the type of soil involved (these terms are not synonymous). For such soils, it is unreasonable to assume that all soil along a failure surface reaches its peak strength simultaneously. In fact, the soil at some points will suffer displacements greater than Δ_a before the soil at other points reaches this deformation. In the limit of a large deformation, all strength at all points will be reduced to the strain-softening limit (point c).

Effect of Rate of Loading

The difference between the rate of loading applied in a laboratory shear test and that experienced in the slope is usually substantial. Most laboratory and in situ tests bring the soil to failure within several hours or at most a few days. For most real structures, the load remains permanently, although in some dynamic situations the peak load may be applied only for short durations. The effect of rate of loading on soil strength, excluding direct drainage effects, may be significant. In general, the undrained strength of soils

Figure 6.3. Typical stress-strain curves for soils.

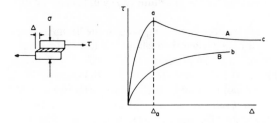

Figure 6.4. Simple test methods for determining soil strength.

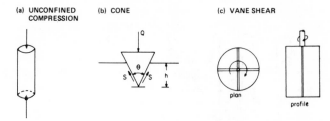

increases as the rate of loading increases; however, this effect depends on the specific material and varies over a wide range (*6.22, 6.73*).

LABORATORY MEASUREMENT OF SHEAR STRENGTH

A variety of methods is available for laboratory measurement of shear strength. The simple methods are designed to determine the shear strength of a sample in a particular condition, such as the water content or void ratio of the soil in situ. These methods are most often used to determine the undrained shear strength (s_u) of saturated cohesive soils. More elaborate laboratory tests are able to establish the shear strength relation defined by Equation 6.4. These methods allow combinations of normal and shear stresses to be employed and pore pressures to be measured or controlled. The more elaborate tests allow more accurate simulation of the field stress or deformation conditions. For example, triaxial compression tests simulate Rankine's active state, and triaxial extension simulates Rankine's passive state. The plane-strain and simple shear tests may be used to provide a better simulation of the actual deformation conditions in a slope.

Simple Tests

Three types of simple tests are discussed below.

1. The unconfined compression test is usually performed on a cylindrical sample with a diameter-to-length ratio of 1:2 or slightly more. The sample is compressed axially (Figure 6.4a) until failure occurs; the shear strength is taken as one-half the compressive strength.

2. In the cone test, a cone with an angle of θ is forced into the soil (Figure 6.4b) under a force (Q), which may be its own weight. The shear strength is obtained from the relation

$$s_u = KQ/h^2 \qquad [6.5]$$

where

    h = penetration, and
    K = constant that depends on the angle θ and the weight
        Q.

Calibration curves for K have been published by Hansbo (*6.34*) and others.

3. In the vane test, a vane is pushed into the soil specimen, and a torque is applied to the stem to produce shear failure over a cylindrical surface (Figure 6.4c). The shear strength

is obtained by equating the torque measured at failure to the moment produced by the shear stresses along the cylindrical surface. According to Cadling and Odenstad (*6.21*), the shear strength for vanes with a diameter-to-height ratio of 1:2 is

$$s = (6/7) (M/\pi D^3) \qquad [6.6]$$

where

    M = torque, and
    D = diameter of the vane.

In the application of the results of these simple tests to the analysis of slopes, consideration should be given to the type of soil and loading conditions in situ. The application of these test results is commonly limited to saturated cohesive soils under undrained conditions. The results are all expressed in terms of total stress because pore pressures are not measured. When the soil is brought to failure rapidly under undrained conditions, the shear strength is defined by $s = s_u$. If the tests are run slowly or if the soil drains during shear, the results are generally not applicable.

It is usually assumed that the measured strength is equal to the in situ strength; however, a major uncertainty is the effect of sampling disturbance on strength. Several studies (*6.15, 6.51*) show that even "good" samples may suffer strength losses as great as 50 percent. The effect of sample disturbance is most severe in soft sensitive soils and appears to become more significant as depth of the sample increases. Other factors to be considered include the state of stress and deformation. The directions of the principal stresses and the orientations of failure surfaces in each of these tests are not the same. They may also be quite different from the directions along the actual slip surface in a slope (*6.50*). Hence, caution should be exercised when the results of these simple strength tests are applied to slope stability problems (*6.15*).

Triaxial Test

The triaxial test is a highly versatile test, and a variety of stress and drainage conditions can be employed (*6.10*). The cylindrical soil specimen is enclosed within a thin rubber membrane and is placed inside a triaxial cell (Figure 6.5a). The cell is then filled with a fluid. As pressure is applied to the fluid in the cell, the specimen is subjected to a hydrostatic compressive stress (σ_3). Drainage from the specimen is provided through the porous stone at the bottom, which is connected to a volume-change gauge. The volume-change gauge is often an enclosed burette so that back pressure can be applied. Volume compressibility can be determined by

141

Figure 6.5. Triaxial test.

(a) EQUIPMENT

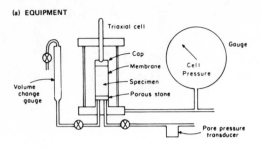

(b) STRESS CONDITIONS

(c) STRESS-STRAIN CURVES

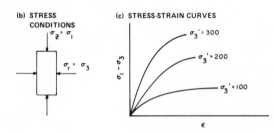

the use of these measurements. Pore pressures in the sample are also measured by a device connected to the porous stone. Several devices may be used. Electric pressure transducers are becoming more common and are replacing the traditional null indicator and hydraulic cylinder system.

The axial stress (σ_z) may be increased by application of a load through the loading ram. From the known stresses at failure ($\sigma_1 = \sigma_z$ and $\sigma_3 = \sigma_2 = \sigma_r$), Mohr circles or other stress plots can be constructed. Several triaxial tests, each using a different value of cell pressure (σ_3), are usually performed on the same material for the definition of the failure envelope. The principal stresses at failure are then used to construct Mohr circles or other stress plots from which a failure envelope can be obtained. Typical plots of the principal stress difference versus axial strain are shown in Figure 6.5c. The stress-strain behavior is influenced by the confining pressure, the stress history, and other factors. Analytical representations of normalized stress-strain relations have been suggested for some soil types. The sample can also be loaded to failure in extension by increasing the radial stress while maintaining the axial stress constant; then, $\sigma_1 = \sigma_2 = \sigma_r$ and $\sigma_3 = \sigma_z$. These two methods of loading simulate the stress states shown in Figures 6.1b and c respectively.

The following tests are commonly performed with the triaxial apparatus.

1. In the consolidated-drained test (sometimes called a drained test or slow test), the soil is allowed to consolidate completely under an effective cell pressure (σ_3') so that at the end of consolidation the excess pore pressure in the soil is zero. The water content of the specimen after consolidation is w_c. In the triaxial compression test, the axial stress is increased at a slow rate, and drainage is permitted. The rate should be slow enough so that water can drain through the soil, and no excess pore pressure should be allowed to build up. A drained test of soils of low permeability often requires several days.

The volume change during shear in a drained test can result in either an increase or a decrease in the water content.

This will depend on the type of soil and the level of stress involved. The water content at failure (w_f) will usually be different from w_c.

Since the excess pore pressure is zero in the drained test, effective stresses are known throughout the test and particularly at failure. In a compression test, the effective radial stress (σ_3') is equal to the cell pressure, and the measured load on the ram can be used to evaluate the effective axial stress (σ_1'). The results of a series of drained tests can be used to evaluate the effective stress strength parameters in Equation 6.4.

2. In the consolidated-undrained test with pore pressure measurement (sometimes called consolidated-quick test), the drainage valves are closed after the initial consolidation of the sample to w_c. Stress changes are applied through the ram, and the excess pore pressure is measured. The pore pressure is subtracted from the total axial and radial stresses to give the effective stresses. The effective stresses at failure from a series of tests are used to define the failure criterion, as in the drained test. Since the test is run in the undrained condition after consolidation, the water content throughout the test and at failure is w_c. Excess pore pressures developed during the test can be either positive or negative, depending on the type of soil and stress level.

Several equations have been proposed to describe the magnitude of the excess pore pressure developed as a result of stress changes in an undrained soil. For soils tested in the triaxial apparatus, or loaded so that $\Delta\sigma_2 = \Delta\sigma_3$, Skempton (6.67) proposed that the excess pore pressure is given by

$$\Delta u = B[\Delta\sigma_3 + A\Delta(\sigma_1 - \sigma_3)] \qquad [6.7]$$

where

B = empirical coefficient related to the soil's compressibility and degree of saturation, and

A = empirical coefficient related to the excess pore pressure developed because of shear of soil.

General relations between pore pressure and applied stresses have been suggested. For example, Henkel (6.37) proposed

$$\Delta u = B(\Delta\sigma_{oct} + \alpha\Delta\tau_{oct}) \qquad [6.8]$$

where

α = empirical coefficient similar to A;

τ_{oct} = octahedral shear stress, equal to $\frac{1}{3}\sqrt{(\sigma_1 - \sigma_2)^2 + (\sigma_2 - \sigma_3)^2 + (\sigma_3 - \sigma_1)^2}$; and

σ_{oct} = octahedral normal stress, equal to $(\sigma_1 + \sigma_2 + \sigma_3)/3$.

The consolidated-undrained test is sometimes performed without pore-pressure measurement. Obviously, effective stresses are not known during this test or at failure. The application of shear strength measured in this test to any field problem involves assumptions of excess pore pressure that are of questionable validity in most cases; thus, this test is not recommended. To relate strength parameters obtained from consolidated undrained tests without pore-pressure measurement to field conditions is difficult.

3. In the unconsolidated-undrained test (sometimes

called undrained test or quick test), no drainage is allowed during any part of the test. When the cell pressure is applied, a pore-water-pressure change (Δu_c) occurs in the soil. When the axial stress is applied, additional pore-pressure changes (Δu_a) occur. These pore-pressure changes are not usually measured, so test results must be interpreted by the use of total stresses. At failure, the undrained shear strength (s_u) is taken to represent the strength at the in situ water content. The unconsolidated-undrained test is, therefore, similar to the simple tests defined in the earlier section on simple tests.

All of the tests described above are usually begun by increasing the cell pressure to the desired stress level. This applies an isotropic or hydrostatic stress to the sample. This initial condition differs from the initial condition in situ (Figure 6.1a) if the vertical and horizontal principal stresses are different. In situ stresses can be simulated in a triaxial test by using an anisotropic stress state during consolidation. This can be accomplished by consolidating the specimen under a cell pressure and an axial load. Experimental results with a wide variety of soils show that effective stress strength parameters determined from isotropically or anisotropically consolidated tests are essentially the same. The stress-strain curves, however, are significantly different. If the stress-strain relation must be determined, anisotropic consolidation should be used.

Plane-Strain Test

The geometry of many geotechnical problems can be approximated by the condition of plane strain, in which the intermediate principal strain (ϵ_2) is zero. To simulate this condition, plane-strain tests have been developed (6.30, 6.39). In plane-strain tests, the sample is consolidated anisotropically with zero lateral strain ($\epsilon_x = \epsilon_y = 0$). After this, the sample is loaded to failure by increasing either σ_z or σ_x and maintaining $\epsilon_y = 0$. The two methods of loading can be used to simulate the stress conditions at points c or a of Figure 6.2. Plane-strain tests can be conducted under undrained, consolidated-undrained, or drained conditions in manners similar to those described for triaxial tests.

Direct Shear Test

The direct shear test is shown in Figure 6.6. The soil specimen is enclosed in a box consisting of upper and lower halves; porous stones on top and bottom permit drainage of water from the specimen. The potential plane of failure is a-a. A normal stress (σ_z') is applied on plane a-a through a loading head, and the shear stress is increased until the specimen fails along plane a-a. A stress-deformation curve is obtained by plotting the shear stress versus the displacement. Because the thickness of shear zone a-a is not precisely known, the shear strain cannot be determined. The test gives the value of τ_{xz} at failure. The vertical stress (σ_z') and the shear stress (τ_{xz}) at point b (Figure 6.6b) are known, but σ_x' is not. The directions of the principal stresses are approximately as shown in Figure 6.6c. Assuming that point a (Figure 6.6d) represents the conditions at failure, a Mohr circle can be constructed. The foregoing represents the common interpretation of the direct shear test. More elaborate analyses have been presented by Hill (6.41) and Morgenstern and Tchalenko (6.57).

Figure 6.6. Direct shear test.

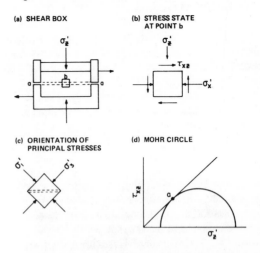

(a) SHEAR BOX

(b) STRESS STATE AT POINT b

(c) ORIENTATION OF PRINCIPAL STRESSES

(d) MOHR CIRCLE

Figure 6.7. Simple shear test.

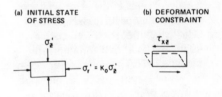

(a) INITIAL STATE OF STRESS

(b) DEFORMATION CONSTRAINT

The failure envelope is obtained from several tests, each using a different effective normal stress, performed on specimens of the same soil. The values of τ_{xz} at failure are plotted against the values of σ_z'. The loading is carried out slowly, so that no excess pore pressure develops; hence, the drained condition is obtained.

In saturated clays, the direct shear test can be performed at a rapid rate so that the time duration is too short for any appreciable amount of water to flow into or out of the sample. This is an undrained condition, and excess pore pressures of unknown magnitude are usually developed in the soil. Consequently, this is essentially a simple test, and the shear stress at failure represents the undrained shear strength (s_u).

Simple Shear Test

Several simple shear tests have been developed, but the one described by Bjerrum and Landva (6.17) is most commonly used for testing undisturbed samples. The cylindrical specimen is enclosed in a rubber membrane reinforced by wire. This allows the shear deformation to be distributed fairly uniformly through the sample, as shown in Figure 6.7b. In the test, the sample is consolidated anisotropically under a vertical stress (Figure 6.7a) and sheared by application of stress τ_{xz} (Figure 6.7b). The simple shear test can be performed under undrained, consolidated-undrained, and drained conditions. Zero volume change during shear in an undrained test can be maintained by adjusting the vertical stress (σ_z) continuously during the test. In the simple shear test, the principal axes are in the vertical and horizontal directions initially. At failure, the horizontal plane becomes the plane of maximum shear strain. This condition approximates that at point b of Figure 6.2.

SHEAR STRENGTH PROPERTIES OF SOME COMMON SOILS

Unlike steel or concrete, whose material properties are known or closely controlled, soil has material properties that are unique at every site. For that reason, the strength properties of the soil should be investigated at every site. Within broad groupings, however, the strength characteristics of many soils are similar. Appreciation of these characteristics can be helpful in planning detailed investigations.

Cohesionless Soils

Granular soils, such as gravel, sand, and nonplastic silts, are called cohesionless soils. The effective stress failure envelope of a cohesionless soil is approximately a straight line passing through the origin. This means that, for those soils, $c' = 0$ in Equation 6.4. The value of ϕ' ranges normally from about 27° to 42° or more and depends on several factors. For a given soil, the value of ϕ' increases as relative density increases. If one considers several soils at the same relative density, the value of ϕ' is affected by particle-size distribution and particle shape. The value of ϕ' for a well-graded soil may be several degrees greater than that for a uniform soil of the same average particle size. The same is true when a soil composed of angular grains is compared with one made up of rounded grains. The effect of moisture on ϕ' is small and amounts to no more than 1° or 2° (6.42).

The failure envelope, which is a straight line at low pressures, cannot be extended to high confining pressures. Tests with effective normal stresses above 700 kPa (1460 lbf/in²) indicate that the failure envelope is curved, as shown in Figure 6.8 (6.7, 6.77). The high normal stresses apparently cause crushing of grain contacts and result in a lower friction angle. Another important factor is the difference in the values of ϕ' as measured by different types of tests. The ϕ' measured in triaxial tests, which permit change in the radial strain, is as much as 4° to 5° smaller than the ϕ' measured in plane-strain tests (6.33). This difference has also been observed in field problems.

In ordinary construction situations, sandy and gravelly soils of high permeability can be considered to be loaded in the drained condition. Volume changes occur rapidly, and no excess pore pressures are sustained. Without excess pore pressures, effective stresses can be estimated from the groundwater levels. Stability analyses can be performed by using the effective stress strength parameters.

For silty soils, the permeability may be sufficiently low that excess pore pressures will develop during construction. When this is the case, the pore pressures must be measured or estimated if an effective stress analysis is to be performed.

The undrained response of sands and gravels is required for only a few situations. Saturated loose sand may fail so rapidly that excess pore pressures are sustained. Similarly, under rapid loading the undrained shear strength may be applicable (see the later section on soil behavior under repeated loads).

Soft Saturated Clays and Clayey Silts

Soils containing significant amounts of clay and silt are called cohesive soils. Because of the low permeability of

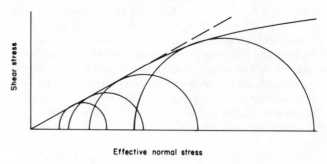

Figure 6.8. Typical failure envelope for cohesionless soils.

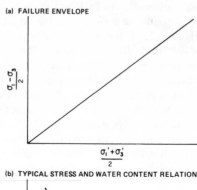

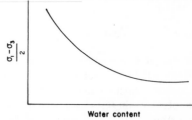

Figure 6.9. Strength behavior of soft saturated clay soils.

(a) FAILURE ENVELOPE

(b) TYPICAL STRESS AND WATER CONTENT RELATION

fine-grained soils, undrained or partially drained situations are common. This is a most important difference between cohesionless soils and cohesive soils. Another important distinction between normally consolidated or lightly overconsolidated clays and heavily overconsolidated clays is based on the kind of excess pore pressures developed in these soils during shear. The general characteristics of normally consolidated clays will be discussed first. Extremely sensitive normally consolidated clays are not discussed here but in the later section on sensitive soils.

A clay soil is considered to be normally consolidated if the consolidation pressures before shear are equal to or greater than the preconsolidation stress (p'_c). When a series of drained triaxial tests is conducted on a normally consolidated soil, the failure envelope is a straight line that passes through the origin (Figure 6.9a); thus, $c' = 0$. A relation between strength and water content is shown in Figure 6.9b. If consolidated-undrained tests are performed on a normally consolidated soil, positive excess pore pressures develop. The effective stresses at failure will define the same failure envelope in consolidated-undrained tests as in drained tests ($c' = 0$ and $\phi' = \phi'_d$). As a result of the positive excess pore pressures, however, the undrained strength ($\sigma_1 - \sigma_3$) will be less than the drained strength of a sample initially consolidated under the same stresses. This characteristic can be applied as a design principle.

If the load or stress change in the field induces positive

excess pore pressures (as in the case of a fill), the undrained strength will be lower than the drained strength. An initially stable design can usually be expected to increase in stability with time as the excess pore pressure dissipates, the water content decreases, and the strength increases. On the other hand, if negative excess pore pressures are induced (as in the case of an excavation), the undrained strength will be larger than the drained strength. Failure may occur sometime after construction, even though the slope is stable in the undrained state. Bishop and Bjerrum (6.9) have described several examples.

A good source of information on the reliability of theoretical models is the failure of real slopes. In a number of careful investigations, the factor of safety of the slope that failed was compared with the measured shear strengths. If the theory and soil properties used are correct, the safety factor of a slope at failure should be unity. The results of these studies show that, for normally consolidated or lightly overconsolidated homogeneous clays of low sensitivity, analysis using the undrained shear strength is reasonably accurate for immediate stability. For long-term stability, the effective stress analysis is also consistently accurate. Several studies on bearing capacity failure likewise show reasonable agreement. Summarized results of several case studies are given in Table 6.1. Eight cases are given of short-term failure immediately after or during construction. These are undrained conditions, and the analyses were made by using the undrained shear strength (s_u). The computed factors of safety are all close to unity and thus show that failure should have occurred according to theoretical predictions.

When a slope is made by excavation, there is a simultaneous increase in shear stress due to the slope and a decrease in mean normal stress due to the general unloading of the excavation (6.9). In a saturated normally consolidated soil, the increase in shear stress produces a positive excess pore pressure, and the decrease in normal stress produces a negative excess pore pressure. The net excess pore pressure in various parts of the slope depends on the relative values of these two effects. If the excess pore pressure is negative, the soil will decrease in strength with time and drainage. In this case, the long-term or drained stability will be critical for a normally consolidated clay. An example of this situation is shown in Figure 6.10. Table 6.1 gives four cases of long-

term failures in soft clay soil and the calculated safety factors.

The examples of field investigations given in Table 6.1 for short-term conditions of soft clay soils all involve undrained behavior of normally consolidated or lightly overconsolidated clays. To date, more than 50 slope failures and foundation failures in such soils have been investigated and reported. For more than 90 percent of those, the discrepancy between calculated and observed safety factors is less than 15 percent. Since most of the clays investigated are fairly uniform deposits, the accuracy would be less for nonuniform clay deposits. Many factors contribute to this uncertainty, but strength anisotropy and rate of loading are probably two of the most important. Bjerrum (6.14, 6.15) reviewed notable cases in which large discrepancies between prediction and performance were observed in highly plastic and organic clays. For those cases, the use of undrained shear strength, as measured by the unconfined compression or vane shear tests, tends to overestimate the safety factor under undrained conditions. Available case studies of drained failure in normally consolidated clays are too few to support a statement about the accuracy of predictions. However, the reliability of these predictions appears to be about the same as that for undrained failure of normally consolidated clays.

Heavily Overconsolidated Clays

Geological and stress histories are important considerations in the behavior of heavily overconsolidated clays. The presence of fissures, which may be due to passive failure under high values of K_0 (6.11, 6.20, 6.68) or other causes such as shrinkage, has an important influence on the strength of soils. The characteristics of some fissures and the strengths along the fissures are described by Skempton and Petley (6.75). The shear strength of laboratory specimens of fissured clays is strongly dependent on the number, shape, and inclination of fissures in the specimen (6.54). The presence of fissures is less likely in small specimens, which are often trimmed from intact soil between the fissures. Hence, the measured strength tends to increase as the size of test specimen decreases (6.52, 6.78). Thus, to extrapolate from the laboratory shear strength to the in situ shear strength is often difficult, and frequently in situ load tests must be conducted.

Most heavily overconsolidated clays show stress-strain relations that suggest general strain softening (Figure 6.3, curve A). Several concepts may be used to explain this strain-softening behavior. Consider the test results from a series of heavily overconsolidated clay specimens. If the peak strength is used to describe failure, an effective stress failure envelope as shown by curve A in Figure 6.11a is obtained. The failure envelope is approximately a straight line and, if extrapolated to the axis of $\sigma' = 0$, there is a cohesion intercept (c'). If the effective stresses at failure are used, results from both drained and undrained tests describe the same envelope. Laboratory tests using normal stresses that are close to the normal stresses in the field should be performed because research (6.11, 6.66) has shown that the failure envelope for the peak strength of heavily overconsolidated clays is curved in the low stress region and passes through the origin.

As time and drainage increase and the negative pore pressure dissipates, a heavily overconsolidated clay will absorb

Figure 6.10. Changes in pore pressure and factor of safety during and after excavation of a cut in clay (6.9).

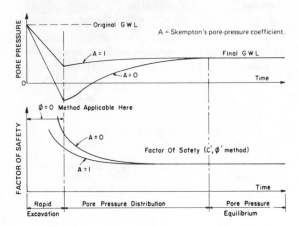

Table 6.1. Examples of stability prediction in soft and overconsolidated clays.

Clay	Condition	Site	LL (%)	PL (%)	w_c (%)	s_u (kPa)	c' (kPa)	ϕ' (deg)	ϕ_r' (deg)	Method of Analysis	Factor of Safety	Remarks	Reference
Soft	Short term	Congress Street, Chicago	32	18	25	40 to 70				$\phi = 0$	1.11 to 0.9	Circular failure	Ireland (**6.43**)
		Welland, Ontario Test cut	53	27	35	60				$\phi = 0$	1.01	Analysis by Bjerrum (**6.15**)	Kwan (**6.48**)
		Channel	40	20	30	29 to 43				Janbu	0.69 to 0.94	Analysis by Bjerrum (**6.15**)	Conlon, Tanner, and Coldwell (**6.26**)
		Portsmouth, New Hampshire, test fill	38	22	50	10 to 18				Simplified Bishop	0.86 to 0.92	Test fill	Ladd (**6.49**)
		Rangsit, Thailand	77	37	65	20 to 30				—	1.08 to 1.26	Loading test	Brand, Muktabhant, and Taecha-thummarak (**6.19**)
		England Chingford Reservoir	145	36	90	14				$\phi = 0$	1.05	Fill	Skempton and Golder (**6.72**)
		Newport	60	26	50	18				$\phi = 0$	1.08	Fill	Skempton and Golder (**6.72**)
		Huntspill	75	28	56	15				$\phi = 0$	0.90	Cut slope	Skempton and Golder (**6.72**)
	Long term	Norway Drammen	35	18	35	—	11	32.5		Simplified Bishop	1.01		Kjaernsli and Simons (**6.47**)
		Lodalen	36	18	31	40 to 60	10	27		Bishop	1.00 to 1.07		Sevaldson (**6.64**)
		Kimola, Finland Upper canal	54	23	44	50	12	27.6		Simplified Bishop	1.16		Kankare (**6.45**)
		Great slide	53	26	53	30	5	27.7		Simplified Bishop	0.97		Kankare (**6.45**)
Overconsolidated	Short term	England Selset	26	13	12		9	30		Bishop	1.05	Peak strength	Skempton and Brown (**6.71**)
		Bradwell 1	95	30	33	77				$\phi = 0$	~1.0	Peak strength corrected for fissures and time to failure	Skempton and LaRochelle (**6.74**)
		Bradwell 2	95	30	33	72				$\phi = 0$	~1.0	Peak strength corrected for fissures and time to failure	Skempton and LaRochelle (**6.74**)
		Valdarno, Italy	75	45	48		100 to 200	27 to 30		Block analysis	~1.0	Failure controlled by orientation of discontinuities	Esu (**6.31**)
	Long term	Caneleira, Brazil	45	30	—		30	40		—	1.0	Peak strength	Vargas and Pichler (**6.76**)
		England Brown London clay (12 cases)	90[a]	30[a]	—		0	20[a]		—	0.8 to 1.0	Fully softened	Skempton (**6.70**)
		Lias clay (12 cases)	60[a]	28[a]	—		0	23 (±1½)		Morgenstern and Price	0.75 to 1.0	Fully softened	Chandler (**6.24**)
		Sudbury Hill	82	28	31		0	20		Morgenstern and Price	1.05	Fully softened	Skempton and Hutchinson (**6.73**)

146

Table 6.1. Continued.

Clay	Condition	Site	LL (%)	PL (%)	w_C (%)	Shear Strength Parameters s_u (kPa)	c' (kPa)	φ' (deg)	φ'_r (deg)	Method of Analysis	Factor of Safety	Remarks	Reference
		Northolt	79	28	30		15[b] 0[c]	20[b]	13[c]	Morgenstern and Price	1.0	R = 0.60	Skempton and Hutchinson (*6.73*)
		Jackfield	44	22	21		0		<19	Morgenstern and Price	1.1	Residual	Henkel and Skempton (*6.38*)
		Sevenoaks	65	26	25		0		16	Fellenius	1.0	Residual	Skempton and Petley (*6.75*)
		River Beas Valley, India	41	25	—		0		15 to 20	Block analysis	1.0	Residual	Henkel and Yudhbir (*6.40*)
		Saskatchewan, Canada	115	23	32				6	Block analysis	0.9	Residual	Bjerrum (*6.12*)
		Balgheim, Germany	61	25	37				17	Block analysis	1.0	Residual	Bjerrum (*6.12*)
		Sadnes, Norway	60	30	36				12 to 18	—	1.0	Residual	Bjerrum (*6.12*)

Note: LL = liquid limit; PL = plastic limit; w_C = water content; s_u = undrained shear strength; c' = cohesion intercept in terms of effective stress; ϕ' = angle of internal friction in terms of effective stress; ϕ'_r = residual angle of internal friction; R = residual factor = (peak strength - average strength at failure)/(peak strength - residual strength); and 1 kPa = 0.145 lbf/in^2.

[a] Average.   [b] Peak.   [c] Residual.

Figure 6.11. Shear strength levels developed by heavily overconsolidated clays.

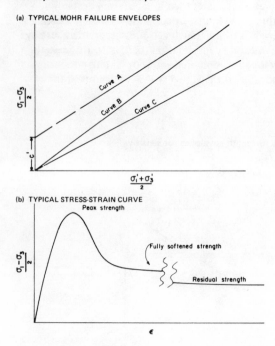

(a) TYPICAL MOHR FAILURE ENVELOPES

(b) TYPICAL STRESS-STRAIN CURVE

water. In landslides, drainage is facilitated by opening of fissures after stress release; this leads to a softening of the clay (*6.13*). The water content, at least in the failure zone, will increase significantly as the strength is reduced to the fully softened strength (Figure 6.11b). For example, several studies (e.g., *6.36*, *6.75*) included data on the differences between soil properties along discontinuities and the same properties for the adjacent "intact" clay. In one case, the water content was 44 percent along the failure surface of London clay and 30 percent in the adjacent intact clay. The failure envelope of a fully softened strength can be defined

by curve B in Figure 6.11a. This fully softened strength is typically the same as the strength of the same soil in the normally consolidated condition. The strain necessary to develop fully softened strength in a heavily overconsolidated clay varies from one soil to another, but is on the order of 10 to 100 percent. The significance of using the fully softened strength in the long-term design of slopes has been discussed by several authors (e.g., *6.24*, *6.70*).

When much larger shear displacements take place within a narrow zone, the clay particles become oriented along the direction of shear, and a polished surface or slickenside is formed (*6.46*, *6.57*). In natural slopes, slickensides may be developed along surfaces of old landslides, bedding planes, or zones of deformation caused by tectonic forces. Along these surfaces, the shear strength may approach the residual strength (*6.69*); this concept has been the subject of extensive study in the field and in the laboratory. The failure envelope, curve C in Figure 6.11a, is typical for the residual strength. The straight line passing through the origin defines the residual angle of internal friction (ϕ'_r). Kenney (*6.46*), as a result of an extensive series of laboratory direct shear tests, concluded that ϕ'_r is dependent on soil mineralogy. As shown in Figure 6.12, massive minerals, like quartz, feldspar, and calcite, have high ϕ'_r values, which are little different from the values of the peak strength parameter (ϕ') for these soils. On the other hand, the various clay mineral groups all show significant differences between ϕ' and ϕ'_r. The largest difference was found in montmorillonitic clays, which have ϕ'_r below 10°.

Both long-term and short-term stability must be considered in the design of slopes in heavily overconsolidated clays. Because of the low permeability, the time required to develop the fully drained condition may be many years (*6.12*). For most slopes in heavily overconsolidated clays, the excess pore pressures immediately after construction are negative. Thus, the undrained strength will be greater than the drained strength. As time and drainage increase, the negative pore pressures dissipate and water is drawn into the

147

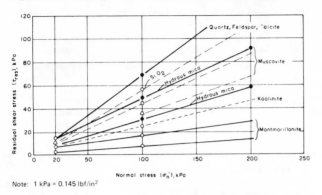

Figure 6.12. Residual shear strengths of soil minerals (6.46).

Note: 1 kPa = 0.145 lbf/in²

sample. As water content increases, strength decreases. Under field conditions, the long-term or drained conditions are critical and there is no assurance that an initially stable slope will remain stable in the long term.

The behavior of heavily overconsolidated clays is obviously much more complicated than that of normally consolidated soils. In the application of shear strength concepts to a slope stability analysis, the most difficult problem is to select the appropriate operating strength. A further complication is that the strength changes are strain-softening (Figures 6.3 and 6.11). As noted in the earlier section on stress-strain characteristics, it is unreasonable to expect all of the soil along a failure surface to reach peak strength at the same time. If the soil at some parts of a failure surface has not yet reached the peak or has passed the peak, the operating strength must be chosen to represent the average of the strengths along the entire failure surface (see Chapter 7). The use of the fully softened strength may be too conservative (6.25).

Another factor that may affect the stability of slopes in heavily overconsolidated clays is the effect of residual stress. Residual stresses released by excavation may be temporarily resisted by bonding within the soil. If these bonds deteriorate with time and weathering, the stress release will occur over a period of time and progressive failure may occur (6.12, 6.25, 6.73).

The various operating strengths of heavily overconsolidated clays are also illustrated by the case histories given in Table 6.1. The undrained failures are all short-term failures of cut slopes. Since slopes in heavily overconsolidated clays will become less stable with time, any design using the undrained strength must be considered temporary and the undrained strength should only be used if the clay is intact. Fissured clays soften and drain so rapidly that undrained conditions should not be assumed in design. The long-term examples are grouped more or less according to the type of strength at the time of failure. The number of examples given is not indicative of the actual distribution of failures in natural slopes.

Sensitive Soils

Sensitivity is defined as the ratio of the peak undrained shear strength to the undrained shear strength of a sample remolded at a constant water content. Causes of clay sensitivity are discussed by Mitchell and Houston (6.55). The

most dramatic landslides in sensitive soils are the flow slides occurring in Pleistocene marine clays in the Scandinavian countries of northern Europe and in the St. Lawrence River valley of eastern North America. The sensitivity of these soils is due to either leaching or natural cementation.

The effective stress failure envelope for both leached and naturally cemented sensitive clays is distinctly different from that of other soft clays, and the strengths of both types of sensitive soil are dominated by structure. Special testing techniques must be used to define these failure envelopes. As shown in Figure 6.13a, the strength envelope for a naturally cemented sensitive clay is a unique curve in stress space (6.59). The failure envelope encloses a low stress region within which the cemented structure remains intact. When stress changes in a slope exceed the limits of the failure curve, the structure is destroyed. Then high positive excess pore pressures are produced, and the soil behaves as a remolded soil. Thus, the failure envelope for the remolded soil (Figure 6.13a) governs the strength only after rupture of the cemented structure and has no influence on initial failure.

The cemented structures of these sensitive soils can be broken down by consolidation, particularly during isotropic consolidation of a triaxial test specimen. The natural state of stress for a sensitive cemented soil lies within its domain of natural cementation. In the design of slopes in these soils, it is usually desirable to keep the stresses within these same limits. Therefore, stresses used in laboratory tests should also lie within these limits. A common mistake in testing these materials is to use a high cell pressure, which destroys the structure during consolidation (6.15). In this case the failure envelope for remolded soil is obtained, and it may not be applicable to the actual stress conditions in the field.

Figure 6.13. Strength envelopes for sensitive soils.

(a) NATURALLY CEMENTED SOILS

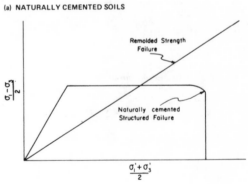

(b) LEACHED SOILS

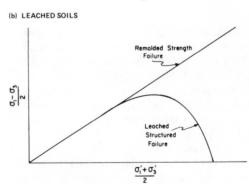

For a leached sensitive soil (Figure 6.13b), the strength envelope falls entirely below the failure envelope for the remolded soil. For both the leached and naturally cemented sensitive soils, the failure envelope is a unique characteristic for each soil. A separate set of tests must be used to determine the strength of each soil. Laboratory studies also show that the strength of these soils is highly anisotropic and that compression, extension, and simple shear tests commonly give different strength envelopes (6.16, 6.56, 6.59).

The difficulties in sampling and testing sensitive soils have resulted in a great deal of emphasis on field measurement of their strengths. The vane test is used extensively; corrections to account for anisotropy and strain rate have been suggested by Bjerrum (6.15).

Partially Saturated Soils

For partially saturated soils, effective stress analysis usually is not possible because excess pore pressures are not known. In this case, the pore pressure consists of two components: pore-air pressure and pore-water pressure. To perform an effective stress analysis of a slope requires that both components and their interaction be known. Coarse granular soils, even when partially saturated, pose little problem because excess pore pressures dissipate rapidly through drainage. On the other hand, partially saturated clays, particularly compacted clays, are a common problem. The process of compaction usually utilizes high pressures, so that the compacted clay resembles in some respects a heavily overconsolidated clay. The strength characteristics of compacted clays are therefore similar to those represented by curve A in Figure 6.11.

Figure 6.14 shows the behavior of air and water pressures

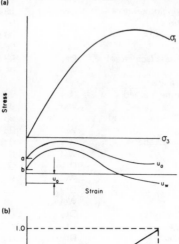

Figure 6.14. Strength of partially saturated soils (6.8).

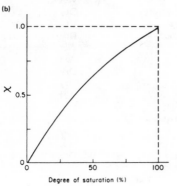

during an undrained triaxial test. In an unsaturated soil, menisci exist at the air-water interfaces and the water is under capillary tension. Hence, the initial pore-water pressures have a negative value equal to u_0. After the all-around pressure (σ_3) is applied, the compression of these air spaces brings about a volume change and a rise in the air and water pressures; these are shown as points a and b in Figure 6.14a. The difference between the pore-air pressure (u_a) and the pore-water pressure (u_w) is equal to the capillary tension. As σ_1 is increased, both u_a and u_w change with strain.

In a system with both air and water under pressure, the effective stress may be written as (6.8)

$$\sigma' = \sigma - u_a + \chi(u_a - u_w) \qquad [6.9]$$

in which χ is a coefficient that depends on the degree of saturation. This relation is evaluated experimentally. To determine the quantity χ for an unsaturated soil requires that both the air and the water pressures be known as well as c' and ϕ', the shear strength parameters of the saturated soil. For an unsaturated soil with given air and water pressures, the value of χ must be such that it satisfies

$$s = c' + [\sigma - u_a + \chi(u_a - u_w)]\,\tan\phi' \qquad [6.10]$$

Equation 6.10 may be rewritten as

$$\chi = [s - c' - (\sigma - u_a)\tan\phi']/[(u_a - u_w)\tan\phi'] \qquad [6.11]$$

In Equation 6.11, c' and ϕ' are known for the soil from tests on saturated specimens, and s, σ, u_w, and u_a are measured for the particular specimen in a triaxial test. The appropriate value of χ can then be calculated directly. The objective is to establish the relation between χ and the degree of saturation (S_r). Therefore, at different degrees of saturation, a curve like the one shown in Figure 6.14b can be obtained. This relation between χ and S_r is a soil property and must be known if the values of c' and ϕ' are to be applied to the unsaturated soil.

To evaluate the shear strength by using Equation 6.10 requires also that the air pressure (u_a) be known. If the soil is compacted at a moisture content on the dry side of optimum, most of the air voids are continuous, and the approximation $u_a = 0$ is commonly used. When this condition does not hold, the air pressure must be estimated by the same method used to evaluate the water pressure. Alternate methods using the total stress to analyze the immediate stability are described by Blight (6.18) and others (6.2).

In dry climates, some partially saturated soils have cemented bonds at the contacts. If these soils are saturated by flooding, the bonds may break leading to collapse of the natural structure. Such soils often are called collapsible soils; when their strengths are evaluated, consideration should be given to the water content under operating conditions (6.1, 6.4).

Residual Soil and Colluvium

The weathering of rock produces residual soil. On flat topography, residual soil remains where formed and is called eluvium. It may develop to great depths, and natural or human-made slopes may be cut into the eluvium. Natural rock slopes also weather to produce residual soil, which will be acted on by forces of gravity. The soil and rock debris that moves down the face of a slope is called colluvium.

Residual soils have a wide range of properties, depending on the parent material and the degree of weathering (6.29, 6.53). In the initial stages of weathering, coarse rock fragments are produced; the ultimate product of weathering is clay. Between these extremes is usually a mixture of grain sizes, and the behavior of residual soils is similar to that described in previous sections for the soils with similar particle size distributions. Some residual soils are also cemented. More detailed information about residual soils and their properties is given by Deere and Patton (6.29).

Colluvium presents some particular slope stability problems (6.27, 6.53). The extremely heterogeneous nature of the material makes sampling and testing difficult. In some cases, loose structure and large grain size result in high values of permeability. This characteristic is often accentuated by the presence of less permeable layers immediately below the colluvium. Under this condition, infiltrating surface water flows through the colluvium, generally parallel to the slope, and the consequence is a decrease in stability. Where large landslides have occurred in the past, slickensides may be found along old slip surfaces. The shear strength along those surfaces may be close to the residual shear strength.

Rocks

The shear strength of rock masses is almost always determined by the configuration of the joints and the nature of the joint surfaces (6.28), as explained in Chapter 9. Numerous theoretical and laboratory studies (6.5, 6.6, 6.32, 6.44) have been conducted to evaluate the apparent strength of rock masses with closed block joints (Figure 6.15a). These studies show that failure of jointed rock depends on several geometrical parameters, including orientation and spacing of joints, joint characteristics, and stress state, and on the strengths of the intact rock and the joint filling (see Chapter 9). In many field problems, the spacing of joints is small relative to the height of slope and the size of the failure surface.

Under these conditions, failure of a jointed rock mass involves the interaction of sliding along the joints, dilatancy, separation and rotation of the blocks, and possible fracture of the intact rock. Most experimental studies indicate that for this type of failure the failure envelope for jointed rocks is nonlinear (Figure 6.15b). Each situation is unique, so such curves must be determined experimentally for the particular situation. With certain joint orientations, failure will occur along a single joint or joint set. This commonly happens in the field when the orientation of joints or other discontinuities is close to that of the slope so that block or wedge sliding becomes possible. To analyze this condition requires that the strength of the rock along the joints be known.

The shear strength of a joint or other discontinuity depends on the characteristics of the joint. The strength of a freshly fractured joint surface will be different from that of a joint that is highly weathered and full of debris. Because fresh rock surfaces may weather rapidly, it is important to recognize the potential for change in the strength of a joint or joint system during the lifetime of an engineering project. Experimental measurements of joint strengths can be made by the use of laboratory or field direct shear tests.

A stress-displacement curve of a rough joint is similar to curve A in Figure 6.3. The peak strength (point a) is obtained when the large projections along the joint are sheared off. Beyond this point, the strength decreases and approaches the residual strength. The residual strength represents the strength of the joint after the projections have been sheared off.

Typical results of drained tests in which effective stresses could be measured are given in Table 6.2. The strength is almost always frictional in character, even when large amounts of debris are present in the joint.

Figure 6.15. Strength of rock masses with closed block joints (6.32).

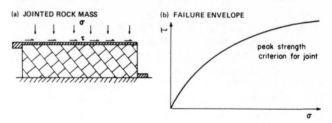

(a) JOINTED ROCK MASS   (b) FAILURE ENVELOPE

peak strength criterion for joint

Table 6.2. Residual angle of friction obtained from sandblasted, rough-sawed, and residual surfaces of various rocks (6.5).

Rock	Moisture	σ' (kPa)	φ'_r (deg)
Amphibolite	Dry	98 to 4 100	32
Basalt	Dry	98 to 8 300	35 to 38
	Wet	98 to 7 700	31 to 36
Conglomerate	Dry	294 to 3 300	35
Chalk	Wet	0 to 390	30
Dolomite	Dry	98 to 7 100	31 to 37
	Wet	98 to 7 100	27 to 35
Gneiss (schistose)	Dry	98 to 7 900	26 to 29
	Wet	98 to 7 700	23 to 26
Granite			
Fine grained	Dry	98 to 7 400	31 to 35
	Wet	98 to 7 300	29 to 31
Coarse grained	Dry	98 to 7 200	31 to 35
	Wet	98 to 7 400	31 to 33
Limestone	Dry	0 to 490	33 to 39
	Wet	0 to 490	33 to 36
	Dry	98 to 7 000	37 to 40
	Wet	98 to 7 000	35 to 38
	Dry	98 to 8 100	37 to 39
	Wet	98 to 8 100	35
Porphyry	Dry	0 to 980	31
	Dry	4 021 to 13 000	31
Sandstone	Dry	0 to 490	26 to 35
	Wet	0 to 490	25 to 33
	Wet	0 to 290	29
	Dry	294 to 2 900	31 to 33
	Dry	98 to 6 900	32 to 34
	Wet	98 to 7 200	31 to 34
Shale	Wet	0 to 290	27
Siltstone	Wet	0 to 290	31
	Dry	98 to 7 400	31 to 33
	Wet	98 to 7 100	27 to 31
Slate	Dry	0 to 1 100	25 to 30

Note: 1 kPa = 0.145 lbf/in².

SOIL BEHAVIOR UNDER REPEATED LOADS

Ground motion during earthquakes subjects slopes to repeated loading. Consider again the slope shown in Figure 6.2. The stresses shown are those acting under static conditions. When subjected to earthquake motions, additional stresses of a cyclic nature are induced in the soil. The nature of the stresses at point b is shown in Figure 6.16. For simplicity, it is assumed here that the ground motion during an earthquake consists only of shear waves propagating vertically through the soil. The cyclic stress (τ^e) during an earthquake usually consists of a series of irregular pulses, as shown by the stress-time plot in Figure 6.16c.

Several methods can be used to calculate the response of a slope to repeated loading (see Chapter 7). Some require use of a constitutive relation for the soil, and others use the soil strength under dynamic loading. When most natural soils are subjected to earthquake or other types of repeated loading, the resulting fluctuations in stress produce irreversible changes in pore pressures. These produce long-term and short-term changes in soil strength, and this must be recognized in the design of slopes to resist earthquakes and other kinds of dynamic loads. Strength changes in soils subjected

to repeated loading have been the subject of extensive research in recent years (*6.3, 6.23, 6.58, 6.60, 6.62*). In some of those studies, pore pressures and effective stresses were determined (*6.23, 6.60, 6.65*). In most cases, however, only total stress analysis was possible because rapid rates of loading did not permit the measurement of pore pressures. The present state of knowledge is based on contributions from both types of study.

Repeated Load Tests

Laboratory tests to measure soil strength under repeated loads can be made with the triaxial cell or the simple shear apparatus (*6.63*). Because of difficulties in interpretation and analysis of random loads, such as those shown in Figure 6.16c, current laboratory tests usually employ regular stress pulses. The shape of the stress pulse may be square or triangular or sinusoidal. In the simple shear test, the stress conditions shown in Figure 6.16a and b can be simulated. The sample is consolidated under the static stresses (σ_z and τ), and the peak shear stress during an earthquake is τ^e. Hence, the cyclic stress ($\tau^e - \tau$) is applied (Figure 6.16b).

The triaxial test cannot simulate the rotation of principal axes under earthquake loading; therefore, this phenomenon must be ignored. The test procedures are similar to those for static tests discussed in the section on the triaxial test. The principal stresses under static loading are σ_1 and σ_3. The sample is consolidated first under the static stresses (Figure 6.17a), after which the cyclic stresses ($\sigma_1^e - \sigma_1$ and $\sigma_3^e - \sigma_3$) are applied (Figure 6.17b). Because of the relatively short duration of earthquakes, the cyclic stresses are usually applied in the undrained condition. If the soil is saturated, the effective stress does not change under an applied hydrostatic stress. Thus, the loading can be simplified to fluctuation in axial stress only (Figure 6.17c).

Model tests of soil slopes and embankments loaded by means of a shaking table have also been used in design (*6.3*).

Stress-Strain Characteristics

Under repeated loading, the strain produced by a given peak stress is usually different from that produced by a static stress of the same magnitude. Strain continues to increase with successive cycles and depends on several factors, but particularly on the duration of the load, the magnitude of the stresses, and the number of load cycles. A typical stress-strain curve is shown in Figure 6.18. At small strain, the cyclic stress produces the hysteresis loop ab. The shear modulus and damping are equal to the slope of ab and the area enclosed by the loop respectively. At large strains, the hysteresis loop is cd. The shear modulus decreases with strain, and the damping increases with strain. Estimates of the shear modulus and damping factor can be made on the basis of available empirical data (*6.35, 6.65*). For dry soils and soils with low degrees of saturation, the modulus tends to increase with cycles of loading. Figure 6.19 compares soil behavior under repeated loading (curves B and C) with that under a monotonically increasing stress (curve A). The stress-strain relation and pore pressure under repeated loading depend on stress level, stress history, type of loading, number of stress cycles, and degree of saturation.

When a saturated soft clay or loose sand is subjected to a

Figure 6.16. Stresses under dynamic loading.

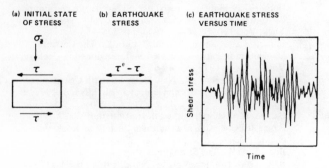

(a) INITIAL STATE OF STRESS   (b) EARTHQUAKE STRESS   (c) EARTHQUAKE STRESS VERSUS TIME

Figure 6.17. Dynamic triaxial test conditions.

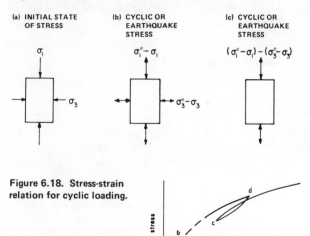

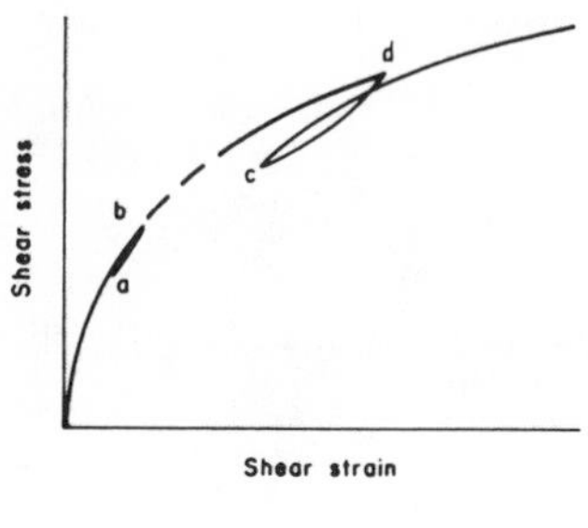

(a) INITIAL STATE OF STRESS   (b) CYCLIC OR EARTHQUAKE STRESS   (c) CYCLIC OR EARTHQUAKE STRESS

Figure 6.18. Stress-strain relation for cyclic loading.

high level of stress (for example, σ_c in Figure 6.19a), positive pore pressures develop (curve C in Figure 6.19b). After a sufficient number of load cycles, the accumulated pore pressure will lead to failure, and the stress-strain curve C (Figure 6.19a) is obtained. Failure occurs as the effective normal stress is reduced (Figure 6.19c). This failure condition is described by several different terms (6.23, 6.62). If the sample is loaded to some stress level lower than σ_c, the pore pressure may build up to a certain value and remain at that level (curve B, Figure 6.19b) and the strain will approach a limiting value (curve B, Figure 6.19a). In this case, no failure occurs.

Failure Under Repeated Loading

Failure caused by high pore pressure, as shown in Figure 6.19, is called liquefaction (6.62), particularly when applied to cohesionless soils. The relation between cyclic stress level (S) and number of cycles (N) necessary to achieve failure or a particular strain is shown in Figure 6.19d. This is a common and useful relation, particularly when only total stresses are known. At lower levels of cyclic stress, failure does not occur, even under a large number of loading cycles. This is shown as the asymptote in Figure 6.19d; it is called the critical level of repeated loading (6.60). Curves such as these make it possible to choose a design stress corresponding to the anticipated number of loading cycles. Alternatively, the critical level of stress may be used for design. In terms of effective stress, the critical level and the corresponding void ratio have been equated (6.60) to the "critical state" of the soil, as defined by Schofield and Wroth (6.61). Soils other than soft clays and loose sands can also experience strength changes as a result of earthquakes or other repeated loading. The fundamental phenomena that control the strength changes and the states that define the critical level of stress have been considered by Sangrey (6.59), and a summary is given in Table 6.3.

REFERENCES

6.1   Aitchison, G. D. Problems of Soil Mechanics and Construction on Soft Clays and Structurally Unstable Soils. Proc., 8th International Conference on Soil Mechanics and Foundation Engineering, Moscow, Vol. 3, 1973, pp. 161-190.

6.2   American Society of Civil Engineers and U.S. Committee on Large Dams. Design Criteria for Large Dams. American Society of Civil Engineers, New York, 1967, 131 pp.

6.3   Arango, I., and Seed, H. B. Seismic Stability and Deformation of Clay Slopes. Journal of Soil Mechanics and Foundations Division, American Society of Civil Engineers, New York, Vol. 100, No. GT2, 1974, pp. 139-156.

6.4   Barden, L., McGown, A., and Collins, K. The Collapse Mechanism in Partly Saturated Soil. Engineering Geology, Vol. 7, No. 1, 1973, pp. 49-60.

6.5   Barton, N. Review of a New Shear-Strength Criterion for Rock Joints. Engineering Geology, Vol. 7, No. 4, 1973, pp. 287-332.

6.6   Barton, N. Estimating the Shear Strength of Rock Joints. Proc., 3rd Congress, International Society of Rock Mechanics, Denver, Vol. 2, Part A, 1974, pp. 219-220.

6.7   Bishop, A. W. The Strength of Soils as Engineering Materials. Geotechnique, Vol. 16, No. 2, 1966, pp. 91-128.

6.8   Bishop, A. W., Alpan, I., Blight, G. E., and Donald, I. B. Factors Controlling the Strength of Partly Saturated Cohesive Soils. Proc., Research Conference on Shear Strength of Cohesive Soils, Boulder, American Society of Civil Engineers, New York, 1960, pp. 503-532.

Figure 6.19. Behavior of soils under repeated loading.

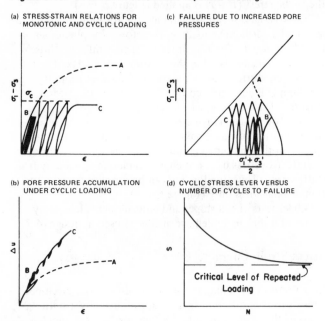

(a) STRESS-STRAIN RELATIONS FOR MONOTONIC AND CYCLIC LOADING

(c) FAILURE DUE TO INCREASED PORE PRESSURES

(b) PORE PRESSURE ACCUMULATION UNDER CYCLIC LOADING

(d) CYCLIC STRESS LEVEL VERSUS NUMBER OF CYCLES TO FAILURE

Critical Level of Repeated Loading

Table 6.3. Design recommendations for strength of soils subjected to earthquake or other repeated loading (6.59).

Soil Type	Critical Level of Stress Under Repeated Loading	Fundamentals of Response
Saturated normally consolidated clay		
Undrained	Strength at critical state; remolded strength	Accumulating positive pore water pressures
Drained	No reduction; strength increases with drainage	Drainage under positive pore-pressure gradient
Saturated overconsolidated clay		
Undrained	Little change from static strength (unless zonal)	Accumulated negative pore water pressure
Drained	Strength decreases with drainage, change in critical state	Negative pore pressures dissipate by increasing water content, decreasing critical state
Extremely sensitive, naturally cemented soils	Reduced strength greater than remolded strength	Fatigue of cementation bonds between particles
Saturated granular materials		
Loose	Strength at critical state or critical void ratio	Accumulation of positive pore pressures
Dense	No reduction from static strength (unless zonal)	Dilation

6.9   Bishop, A. W., and Bjerrum, L. The Relevance of the Tri-axial Test to the Solution of Stability Problems. Proc., Research Conference on Shear Strength of Cohesive Soils, Boulder, American Society of Civil Engineers, New York, 1960, pp. 437-501.

6.10   Bishop, A. W., and Henkel, D. J. The Measurement of Soil Properties in the Triaxial Test. Edward Arnold Publishers, London, 1962, 227 pp.

6.11   Bishop, A. W., Webb, D. L., and Lewin, P. I. Undisturbed Samples of London Clay From the Ashford Common Shaft. Geotechnique, Vol. 15, No. 1, 1965, pp. 1-31.

6.12   Bjerrum, L. Progressive Failure in Slopes of Overconsolidated Plastic Clay and Clay Shales. Journal of Soil Mechanics and Foundations Division, American Society of Civil Engineers, New York, Vol. 93, No. SM5, 1967, pp. 3-49.

6.13   Bjerrum, L. Closure to Main Session 5. Proc., 7th International Conference on Soil Mechanics and Foundation Engineering, Mexico City, Vol. 3, 1969, pp. 410-414.

6.14   Bjerrum, L. Embankments on Soft Ground. Proc., Specialty Conference on Performance of Earth and Earth-Supported Structures, Purdue Univ., Lafayette, Ind., American Society of Civil Engineers, New York, Vol. 2, 1972, pp. 1-54.

6.15   Bjerrum, L. Problems of Soil Mechanics and Construction on Soft Clays and Structurally Unstable Soils. Proc., 8th International Conference on Soil Mechanics and Foundation Engineering, Moscow, Vol. 3, 1973, pp. 111-159.

6.16   Bjerrum, L., and Kenney, T. C. Effect of Structure on the Shear Behavior of Normally Consolidated Quick Clays. Proc., Geotechnical Conference on Shear Strength Properties of Natural Soils and Rocks, Norwegian Geotechnical Institute, Oslo, 1967, Vol. 2, pp. 19-27.

6.17   Bjerrum, L., and Landva, A. Direct Simple-Shear Tests on a Norwegian Quick Clay. Geotechnique, Vol. 16, No. 1, 1966, pp. 1-20.

6.18   Blight, G. E. Effective Stress Evaluation for Unsaturated Soils. Journal of Soil Mechanics and Foundations Division, American Society of Civil Engineers, New York, Vol. 93, No. SM2, 1967, pp. 125-148.

6.19   Brand, E. W., Muktabhant, C., and Taechathummarak, A. Load Tests on Small Foundations in Soft Clay. Proc., Specialty Conference on Performance of Earth and Earth-Supported Structures, Purdue Univ., Lafayette, Ind., American Society of Civil Engineers, New York, Vol. 1, Part 2, 1972, pp. 903-928.

6.20   Brooker, E. W., and Ireland, H. O. Earth Pressures at Rest Related to Stress History. Canadian Geotechnical Journal, Vol. 2, No. 1, 1965, pp. 1-15.

6.21   Cadling, L., and Odenstad, S. The Vane Borer: An Apparatus for Determining the Shear Strength of Clay Soils in the Ground. Proc., Royal Swedish Geotechnical Institute, No. 2, 1950, 88 pp.

6.22   Casagrande, A., and Wilson, S. D. Effect of Rate of Loading on the Strength of Clays and Shales at Constant Water Content. Geotechnique, Vol. 2, No. 3, 1951, pp. 251-263.

6.23   Castro, G. Liquefaction and Cyclic Mobility of Saturated Sands. Journal of Geotechnical Engineering Division, American Society of Civil Engineers, New York, Vol. 101, No. GT6, 1975, pp. 551-569.

6.24   Chandler, R. J. Lias Clay: The Long-Term Stability of Cutting Slopes. Geotechnique, Vol. 24, No. 1, 1974, pp. 21-38.

6.25   Chandler, R. J., and Skempton, A. W. The Design of Permanent Cutting Slopes in Stiff Fissured Clays. Geotechnique, Vol. 24, No. 4, 1974, pp. 457-466.

6.26   Conlon, R. J., Tanner, R. G., and Coldwell, K. L. The Geotechnical Design of the Townline Road-Rail Tunnel. Canadian Geotechnical Journal, Vol. 8, No. 2, 1971, pp. 299-314.

6.27   D'Appolonia, E., Alperstein, R., and D'Appolonia, D. J. Behavior of a Colluvial Slope. Journal of Soil Mechanics and Foundations Division, American Society of Civil Engineers, New York, Vol. 93, No. SM4, 1967, pp. 447-473.

6.28   Deere, D. U., Hendron, A. J., Jr., Patton, F. D., and Cording, E. J. Design of Surface and Near-Surface Construction in Rock. In Failure and Breakage of Rock (Fairhurst, C., ed.), Proc., 8th Symposium on Rock Mechanics, American Institute of Mining, Metallurgy and Petroleum Engineers, New York, 1967, pp. 237-302.

6.29   Deere, D. U., and Patton, F. D. Slope Stability in Residual Soils. Proc., 4th Pan-American Conference on Soil Mechanics and Foundation Engineering, San Juan, American Society of Civil Engineers, New York, Vol. 1, 1971, pp. 87-170.

6.30   Duncan, J. M., and Seed, H. B. Strength Variation Along Failure Surfaces in Clay. Journal of Soil Mechanics and Foundations Division, American Society of Civil Engineers, New York, Vol. 92, No. SM6, 1966, pp. 81-104.

6.31   Esu, F. Short-Term Stability of Slopes in Unweathered Jointed Clays. Geotechnique, Vol. 16, No. 4, 1966, pp. 321-328.

6.32   Goodman, R. E. Methods of Geological Engineering in Discontinuous Rocks. West Publishing, St. Paul, 1976, 472 pp.

6.33   Green, G. E., and Bishop, A. W. A Note on the Drained Strength of Sand Under Generalized Strain Conditions. Geotechnique, Vol. 19, No. 1, 1969, pp. 144-149.

6.34   Hansbo, S. A New Approach to the Determination of the Shear Strength of Clay by the Fall-Cone Test. Proc., Royal Swedish Geotechnical Institute, No. 14, 1957, 47 pp.

6.35   Hardin, B. O., and Drnevich, V. P. Shear Modulus and Damping in Soils: Measurement and Parameter Effects. Journal of Soil Mechanics and Foundations Division, American Society of Civil Engineers, New York, Vol. 98, No. SM6, 1972, pp. 603-624.

6.36   Henkel, D. J. Investigations of Two Long-Term Failures in London Clay Slopes at Wood Green and Northolt. Proc., 4th International Conference on Soil Mechanics and Foundation Engineering, London, Vol. 2, 1957, pp. 315-320.

6.37   Henkel, D. J. The Shear Strength of Saturated Remolded Clays. Proc., Research Conference on Shear Strength of Cohesive Soils, Boulder, American Society of Civil Engineers, New York, 1960, pp. 533-554.

6.38   Henkel, D. J., and Skempton, A. W. A Landslide at Jackfield, Shropshire, in a Heavily Overconsolidated Clay. Geotechnique, Vol. 5, No. 2, 1955, pp. 131-137.

6.39   Henkel, D. J., and Wade, N. H. Plane Strain Tests on a Saturated Remolded Clay. Journal of Soil Mechanics and Foundations Division, American Society of Civil Engineers, New York, Vol. 92, No. SM6, 1966, pp. 67-80.

6.40   Henkel, D. J., and Yudhbir. The Stability of Slopes in the Siwalik Rocks in India. Proc., 1st Congress, International Society of Rock Mechanics, Lisbon, Vol. 2, 1966, pp. 161-165.

6.41   Hill, R. The Mathematical Theory of Plasticity. Oxford Univ. Press, London, 1950, 355 pp.

6.42   Holtz, W. G., and Gibbs, H. J. Triaxial Shear Tests on Pervious Gravelly Soils. Journal of Soil Mechanics and Foundations Division, American Society of Civil Engineers, New York, Vol. 82, No. SM1, 1956, pp. 1-22.

6.43   Ireland, H. O. Stability Analysis of the Congress Street Open Cut in Chicago. Geotechnique, Vol. 4, No. 4, 1954, pp. 163-168.

6.44   Jaeger, J. C. Friction of Rocks and Stability of Rock Slopes. Geotechnique, Vol. 21, No. 1, 1971, pp. 97-134.

6.45   Kankare, E. Failures at Kimola Floating Canal in Southern Finland. Proc., 7th International Conference on Soil Mechanics and Foundation Engineering, Mexico City, Vol. 2, 1969, pp. 609-616.

6.46   Kenney, T. C. The Influence of Mineral Composition on the Residual Strength of Natural Soils. Proc., Geotechnical Conference on Shear Strength Properties of Natural Soils and Rocks, Norwegian Geotechnical Institute, Oslo, 1967, Vol. 1, pp. 123-129.

6.47   Kjaernsli, B., and Simons, N. Stability Investigations of the North Bank of Drammen River. Geotechnique, Vol. 12, No. 2, 1962, pp. 147-167.

6.48   Kwan, D. Observations of the Failure of a Vertical Cut in Clay at Welland, Ontario. Canadian Geotechnical Journal, Vol. 9, No. 3, 1971, pp. 283-298.

6.49   Ladd, C. C. Test Embankment on Sensitive Clay. Proc., Specialty Conference on Performance of Earth and Earth-

Supported Structures, Purdue Univ., Lafayette, Ind., American Society of Civil Engineers, New York, Vol. 1, Part 1, 1972, pp. 101-128.

6.50 Ladd, C. C., and Foott, R. New Design Procedure for Stability of Soft Clays. Proc., Journal of Geotechnical Engineering Division, American Society of Civil Engineers, New York, Vol. 100, No. GT7, 1974, pp. 763-786.

6.51 Ladd, C. C., and Lambe, T. W. The Strength of "Undisturbed" Clay Determined From Undrained Tests. Proc., Symposium on Laboratory Shear Testing of Soils, Ottawa, American Society for Testing and Materials, Philadelphia, Special Technical Publ. 361, 1964, pp. 342-371.

6.52 Lo, K. Y. The Operational Strength of Fissured Clays. Geotechnique, Vol. 20, No. 1, 1970, pp. 57-74.

6.53 Lumb, P. The Properties of Decomposed Granite. Geotechnique, Vol. 12, No. 3, 1962, pp. 226-243.

6.54 Marsland, A. The Shear Strength of Stiff Fissured Clays. In Stress-Strain Behavior of Soils (Foulis, G. T., ed.), Proc., Roscoe Memorial Symposium, Cambridge Univ. Press, 1971, pp. 59-78.

6.55 Mitchell, J. K., and Houston, W. N. Causes of Clay Sensitivity. Journal of Soil Mechanics and Foundations Division, American Society of Civil Engineers, New York, Vol. 95, No. SM3, 1969, pp. 845-871.

6.56 Mitchell, R. J., and Wong, K. K. The Generalized Failure of an Ottawa Valley Champlain Sea Clay. Canadian Geotechnical Journal, Vol. 10, No. 4, 1973, pp. 607-616.

6.57 Morgenstern, N. R., and Tchalenko, J. S. Microscopic Structures in Kaolinite Subjected to Direct Shear. Geotechnique, Vol. 17, No. 4, 1967, pp. 309-328.

6.58 Sangrey, D. A. Changes in Strength of Soils Under Earthquake and Other Repeated Loading. In Earthquake Engineering (Cherry, S., ed.), Proc., 1st Canadian Conference on Earthquake Engineering, Univ. of British Columbia, Vancouver, 1972, pp. 82-96.

6.59 Sangrey, D. A. Naturally Cemented Sensitive Soils. Geotechnique, Vol. 22, No. 1, 1972, pp. 139-152.

6.60 Sangrey, D. A., Henkel, D. J., and Esrig, M. I. The Effective Stress Response of a Saturated Clay Soil to Repeated Loading. Canadian Geotechnical Journal, Vol. 6, No. 3, 1969, pp. 241-252.

6.61 Schofield, A., and Wroth, P. Critical State Soil Mechanics. McGraw-Hill, New York, 1968, 310 pp.

6.62 Seed, H. B. Landslides During Earthquakes Due to Soil Liquefaction. Journal of Soil Mechanics and Foundations Division, American Society of Civil Engineers, New York, Vol. 94, No. SM5, 1968, pp. 1055-1122.

6.63 Seed, H. B., and Peacock, W. H. Test Procedures for Measuring Soil Liquefaction Characteristics. Journal of Soil Mechanics and Foundations Division, American Society of Civil Engineers, New York, Vol. 97, No. SM8, 1971, pp. 1099-1119.

6.64 Sevaldson, R. A. The Slide in Lodalen, October 6, 1954. Geotechnique, Vol. 6, No. 4, 1956, pp. 167-182.

6.65 Silver, M. L., and Seed, H. B. Deformation Characteristics of Sands Under Cyclic Loading. Journal of Soil Mechanics and Foundations Division, American Society of Civil Engineers, New York, Vol. 97, No. SM8, 1971, pp. 1081-1098.

6.66 Singh, R., Henkel, D. J., and Sangrey, D. A. Shear and K_O Swelling of Overconsolidated Clay. Proc., 9th International Conference on Soil Mechanics and Foundation Engineering, Moscow, Vol. 1, Part 2, 1973, pp. 367-376.

6.67 Skempton, A. W. The Pore-Pressure Coefficients A and B. Geotechnique, Vol. 4, No. 4, 1954, pp. 143-147.

6.68 Skempton, A. W. Horizontal Stresses in an Over-Consolidated Eocene Clay. Proc., 5th International Conference on Soil Mechanics and Foundation Engineering, Paris, Vol. 1, 1961, pp. 351-357.

6.69 Skempton, A. W. Long-Term Stability of Clay Slopes. Geotechnique, Vol. 14, No. 2, 1964, pp. 77-101.

6.70 Skempton, A. W. First-Time Slides in Over-Consolidated Clays. Geotechnique, Vol. 20, No. 3, 1970, pp. 320-324.

6.71 Skempton, A. W., and Brown, J. D. A Landslide in Boulder Clay at Selset, Yorkshire. Geotechnique, Vol. 11, No. 4, 1961, pp. 280-293.

6.72 Skempton, A. W., and Golder, H. Q. Practical Examples of the $\phi = 0$ Analysis of Stability of Clays. Proc., 2nd International Conference on Soil Mechanics and Foundation Engineering, Rotterdam, Vol. 2, 1948, pp. 63-70.

6.73 Skempton, A. W., and Hutchinson, J. N. Stability of Natural Slopes and Embankment Foundations. Proc., 7th International Conference on Soil Mechanics and Foundation Engineering, Mexico City, State-of-the-Art Vol., 1969, pp. 291-340.

6.74 Skempton, A. W., and La Rochelle, P. The Bradwell Slip: A Short-Term Failure in London Clay. Geotechnique, Vol. 15, No. 3, 1965, pp. 221-242.

6.75 Skempton, A. W., and Petley, D. J. The Strength Along Structural Discontinuities in Stiff Clays. Proc., Geotechnical Conference on Shear Strength Properties of Natural Soils and Rocks, Norwegian Geotechnical Institute, Oslo, Vol. 2, 1967, pp. 29-46.

6.76 Vargas, M., and Pichler, E. Residual Soil and Rock Slides in Santos (Brazil). Proc., 4th International Conference on Soil Mechanics and Foundation Engineering, London, Vol. 2, 1957, pp. 394-398.

6.77 Vesíc, A. S., and Clough, G. W. Behavior of Granular Materials Under High Stresses. Journal of Soil Mechanics and Foundations Division, American Society of Civil Engineers, New York, Vol. 94, No. SM3, 1968, pp. 661-668.

6.78 Ward, W. H., Marsland, A., and Samuels, S. G. Properties of the London Clay at the Ashford Common Shaft: In Situ and Undrained Strength Tests. Geotechnique, Vol. 15, No. 4, 1965, pp. 321-344.

Chapter 7

Methods of Stability Analysis

N. R. Morgenstern and Dwight A. Sangrey

The functional design of a slope requires that the deformations of the earth mass constituting the slope be limited to amounts governed by the use of the ground on or adjacent to the slope. If there is no particular development near the slope, the allowable deformations can be large, provided that the earth mass does not fail with resulting uncontrolled deformations. On the other hand, if there are structures close to the slope or services buried within the soil beneath the slope, smaller deformations may result in unacceptable performance, and functional design would require that the average stress level in the earth mass be lower in this case than in the case in which no development is near the slope.

Analysis is an important component of design, which also includes other equally important components, such as drainage considerations and construction control. In analysis, the mechanical properties of the earth materials are evaluated quantitatively to arrive at a configuration consistent with the performance requirements of the slope. In this chapter, the relative roles of limit equilibrium and deformation analyses are discussed. The principles of limit equilibrium analysis are outlined, and details of the techniques commonly used in practice are given with examples. Both soil and rock slopes are discussed. The chapter concludes with a brief discussion of deformation analysis as it is currently used. Other aspects of design are discussed in Chapters 8 and 9.

ROLES OF LIMIT EQUILIBRIUM AND DEFORMATION ANALYSES

Although the performance of a slope often is dictated by allowable deformations, quantitative prediction of displacements of the slope is seldom undertaken routinely. Instead, analyses are made by the use of limit equilibrium methods, and the performance of a slope is evaluated in terms of its factor of safety (F). There are several reasons for this.

For realistic configurations commonly encountered in design, particularly in the design of cuts and natural slopes,

a method of analysis that predicts deformations to an acceptable degree of accuracy would likely have to include

1. Representative stress-strain relations, including behavior from peak to residual shear strengths;
2. Anisotropy;
3. Variable pore pressure distributions;
4. Nonhomogeneity arising from variation of material properties with depth, layering, and discontinuities;
5. Influence of initial stress; and
6. Construction sequence.

Most of these factors have a major influence on the behavior of a slope; and, even if proven analytical techniques were readily available to predict deformations, the determination of all of these factors in a manner suitable for a deformation analysis would seldom be a practical undertaking. Appropriate analytical techniques are not yet at hand, although impressive progress is being made by means of finite element methods.

The situation is not so pessimistic with regard to fills. Representative stress-strain relations within the working range can be specified with a greater degree of reliability for embankments than for cuts or natural slopes. Moreover, since the soil mass is a processed material, it is generally more uniform and homogeneous than naturally occurring deposits. Hence, deformation analysis has a role in the design of embankment slopes, and that role will be reviewed in greater detail later in this chapter.

The stabilities of natural slopes, cut slopes, and fill slopes are commonly analyzed by limit equilibrium methods. These methods take into account all of the major factors that influence the shearing resistance of a soil or rock mass; this is one of their significant advantages. In addition, they are simpler than deformation analyses. However, because the actual stress-strain relations are not used in the analysis, limit equilibrium methods do not result in calculation of expected deformations. Deformations are controlled by designing for an appropriate factor of safety. At least to this extent, the use of limit equilibrium methods of analysis is semiempirical. The factor of safety cannot be measured except when the slope is failing, in which case it is known to be unity. As noted previously, there are many instances in which the precise deformations in a slope are of little concern, provided the material stays in place. In such cases, it is entirely appropriate to undertake analyses with limit equilibrium methods; however, regardless of the complexity of the method employed, the empirical nature of the design criteria should be borne in mind.

LIMIT EQUILIBRIUM ANALYSIS

Limit equilibrium methods for slope stability analysis have been used for several decades, and numerous techniques have been developed. Limit equilibrium analysis is used in design to determine the magnitude of the factor of safety. When a slope has failed, however, the factor of safety is unity, and the analysis can then be used to estimate the average shearing resistance along the failure surface or along part of the failure surface if the shearing resistance is assumed to be known along the remainder. Regardless of the specific procedure for carrying out the computations, the following principles are common to all methods of limit equilibrium analysis.

1. A slip mechanism is postulated. This is done without any major kinematic restriction except that the mechanism be feasible and sensible. In the simpler configurations, the slopes are assumed to fail along planes or circular sliding surfaces. When conditions are not uniform, more complex shapes are known to be appropriate, and analyses have been developed to handle surfaces of arbitrary shape.

2. The shearing resistance required to equilibrate the assumed slip mechanism is calculated by means of statics. The physical concepts used here are that the potential slip mass is in a state of limiting equilibrium and that the failure criterion of the soil or rock is satisfied everywhere along the proposed surfaces. Various methods differ in the degree to which the conditions for equilibrium are satisfied, and some common methods of analysis violate conditions of static equilibrium. This is an important factor in evaluating the rigor of any method.

3. The calculated shearing resistance required for equilibrium is compared with the available shear strength. This comparison is made in terms of the factor of safety, which will be defined more precisely below.

4. The mechanism with the lowest factor of safety is found by iteration. For example, if it is assumed that the slip surface is circular, a search is made for the critical slip circle. When the position of the slip surface is dictated by a dominant weakness, such as sheared clay at residual strength, other trials are unnecessary.

The factor of safety should be defined as clearly as possible and its role understood. The definition used here is as follows:

The factor of safety is that factor by which the shear strength parameters may be reduced in order to bring the slope into a state of limiting equilibrium along a given slip surface.

According to this definition, the factor of safety relates to the strength parameters and not to the strength itself, which in the case of an effective stress analysis depends on the effective normal stress as well. Moreover, this definition implies that the factor of safety is uniform along the entire slip surface.

Other definitions of the factor of safety are also used in geotechnical engineering. For example, the factor of safety is commonly defined as the ratio of a disturbing moment to a restoring moment when the moment equilibrium about the center of rotation of a slip circle is under consideration. This definition is not convenient when slip surfaces are noncircular and therefore is of limited applicability. It is equivalent to the definition given above in the case of a $\phi = 0$ analysis. When the bearing capacity of a foundation is calculated, the factor of safety is usually determined as the ratio of the maximum bearing capacity to the allowable bearing capacity. Used in this way, the factor of safety is a ratio of loads and differs from the factor of safety applied to strength parameters. This is one reason why the magnitude of the factor of safety commonly adopted in design for bearing capacity differs substantially from the values used in slope stability analyses. Occasionally, the factor of safety is also taken as the ratio of the calculated stress at a point to the allowable stress. This definition can only be applied when the stress distribution in the slope is evaluated in detail, and it bears no simple relation to the definition adopted in limit equilibrium analysis. As noted by Barron (*7.2*), even the definition given here is restrictive because it is based on an assumption for the likely stress path to failure. However, the implications of whether failure might occur under drained or undrained conditions and whether the soil may have a metastable structure can be taken into account by adjusting the factor of safety accordingly.

Partial safety factors are sometimes used in stability analysis. A simple case is the application of separate factors of safety to two strength parameters, for example, c'/F_1 and $\tan \phi'/F_2$, where F_1 and F_2 have different values. Uncertainty about other terms in the stability equation might also be included by application of partial safety factors or coefficients to loads applied to the slope, water pressures, and other parameters. This practice has been common in some parts of Europe (*7.29*) and recently has been used in probabilistic methods of design.

An understanding of the role of the factor of safety is vital in the rational design of slopes. One well-recognized role is to account for uncertainty and to act as a factor of ignorance with regard to the reliability of the items that enter into analysis. These include strength parameters, pore-pressure distribution, and stratigraphy. In general, the lower the quality of the site investigation is, the higher the factor of safety should be, particularly if the designer has only limited experience with the materials in question. Another use of the factor of safety is to provide a measure of the average shear stress mobilized in the slope. This should not be confused with the actual stresses. A realistic stress distribution cannot, of course, be calculated from a limit equilibrium analysis because stress-strain relations are not used in the analysis. As indicated earlier, a major role of the factor of safety is that it constitutes the empirical tool whereby deformations are limited to tolerable amounts within economic restraints. In this way, the choice of the factor of safety is greatly influenced by the accumulated experience with a particular soil or rock mass. Since the degree of risk that can be taken is also much influenced by experience, the actual magnitude of the factor of safety used in design will vary with material type and performance requirements. These and related topics influencing the design of stable slopes are discussed in Chapter 8.

Total and Effective Stress Analyses

When a fully saturated soil is sheared to failure without permitting drainage, the soil behaves as though it is purely cohesive (i.e., $\phi = 0$ and $s = s_u$) and the results are interpreted in terms of total stresses. This behavior is more properly viewed as a response to the type of test and method of interpretation than as a reflection of basic strength properties. Nevertheless, simple, practical analyses result when $\phi = 0$ behavior is assumed, and, provided the available shearing resistance is selected with care, reasonable results can be obtained. Undrained total stress analysis is limited to slopes where the pore pressures are governed by total stress changes and little time has elapsed so that no significant dissipation has occurred. This is the so-called end-of-construction class of problems. Even in the case of normally or lightly overconsolidated soils, the factor of safety under undrained conditions may not be critical during the life of a slope, as Bishop and Bjerrum (7.4) point out. The case of the failure along the slope of the Kimola Canal in Finland is particularly instructive in this regard (7.45, 7.48).

When pore pressures are governed by steady seepage configurations, stability analysis should be performed in terms of effective stresses. This is the usual condition for natural slopes in both soil and rock. Moreover, all permanent cuts or fills should be analyzed for long-term conditions to see whether these control the design. Some jointed and fissured clays respond to drainage along dominant discontinuities so quickly that it is doubtful that a $\phi = 0$ idealization is ever representative (7.24). These materials should be analyzed in terms of effective stresses regardless of the time of loading or unloading.

It is also possible to analyze slopes under undrained conditions in terms of effective stresses, but the pore pressures must be obtained either by prediction in terms of total stress changes and pore pressure parameters or from field observations. Embankments on soft clay and silt foundations are commonly designed by the use of a $\phi = 0$ analysis and monitored during construction in terms of effective stress. If design has been based on a factor of safety of 1.5 in terms of a total stress analysis, it does not follow that the factor of safety during construction should also be 1.5 in terms of an effective stress analysis. For a given problem the two coincide only at failure, in which case the factor of safety is unity.

A $\phi = 0$ analysis requires knowledge of only the undrained strength, while an effective stress analysis is based on c', ϕ', and the pore-pressure distribution. The selection of strength parameters is a key factor in analysis, and this is made difficult by the fact that the relevant parameters are often anisotropic, depend on rate of test and size, and decay with time. Some of the problems associated with selecting appropriate parameters are discussed by Skempton and Hutchinson (7.66) and in Chapter 6. The analytical capabilities discussed

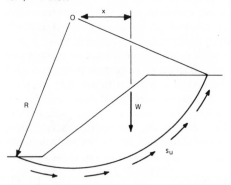

Figure 7.1. Geometry used in slip circle analysis for $\phi = 0$ case.

in the following sections, particularly when facilitated by computer programs, invariably exceed our ability to identify the appropriate slip mechanism, determine representative shear strength parameters, and specify the correct water pressure distribution.

Total Stress Analysis of Soil Slopes ($\phi = 0$)

Homogeneous Isotropic

Figure 7.1 shows a uniform isotropic purely cohesive slope. From considerations of moment equilibrium it can be shown that

$$F = R \int s_u \, ds / W x \qquad [7.1]$$

where

  F = factor of safety,
  ds = increment of arc length,
  s_u = undrained shear strength,

and the other terms are specified in Figure 7.1. This problem is statically determinate, and it is possible to solve for the critical circle as a function of the slope angle and geometry (7.68). The solution takes the form

$$F = N(s_u / \gamma H) \qquad [7.2]$$

where

  γ = unit weight of the soil,
  H = height of the slope, and
  N = stability factor.

N is a function only of the slope angle (β) and the depth factor (n_d). Charts presenting the results of analytical studies are given by Taylor (7.68, 7.69) and Terzaghi and Peck (7.71).

Nonhomogeneous Isotropic

Normally consolidated clays characteristically exhibit a linear increase of strength with depth. This often is expressed as the ratio of the undrained strength (s_u) to the vertical effective stress (p) under which the material has been consolidated. Although the ground close to the surface may have been overconsolidated by desiccation, this will be neglected in the analysis presented here. A linear increase in undrained strength with depth can be treated in analysis by assuming that the soil consists of a sequence of layers chosen to represent the variation of strength with depth to an appropriate degree of accuracy. Since this is cumbersome, Gibson and Morgenstern (7.27) sought an analytical solution to the problem. They showed that the factor of safety for a cut in a normally consolidated soil may be found from the following equation, which is analogous to Equation 7.2:

$$F = N(s_u/p)_n [1 - \gamma_w/\gamma)] \qquad [7.3]$$

where

γ_w = unit weight of water, and
N = stability factor that depends only on the slope angle (β).

The variation of N with β is shown in Figure 7.2. The factor of safety is independent of the height of the slope. In theory, failure occurs along an infinite number of slip circles that emerge above the toe, and the need to consider a depth factor is obviated. In practice, the position of the actual slip circle will be dictated by some small variation from the assumed strength-depth relation. Further theoretical support for the use of slip circles for this class of problems has been provided by Booker and Davis (7.6), who show, from a more exact analysis based on the theory of plasticity, that the slip circle analysis provides acceptable results for slopes steeper than 5°, i.e., within the usual working range.

Hunter and Schuster (7.38) considered the stability of a cut in soil with a strength profile shown in Figure 7.3. The undrained shear strength increases with depth, but there is a finite intercept at the surface. Both finite and infinite depths of soil were taken into consideration and showed that

$$F = (s_u/p) (\gamma'/\gamma)N \qquad [7.4]$$

where γ' is the submerged unit weight of the soil and N depends on β and M. M is a parameter used to specify the strength of the soil at the surface of the ground and is given by

$$M = (h/H)(\gamma_w/\gamma') \qquad [7.5]$$

where H is the height of slope, and the other terms are specified in Figure 7.3. The results for an unlimited depth of clay are shown in Figure 7.4, and charts describing the influence of the depth factor are included in the original reference.

Design charts have also been constructed to facilitate the stability analysis of embankments on soils subjected to undrained loading. Nakase (7.60) summarizes the development of various solutions and presents in a concise form the results of the analysis for the case shown in Figure 7.5. This is particularly useful in that it embraces a variety of possible embankment loadings on a soil whose strength increases with depth from a finite intercept at the surface. Both finite and infinite depths of subsurface clay are included in the

analysis. Although the strength of the embankment material is neglected, this assumption is often realistic for reasons of strain compatibility between the fill and the foundation materials. In any case, the assumption leads to conservative results. The final design charts are too numerous to be included here.

A comprehensive set of design charts for the stability of an embankment on a uniform homogeneous purely cohesive foundation was prepared by Pilot and Moreau (7.62). In this case the strength of the embankment material is taken into account, but the foundation is characterized by a uniform undrained strength. The depth to a rigid stratum is varied as well. An extensive analysis over the range of all the parameters of interest has been undertaken, and the factor of safety is presented directly in terms of these parameters.

Anisotropic

The undrained strength of a soil is commonly anisotropic.

Figure 7.2. Variation of stability number with slope angle for cuts in normally consolidated clays (7.27).

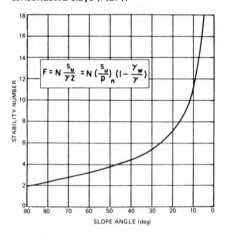

Figure 7.3. Idealized strength-depth relation for normally consolidated clay (7.38).

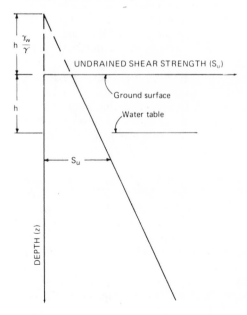

This can arise from the development of an anisotropic fabric during deposition and subsequent consolidation. Even if anisotropic structure can be ignored, anisotropic states of stress during consolidation result in anisotropy in the undrained strength (7.30). Casagrande and Carrillo (7.8) proposed that the undrained strength varies with direction according to

$$C_i = C_2 + (C_1 - C_2) \cos^2 i \qquad [7.6]$$

where

    i = angle of deviation from vertical of major principal stress at failure,
    C_i = directional strength in i direction,
    C_1 = strength of vertical specimen, and
    C_2 = strength of horizontal specimen.

There is some experimental support for Equation 7.6 and, since the major principal stress rotates from the vertical toward the horizontal around a slip circle, design charts for the stability of slopes in anisotropic soils are of practical interest. Based on Equation 7.2,

$$F = N(C_1/\gamma H) \qquad [7.7]$$

where N depends on C_2/C_1 and the slope angle (β). Lo (7.53) undertook the evaluation of N for uniform soils and for profiles where the strength increases with depth. For soils conforming to Equation 7.6, the factor of safety obtained by taking anisotropy into account is significantly lower than that from conventional analysis, except for steep slopes. The design charts presented by Lo allow this to be assessed readily.

Method of Slices

When the surface profile or the stratigraphy is irregular, it is not convenient to seek an analytical solution. Instead, Equation 7.1 is evaluated numerically by the method of slices. Figure 7.6 shows the geometry of a circular surface of sliding as used in that method. F is found from

$$F = (\Sigma s_{u_i} \ell_i)/(\Sigma W_i \sin \alpha_i) \qquad [7.8]$$

where i denotes each slice in turn, and Σ is the summation of all slices. By use of Equation 7.8, an almost arbitrary degree of complexity can be considered, including anisotropy and variation of strength with depth. It is common to undertake these analyses with a digital computer and many programs exist. For illustrative purposes, the factor of safety is calculated below for a sample problem. The assumptions for the problem are that

1. The soil is saturated,
2. No volume change occurs,

Figure 7.4. General solution for cuts in unlimited depths of normally consolidated clays (7.38).

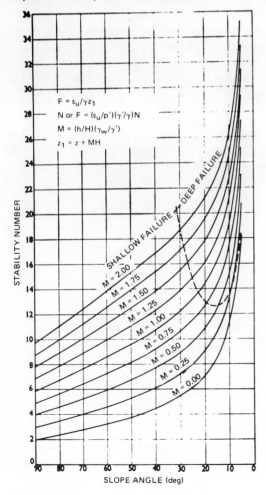

$F = s_u/\gamma z_1$
N or $F = (s_u/p')(\gamma'/\gamma)N$
$M = (h/H)(\gamma_w/\gamma')$
$z_1 = z + MH$

SHALLOW FAILURE — DEEP FAILURE

M = 2.00
M = 1.75
M = 1.50
M = 1.25
M = 1.00
M = 0.75
M = 0.50
M = 0.25
M = 0.00

STABILITY NUMBER

SLOPE ANGLE (deg)

Figure 7.5. Idealized problem of embankment loading on cohesive soils. Stability charts are available for this problem (7.60).

(a) EMBANKMENT AND POTENTIAL FAILURE SURFACE

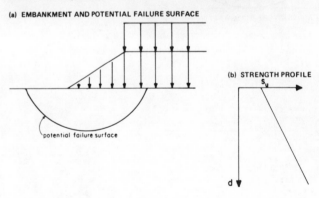

(b) STRENGTH PROFILE

potential failure surface

Figure 7.6. Geometry pertaining to a particular circular surface of sliding as used in the method of slices.

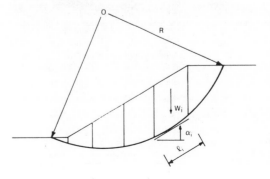

3. No dissipation of pore pressure occurs,
4. The mode of failure is slip circle, and
5. The shear strength along the slip surface is the only factor contributing to resistance.

Figure 7.7 shows the cross section of the failure and the force diagram.

The factor of safety is determined by the use of Equation 7.8; i.e., F is the resisting moment divided by the driving moment. The resisting moment is the sum of the mobilized strength (s_u) in each layer times the radius of the circle, and the driving moment is the sum of the weight of each slice times its moment arm from center of rotation. F can be defined this way only because all terms in the resisting moment have s_u.

The calculated dimensions of the slices shown in Figure 7.7 are given in Table 7.1. The remaining calculations are

1. Sum of the driving moments = 29 507 kN·m;
2. Resisting moments = 24 (25 x AB + 15 x BC + 35 x CD) = 24 (25 x 7.5 + 15 x 4 + 35 x 34) = 34 500 kN·m; and
3. For this example, F = 34 500/29 507 = 1.17.

Figure 7.7. Cross section and force diagram of example slip surface failure.

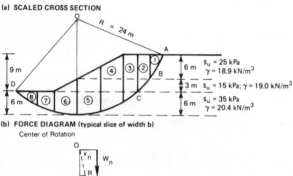

(a) SCALED CROSS SECTION

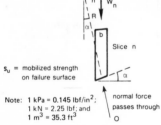

(b) FORCE DIAGRAM (typical slice of width b)

s_u = mobilized strength on failure surface

Note: 1 kPa = 0.145 lbf/in²;
1 kN = 2.25 lbf; and
1 m³ = 35.3 ft³

Table 7.1. Calculated dimensions of slices in example.

Slice	Width (m)	Height (m)	Force (kN)	Lever Arm (m)	Driving Moment (kN·m)
1	3	3	170	20	3 400
2	3	7.5	426	17.5	7 455
3	3.5	10.5	703	14	9 842
4	6	12	1400	9	12 600
5	6	10	1182	3	3 546
6	6	7	844	-3	-2 532
7	4	4	326	-8	-2 608
8	6	1.5	183	-12	-2 196
Total					29 507

Note: 1 m = 3.3 ft; 1 kN = 2.25 lbf; and 1 kN·m = 0.74 lbf·ft.

The critical slip circle and the minimum F must be found by iteration. Convergence is fastest if O is moved horizontally.

Effective Stress Analysis of Soil Slopes

Planar Slip Surface

When a soil mass moves predominantly in a translational manner, there is little internal distortion and an infinite slope analysis is often representative. In this analysis, the slip surface is assumed to be a plane parallel to the ground surface and the end effects can be neglected. When sliding takes place in the mantle of weathered material with small ratios of depth to length, this type of analysis is often appropriate. The mechanics of sliding along a planar slip surface will be presented in detail, for they are instructive with regard to all effective stress analyses.

Consider the slope and slip surface shown in Figure 7.8. In an effective stress analysis the pore pressure distribution is assumed to be known, and in this case steady seepage is assumed to be parallel to the surface with the groundwater table as indicated. The forces acting on a slice of width b are shown. Since there is no internal distortion and end effects are neglected,

$$dE = dX = 0 \qquad [7.9]$$

The weight of a slice W is given by

$$W = \gamma bd \qquad [7.10]$$

The total normal force (N) is related to the effective force (N') and pore pressure (u) by

$$N = N' + u\ell \qquad [7.11]$$

and

$$u\ell = (\gamma_w h) \, b \sec \alpha \qquad [7.12]$$

where h is the piezometric head acting on the slip surface. The shear force acting along the base of the slice (S) is

$$S = [(c'b \sec \alpha)/F] + (N - u\ell)(\tan \phi'/F) \qquad [7.13]$$

where

c' = cohesion intercept in terms of effective stress, and
ϕ' = angle of internal friction in terms of effective stress.

By resolving W into components parallel and perpendicular to the slope, we get

$$S = W \sin \alpha \qquad [7.14]$$

and

$$N = W \cos \alpha \qquad [7.15]$$

The S and N forces can be eliminated by combining Equations 7.13, 7.14, and 7.15, and an explicit relation for F can be found as

$$F = (c'/\gamma d) \sec \alpha \csc \alpha$$
$$+ (\tan \phi'/\tan \alpha)\,[1 - (\gamma_w h/\gamma d) \sec^2 \alpha] \qquad [7.16]$$

F is a function of two dimensionless ratios, $(c'/\gamma d)$ and $(\gamma_w h/\gamma d)$. The latter, which expresses the relation of the magnitude of the pore pressure at a point and the overburden pressure at that point, has been called the pore-pressure ratio (r_u) by Bishop and Morgenstern (7.5). When r_u is used, Equation 7.16 can be rewritten as

$$F = (c'/\gamma d) \sec \alpha \csc \alpha$$
$$+ (\tan \phi'/\tan \alpha)\,(1 - r_u \sec^2 \alpha) \qquad [7.17]$$

which in the case of zero cohesion becomes

$$F = (\tan \phi'/\tan \alpha)\,(1 - r_u \sec^2 \alpha) \qquad [7.18]$$

Analyses using Equation 7.17 or 7.18 are readily performed by hand. Several cases exist to illustrate the utility of this analysis in practice (7.39, 7.66).

Circular Slip Surface

Rotational slides occur in many types of soil, and to analyze slopes in these materials by assuming a circular slip surface is common practice. Taylor (7.68) presents stability charts for uniform slopes characterized by both friction and cohesion. However, since these charts are in terms of total stresses, they are of limited value. For heterogeneous conditions, the friction circle method of analysis used by Taylor is less practical than methods that use slices. Therefore, only various methods of slices will be discussed here.

One method of analysis that is accurate for most purposes is that advanced by Bishop (7.3). The forces acting on a typical slice are shown in Figure 7.9. Moment equilibrium about the center of rotation gives

$$\Sigma W x = \Sigma S R \qquad [7.19]$$

Substituting for S in Equation 7.19 and using the Mohr-Coulomb failure criterion in terms of the factor of safety give

$$F = (R/\Sigma W x)\,\Sigma [c'\ell + (P - u\ell)\tan \phi'] \qquad [7.20]$$

The equilibrium of a slip circle in a frictional medium is statically indeterminate. Attempts to eliminate P from Equation 7.20 by using force equilibrium conditions introduce E or X terms, and an assumption must be made to render the problem statically determinate. In what is usually termed the simplified Bishop method, force equilibrium of a slice in the vertical direction is taken, and the variation in X forces across a slice is ignored. This is tantamount to assuming zero shear between slices and ignoring the requirement for equilibrium in the horizontal direction. The resulting equation for the factor of safety is

$$F = \Sigma \{[c'b + (W - ub)\tan \phi']\,(1/m_\alpha)\}$$
$$\div (\Sigma W \sin \alpha) \qquad [7.21]$$

where

$$m_\alpha = \cos \alpha\,[1 + (\tan \alpha \tan \phi'/F)]. \qquad [7.22]$$

Since F appears on both sides of Equation 7.21, it is most convenient to solve for F in an iterative manner. Initially a value of F is assumed and the right side of the equation is evaluated to find a new value. The magnitudes of m_α are changed accordingly, and another iteration is undertaken. The process converges rapidly. Hand calculations are made convenient by the use of tabular forms, such as that shown in Figure 7.10. It is useful to have m_α plotted for any assumed value of F, as shown in Figure 7.11. Several versions of efficient computer programs are available for the solution of Equation 7.21 (7.52, 7.73).

A simpler but less satisfactory method of analysis may be derived by neglecting the internal forces and assuming that

$$N = W \cos \alpha \qquad [7.23]$$

Substituting this relation into the equation for overall moment equilibrium results in the following expression for F:

$$F = \Sigma [c'\ell + (W \cos \alpha - u\ell)\tan \phi']/(\Sigma W \sin \alpha) \qquad [7.24]$$

Figure 7.8. Forces acting on idealized slice of infinite slope.

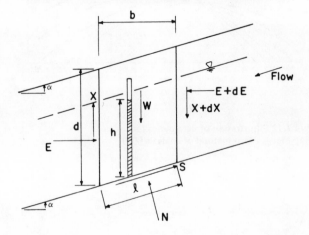

Figure 7.9. Forces involved in effective stress slip circle analysis.

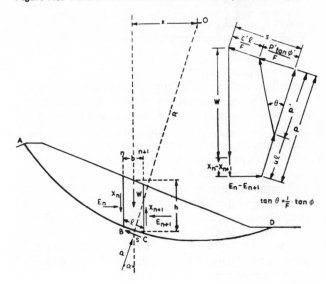

161

Figure 7.10. Tabular form for computation of factor of safety if surface of sliding is circular and interslice forces are neglected (7.71).

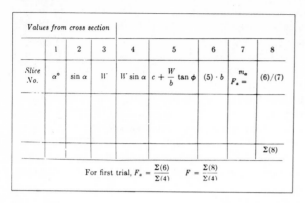

Values from cross section								
	1	2	3	4	5	6	7	8
Slice No.	α°	$\sin \alpha$	W	$W \sin \alpha$	$c + \dfrac{W}{b} \tan \phi$	$(5) \cdot b$	$\begin{array}{c} m_\alpha \\ F_a = \end{array}$	$(6)/(7)$
								$\Sigma(8)$

For first trial, $F_a = \dfrac{\Sigma(6)}{\Sigma(4)}$ $\qquad F = \dfrac{\Sigma(8)}{\Sigma(4)}$

Figure 7.11. Values of m_α for use in calculation of factor of safety if surface of sliding is circular and interslice forces are neglected (7.71).

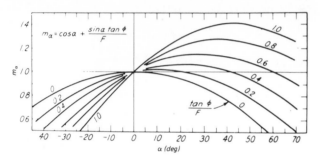

Figure 7.12. Illustration of parameters for use in Bishop-Morgenstern method of analysis (7.5).

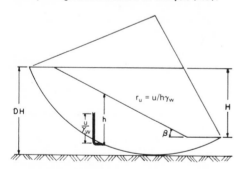

This method has been called the Fellenius method, the U.S. Bureau of Reclamation method, the common method of slices, and the ordinary method. As confirmed in practice (7.64), it results in factors of safety that are too low. Although simplicity may have been a reasonable argument at one time for using Equation 7.24, the widespread use of computers precludes algebraic complexity as an excuse for not adopting methods of analysis that are superior from the point of view of mechanics.

Bishop and Morgenstern (7.5) sought to develop charts that would facilitate the use of Equation 7.21 for problems formulated in terms of effective stress. For the uniform slope shown in Figure 7.12, they showed that Equation 7.21 could be cast into the dimensionless form

Figure 7.13. Idealized cases of seepage in slopes for which chart solutions are available (7.37).

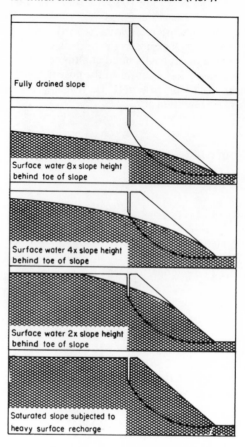

Fully drained slope

Surface water 8x slope height behind toe of slope

Surface water 4x slope height behind toe of slope

Surface water 2x slope height behind toe of slope

Saturated slope subjected to heavy surface recharge

$$F = \Sigma \{ [(c'/\gamma H)(b/H) + (b/H)(h/H)$$
$$\times (1 - r_u) \tan \phi'] (1/m_\alpha) \}$$
$$\div [\Sigma(b/H)(h/H) \sin \alpha] \qquad [7.25]$$

Hence, given $(c'/\gamma H)$, ϕ', and r_u, F depends only on geometry. They further showed that, for a given geometry expressed by $\cot \beta$ and D and for given soil properties expressed by $(c'/\gamma H)$ and ϕ', the factor of safety (F) decreased in a linear manner with an increase in the pore-pressure ratio (r_u). This relation is given by

$$F = m - n r_u \qquad [7.26]$$

where m and n are called stability coefficients and are presented as functions of the parameters and ratios noted above. The stability coefficients were evaluated over a wide range of values for $(c'/\gamma H)$, ϕ', β, and D, and the results were presented in both chart and tabular form for ease of use. For a given problem, reference to the appropriate chart or linear interpolation between charts allows one to extract the appropriate values of m and n. The factor of safety is given directly by Equation 7.26. In many cases that arise in practice, r_u is not uniform, and an average value has to be evaluated in order to use the stability coefficients. Guidance in calculating an average value of r_u is given (7.5). When r_u is zero, stability coefficient m gives the same factor of safety as Taylor's charts for a material without cohesion.

Other stability charts that may be found useful have been prepared by Hoek and Bray (7.37) for each of the cases

Figure 7.14. Slip surface dimensions and slice forces for use in Spencer's analysis (7.67).

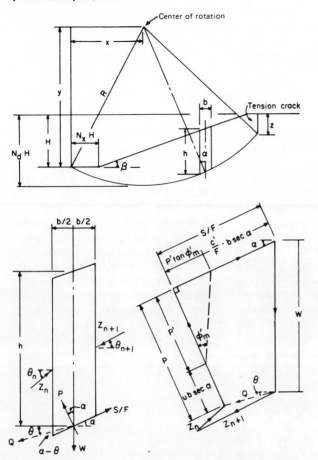

Figure 7.15. Examples of variation of F_m and F_f with θ in Spencer's analysis (7.67).

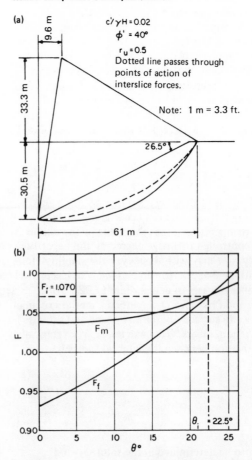

shown in Figure 7.13. These cases represent failure along a circular slip surface in a soil mass subjected to steady seepage and containing a tension crack. The magnitudes of $(c'/\gamma H)$, ϕ', and β are needed to enter a particular chart representing one of the cases shown in Figure 7.13. The factor of safety is then readily found. The charts are based on a form of friction circle analysis that tends to underestimate the factor of safety. Although they are convenient to use for preliminary design, final design should be checked with a more rigorous analysis.

As pointed out previously, the simplified Bishop method of analysis is adequate for circular slip surfaces, but it assumes that no shear is mobilized between slices and it does not satisfy the condition of force equilibrium in the horizontal direction. Why the method works can be demonstrated by reference to the method presented by Spencer (7.67). Spencer followed the usual procedures in the method of slices (Figure 7.14) and argued that the action of the internal forces acting on a slice can be replaced by their resultant (Q) inclined at θ to the horizontal and passing through the midpoint of a thin slice in order to satisfy moment equilibrium. In general, both Q and θ will vary from slice to slice, and from force equilibrium it can be shown that

$$Q = [(c'b \sec \alpha/F) + (\tan \phi'/F)$$
$$\times (W \cos \alpha - ub \sec \alpha) - W \sin \alpha]$$
$$\div \cos (\alpha - \theta)[1 + (\tan \phi'/F) \tan (\alpha - \theta)] \quad [7.27]$$

For overall equilibrium, the sum of the horizontal components and the vertical components of the interslice forces must each be zero; that is,

$$\Sigma [Q \cos \theta] = 0 \quad [7.28]$$

and

$$\Sigma [Q \sin \theta] = 0 \quad [7.29]$$

Moreover, if the sum of the moments of the external forces about the center of rotation is zero, the sum of the moments of the interslice forces about the center of rotation is also zero. For a circular slip surface,

$$\Sigma [Q \cos (\alpha - \theta)] = 0 \quad [7.30]$$

If θ is assumed constant in order to make the problem determinate, Equations 7.28 and 7.29 reduce to

$$\Sigma Q = 0 \quad [7.31]$$

In either Equation 7.30 or 7.31, the value of F will depend on the assumed constant value of θ. For all but one value of θ, the safety factor determined (Equation 7.31) by force equilibrium (F_f) will not equal the safety factor determined (Equation 7.30) by moment equilibrium (F_m). This is shown in Figure 7.15 by the intersection of curves of F_f and

163

F_m plotted against values of θ. This intersection gives the pair of values of θ and F that satisfy both equations simultaneously. If θ is set equal to zero and substituted into Equation 7.30, the equation governing Bishop's simplified method (Equation 7.21) is recovered. Therefore, Bishop's simplified method represents the solution for F_m with θ equal to zero. In general, the equation governing moment equilibrium is rather insensitive to variations in θ, and errors of only a few percent result. For the assumption that θ equals zero used in the Bishop simplified method, these errors are not conservative with respect to actual values of θ greater than zero.

Noncircular Slip Surface

When the distribution of shearing resistance within an earth mass becomes nonuniform, slip can occur along surfaces more complex than a circle. Since the shape of the failure surface will be controlled to a large degree by the departure from uniformity, one may wish to analyze the stability along surfaces of arbitrary shapes. Noncircular analysis has proved useful in a number of cases of actual slides (7.22, 7.40).

Morgenstern and Price (7.57) developed a method of analysis that treats a slip surface of arbitrary shape and satisfies all equilibrium requirements. To make the analysis statically determinate, a relation between the internal forces is assumed of the form

$$X = \lambda \, f(x) \, E \qquad\qquad [7.32]$$

where

λ = factor to be determined in the solution, and
$f(x)$ = arbitrary function concerning the distribution of internal forces.

The choice of $f(x)$ is limited by conditions of physical admissibility, which require that no tension be developed in the earth mass and that the failure criterion not be violated. Subject to these conditions, the factor of safety is rather insensitive to variations in $f(x)$. A digital computer is needed to solve resulting Equations 7.30 and 7.31. The Spencer analysis of a slip circle discussed previously is equivalent to setting $f(x)$ equal to unity.

A convenient approximate method of analysis suitable for hand calculations and sufficiently accurate for many practical purposes is described by Janbu, Bjerrum, and Kjaernsli (7.43) and Janbu (7.42). This method follows the method of slices and is based on a more elaborate procedure described in detail in the latter reference. The factor of safety is given by

$$F = f_0 \, \Sigma \, \{ [c'b + (W - ub) \tan \phi'] / \cos \alpha \, m_\alpha \,] \}$$
$$\div (\Sigma W \tan \alpha) \qquad [7.33]$$

where m_α is defined in Equation 7.22 and used as shown in Figure 7.10. The term f_0, which is a correction factor for the role of the internal forces, is a function of the curvature of the slip surface and the type of soil; recommended values for this factor are given in Figure 7.16.

Another approximation procedure for analyzing slip along a noncircular sliding surface is the wedge analysis. In this method the potential sliding mass is separated into a series of wedges, and the equilibrium of each wedge is considered in turn. Only the conditions of horizontal and vertical equilibrium are used in the analysis, which is often performed in a graphical manner. Moreover, an assumption must be made regarding the inclination of the force transmitted across the interface between any two wedges. The forces utilized in this method of analysis are shown in Figure 7.17 for a rock mass separated into two wedges; the notation is defined below.

W_1, W_2 = weight of wedge,
U_1, U_2 = resultant water pressure acting on base of wedge,
N_1, N_2 = effective force normal to base,
T_1, T_2 = shear force acting along base of wedge,
L_1, L_2 = length of base,
α_1, α_2 = inclination of base to horizontal,
P_{w12} = resultant water pressure at interface,
P_{12} = effective force at interface, and
δ = inclination of P_{12} to horizontal.

Although the rock mass shown in Figure 7.17 is separated into two wedges, the analysis can be extended readily to any number of wedges. The factor of safety is varied so that the force polygons constructed for each of the wedges satisfy horizontal and vertical equilibrium. A value of δ is

Figure 7.16. Curves for use in Janbu's simplified analysis (7.42, 7.43).

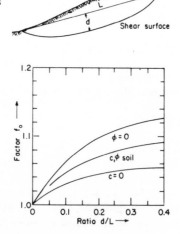

Figure 7.17. Forces used in typical wedge analysis.

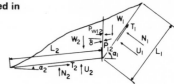

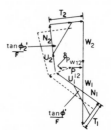

assumed, and an initial value of F is taken; this allows the force polygon for the first wedge to be constructed, as shown by the dashed line in Figure 7.17. With the value of P_{12} obtained in this manner, the force polygon for the second wedge is readily constructed. This will not close, in general, and the value of F must be varied until it does. The final solution also is given in Figure 7.17.

The factor of safety is sensitive to the assumed value of δ. Conservative values of the factor of safety may be found by setting δ equal to zero. The maximum possible value of δ is that value which is compatible with failure in the direction of the interface, but this usually leads to an overestimate of the factor of safety. Experience indicates that putting δ between 10° and 15° usually gives reasonable results.

Pore-Pressure Distribution

To perform a stability analysis in terms of effective stress requires that the pore pressures be specified. A distinction is made between problems of undrained and drained loading. In the first case, the magnitude of the pore pressure is influenced to a large degree by the changes in total stress to which the earth mass is subjected; in the second case, the pore-pressure distribution is controlled by steady seepage conditions. When an analysis of a soil loaded under undrained conditions is made in terms of effective stress, the pore pressures must be either predicted or measured. The confidence level for predicting pore pressures in cuts is not high; however, it is somewhat better in the case of foundations beneath embankments because the total stress change is less influenced by the initial state of stress in the ground. For a clay soil, the initial stresses in the ground are not usually known to a high degree of accuracy. In general, the pore pressure (u) that is used in a stability analysis is given by

$$u = u_0 + \Delta u \qquad [7.34]$$

where

u_0 = pore pressure before any stress change, and
Δu = change in pore pressure due to change in total stress.

The change in pore pressure (Δu) can be estimated approximately if the pore-pressure coefficients and the changes in total stresses are known. In turn, the change in total stress at a point may be estimated from stress analyses. Linear elasticity is commonly assumed; therefore, the results are of limited applicability, particularly if the factor of safety approaches unity and a substantial volume is permitted to yield. In the case of cuts, the total stresses decrease so that with time the pore pressures increase from their end-of-construction values, thereby resulting in reduction of the factor of safety.

Although it is not common to predict pore-pressure changes in slopes due to changes in total stress, it is common to measure them on major projects and to use the data directly in stability analyses. Instrumentation for measuring pore pressures is discussed in Chapter 5. In the case of embankments on soft impervious foundations, monitoring pore pressures is almost routine. However, this monitoring may not be adequate to avert a slip (7.1).

Steady seepage through a soil is governed by the Laplace equation, for which a variety of methods are available to find solutions subject to appropriate boundary conditions (7.9, 7.31). These techniques are equally useful in the design of drainage measures if the influence of the design on the magnitude and distribution of the pore pressures must be ascertained. In some cases, analytical methods will have the capability to provide solutions in closed form, but more generally the problems that arise in practice are analyzed by constructing flow networks graphically by the use of electric analogue techniques or numerically by means of the finite element method (7.33, 7.46, 7.70).

An understanding of hydrogeologic controls is important in the solution, either by analysis or by observation, of any steady seepage problem. In particular, it is essential to recognize the control that regional groundwater systems have on the local pore-pressure distribution in a natural slope or excavation. The recognition of groundwater discharge areas and an identification of the recharge areas are a starting point for almost all seepage studies, either computational or observational. High pore pressures are often found in discharge areas associated with topographic lows, and they may have a controlling influence on the stability of slopes. Analyses of regional groundwater systems have been made by Toth (7.72) and Freeze and Witherspoon (7.25, 7.26), and their significance for slope stability problems is noted by Patton and Deere (7.61).

Heterogeneity of soil deposits has a marked influence on the pore-pressure distribution under steady seepage conditions. Kenney and Chan (7.47) describe a detailed field study of the anisotropy of the permeability in a varved soil. The ratio of the permeability of flow parallel to the varves to that perpendicular to the varves was found to be less than five. Although this is less than might be expected, it still exerts a significant influence on the pressure distribution that develops during steady seepage conditions. Eigenbrod and Morgenstern (7.22) report a striking example of the influence of geologic detail on the pore-pressure distribution within a slope about 20 m (70 ft) high in Cretaceous bedrock overlain by till. This slope was shown to possess at least three water tables, the upper two of which were perched on bentonitic layers. Moreover, no pore pressures acted along much of the failure surface, since it lay on a fractured free-draining coal layer. Careful field observations guided by an appreciation for the role of geologic details are needed in cases like this if misleading results are to be avoided.

Analysis of Rock Slopes

The analysis of the stability of a rock slope is normally undertaken in terms of effective stress because the permeability of rock masses is usually so high that undrained loading does not arise. To perform an analysis requires that the shear strength parameters, the pore-water pressures, and the sliding mechanism be specified. In hard rocks, the sliding mechanism is controlled by structural features, and a detailed study of those features is a prerequisite for rational design of a rock slope. Further guidance on the evaluation of rock structure characteristics is given in Chapter 9.

The analysis of a rock slope in terms of a factor of safety is a subordinate activity to achieving a clear understanding of the controlling geology and water-pressure configuration. In the case of slopes in soil, the confidence limits attached to the application of the total process of investigation, testing, and analysis are determined by the successful explanation of well-documented case histories (7.66). However, in the case of rock slides, few such case histories report completely the geologic configuration, give good quality test data, and have been subjected to an acceptable analysis. Moreover, the shearing resistance in rocks is often sensitive to movements. Small movements can result in substantial decreases in available shearing resistance and hence in the factor of safety. Therefore, the confidence limits of design for a given factor of safety are generally lower for rock slopes than for soil slopes. In other words, rock slopes designed for a given factor of safety have a higher degree of risk associated with them than most soil slopes designed for the same factor of safety. If the degree of risk is to be the same, it is prudent to design rock slopes for higher factors of safety, unless only the residual shearing resistance is being mobilized. However, design for a higher factor of safety is not always economically feasible.

The shearing resistance of rock is more variable than that of soil, and the same is true of the water-pressure distribution. Both the strength and the water-pressure distribution are dominated to a large degree by the pattern of discontinuities in the rock mass. In the case of strength, the usual procedures for investigation involve in situ testing or sampling and testing, as discussed in Chapter 4. The water-pressure distribution can only be evaluated by field investigations coupled with an appreciation of the hydrogeologic constraints.

Morgenstern (7.55) reviews the influence of groundwater on the stability of rock slopes and shows how flow through a discrete network of discontinuities can be replaced by flow through an imaginary medium of equivalent permeability. If the crack spacing is small compared to the size of the discontinuity, conventional procedures for solving problems of flow through porous media are applicable. If this is not the case, the attitudes and hydraulic conductivities of the actual discontinuities should be considered (7.54, 7.74). This is difficult to do, and considerable judgment is required to properly evaluate the water pressure distribution. Patton and Deere (7.61) stress the large local differences in water pressures that can arise in a jointed rock mass and the large fluctuations that can develop as a consequence of rainfall. They also note the significant effects that faults can have on groundwater conditions. Those effects can be either beneficial or deleterious, depending on the attitude and characteristics of the fault gouge. Groundwater flow around faults and other similar features requires special study in slope-stability analyses. The evaluation of water pressures in a rock mass is made more difficult by the sensitivity of the hydraulic conductivity to small deformations. A small amount of slip along a discontinuity can result in a disproportionate change in hydraulic conductivity (7.65). Although methods of analysis can be developed to take these effects into consideration (7.7, 7.56), the information needed to conduct these analyses is only rarely available.

Stability analyses are usually concerned with sliding. However, discrete blocks of rock can also rotate outward or topple. A tendency to topple depends on the shape of the block and is usually readily discernible in the field. The condition for toppling is dictated by the position of the weight vector of the block relative to the base of the block. If the weight vector passing through the center of gravity of the block falls outside the base of the block, toppling will occur; this is shown in Figure 7.18 (7.37).

A common case of instability in rock slopes is sliding along a planar surface. The analysis of this case is a simple extension of the stability of a rough rigid block on an inclined plane, and, for the case shown in Figure 7.19a, we have

$$F = (2c/\gamma H) \, P + [Q \cot \psi_p - R(P + S)] \, \tan \phi$$
$$\div (Q + RS \cot \psi_p) \qquad [7.35]$$

where

$$P = [1 - (Z/H)] \, \operatorname{cosec} \psi_p , \qquad [7.36]$$

$$Q = \{[1 - (Z/H)]^2 \cot \psi_p - \cot \psi_f\} \sin \psi_p , \quad [7.37]$$

$$R = (\gamma_w/\gamma) \, (Z_w/Z) \, (Z/H) , \text{ and} \qquad [7.38]$$

$$S = (Z_w/Z) \, (Z/H) \sin \psi_p . \qquad [7.39]$$

But if the tension crack is in the surface above the slope (Figure 7.19h),

$$Q = \{[1 - (Z/H)]^2 \cos \psi_p$$
$$(\cot \psi_p \tan \psi_f - 1)\} \qquad [7.40]$$

should be used. All other terms are as shown in the figure. Hoek and Bray (7.37) undertook a systematic study of the influence of the various parameters that enter into Equation 7.35 and prepared design charts to facilitate its use. If a planar mechanism is appropriate, it is also possible to assess the influence of drainage and other stabilizing measures, such as berms or cables, by using simple extensions of Equation 7.35.

Often two or three sets of discontinuities intersect, and sliding of a wedge of rock becomes possible. In general, differing amounts of friction and cohesion can act on the planes, and the water pressure acting normal to the surfaces bounded by the planes should be taken into consideration. A comprehensive analysis of the mechanics of wedge failure is given by Hendron, Cording, and Aiyer (7.32). Figure 7.20 shows a tetrahedron bounded by two base planes, which may be intersecting joint sets. Failure may occur by sliding along the line of intersection of the two planes or by sliding on either one of the two planes. General procedures have been developed for determining the factor of safety graphically by vector analysis by the use of stereo networks and numerically by hand or by computer (7.32, 7.37). The following steps are included in the analysis.

1. The intersections of the various joint sets with one another and with the slope face are inspected to determine the tetrahedra that may be potential failure wedges; these are then analyzed in detail.

2. The forces tending to disturb the equilibrium of the wedge are added vectorially to give the resultant driving force. These disturbing forces arise from the weight of the

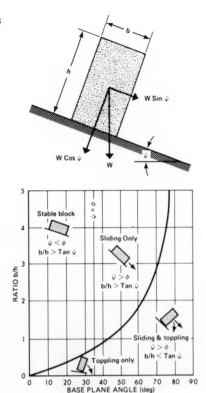

Figure 7.18. Conditions for sliding and toppling of rock block on inclined surface (7.37).

Figure 7.19. Rock-slope failure due to sliding along single discontinuity (7.37).

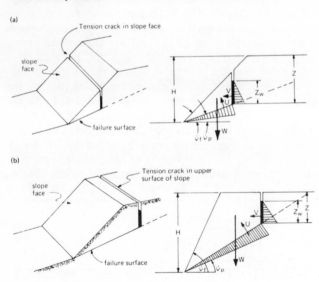

(a)

(b)

Figure 7.20. Rock-slope failure by sliding on intersecting joint sets.

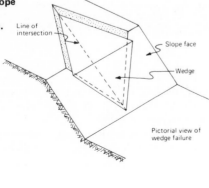

wedge, any external load applied to the wedge, and the pore-water forces acting on the various faces.

3. The mode of failure is then determined. For example, a wedge supported on two base planes can either slide along the line of intersection of the two planes, slide on either plane, or rotate on either plane. The kinematics of failure will depend on the orientation of the disturbing force in relation to the orientation of the supporting planes. Various kinematic tests can be defined (7.32).

4. After the mode of failure is determined, the maximum shearing resistance that can be mobilized in the direction of movement is compared to the shearing forces necessary for equilibrium in order to compute a factor of safety.

Hendron, Cording, and Aiyer (7.32) point out that, in the common case of a wedge resting on two base planes and acted on only by its own weight, sliding will occur along the line of intersection of the two planes if a line drawn down the dip in both planes tends to intersect the line of intersection. If in either one of the planes a line drawn down the dip is directed away from the line of intersection, sliding will occur on that plane only and the wedge will move away from the line of intersection. If a wedge is acted on only by its own weight, it will slide down the maximum dip if sliding occurs on one plane, and the factor of safety can be readily computed. If sliding occurs along the line of intersection in the absence of pore pressures, the angle of shearing resistance required for equilibrium will always be equal to or less than the slope of the line of intersection. The actual factor of safety will depend on the relative attitudes of bounding planes. Stability charts for some frictional wedges without water-pressure loading and with sliding along the line of intersection have been prepared by Hoek and Bray (7.37). Stereographic projection is a most useful tool for the analysis of the stability of rock wedges. Both friction and cohesion as well as water pressures can be taken into account. Designers concerned with rock slopes are well advised to become familiar with these techniques, which are described in the references cited earlier, as well as by John (7.44), Heuze and Goodman (7.34), and Hoek (7.36).

Some rock slides can be treated as two-dimensional problems on noncircular surfaces. Figure 7.21 shows the failure surface of the famous Frank slide (7.13), an example of this type of slide; the Vaiont slide is another example (7.59).

Figure 7.21. Cross section of Frank slide (7.13).

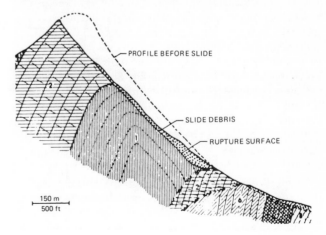

167

These cases are amenable to analysis by the use of the same methods adopted for noncircular slides in soil.

DEFORMATION ANALYSIS

The use of deformation analysis in the design of slopes is an attractive concept; however, practical applications of deformation analysis are limited at present. Until the last decade, this limitation was imposed by the absence of suitable analytical methods to perform deformation analysis. Few elastic solutions were available for boundary value problems similar to slopes, and the solutions available were not well suited to practical design. The development of the finite element method has largely eliminated this impediment. As engineers have taken advantage of this analytical tool, however, other equally vexing problems have been recognized, and these still severely limit the practical application of the finite element method in most slope design problems.

In principle, a deformation analysis, particularly an analysis utilizing the finite element method, must include the following characteristics.

1. The stress field must satisfy equilibrium at every point. The prediction of this equilibrium stress field is usually done by use of elastic theory to describe the stresses and deformations, but plasticity or other material models might also be used. To predict a stress field requires that the stress-strain relation for the soil be known.

2. Boundary conditions of stress and deformation must be satisfied. These characteristics can be contrasted to the principles of limit equilibrium noted in the earlier section on principles of limit equilibrium analysis.

The lack of success in applying deformation analysis in the design of slopes can be attributed directly to the difficulties associated with the stress-strain relation noted above. Soil is generally nonlinear, nonuniform, inelastic, and anisotropic. Each of these characteristics must be idealized and used in a deformation analysis, and the difficulty in describing natural soil deposits in these terms is the major factor that limits application of the finite element method in the analysis of slopes. The problem is aggravated by our limited knowledge of the in situ stresses before excavation. Under special conditions, particularly in analysis of dams and embankments of compacted earth fill, the method is being used successfully (7.51); but the deformation analysis of natural soils is primarily a research activity at present and has little practical application. A review of the principles of the finite element method is nevertheless appropriate because some practical applications are being made at present and progress in developing practical applications in the future is anticipated.

The finite element method has been applied to geotechnical engineering problems since the late 1960s, having been developed a decade earlier for applications in structural engineering and continuum mechanics. Since that time, extensive literature has developed, including several texts. An important and useful reference and summary of this work was compiled by Desai (7.16); this volume is recommended as a starting point for those interested in the method.

The underlying principle of the finite element method is that the behavior of a complex continuum can be approximated by the collective behavior of parts representing the continuum. The parts are selected so that their individual behaviors can be described simply. By requiring adjacent elements to behave similarly at selected points of contact or nodes, the overall continuum is modeled.

In a typical application (Figure 7.22a), an embankment deformation analysis begins by using a group of simple elements to approximate the fill. In a two-dimensional analysis, either triangular or quadrilateral elements are commonly used. For a two-dimensional analysis, equations describing the behavior of these elements in plane strain are used in contrast to the plane stress formulation often used for problems in structural engineering (7.18). A three-dimensional problem can be modeled by using solid elements. For each element in the model, a series of equations and an appropriate stress-strain relation, such as elasticity, are used to relate the application of loads to deformation at the nodes. In the direct method of element formulation (Figure 7.22b), a triangular element with nodes only at the corners can be used. Other elements in the problem are formulated in a similar fashion. The deformation of adjacent elements is then constrained to be identical at the nodal points. There

Figure 7.22. Finite element idealization of embankment.

(a) FINITE ELEMENT REPRESENTATION OF EMBANKMENT CROSS SECTION

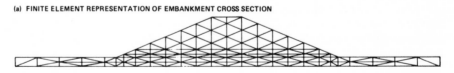

(b) TRIANGULAR ELEMENT OF NETWORK ABOVE
    SHOWING FORCES AT NODAL POINTS

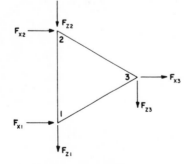

is no requirement that deformations or stress be compatible between other than nodal points on adjacent elements in this example. As a result, this solution would be only an approximation of the actual system behavior. A more accurate approximation might result by adding nodes at the midpoint of each side of the element. Variational methods of element formulation also add to the power, accuracy, and flexibility of the procedure (7.17).

After a model of the embankment is formulated, appropriate loading and deformation boundary conditions are imposed. In geotechnical engineering problems, one of the important loads is the self-weight of the material, which can be represented by applying a vertical force at each node (7.21). Several studies have shown, however, that this method has limitations in predicting deformations. An alternate approach is to perform a sequential analysis in which the problem is solved in successive stages to represent a series of steps in construction or excavation (7.20).

The procedures described above lead to the assembly of a large number of simple equations, and the solution of these equations requires use of a digital computer. Finite element programs applicable to geotechnical engineering problems are readily available from several sources and the literature. Because continuing improvements are made in these programs, no specific references are given here.

The use of finite element methods to predict the stability of slopes does not offer any improvement over accurate limit equilibrium methods (7.19). Its usefulness lies in deformation analysis, in which accuracy is directly related to the stress-strain behavior of the soil. Active research continues to be directed toward this subject. Modeling the complex behavior of soil has progressed from initial efforts using linear elasticity to include many of the more complicated elements of the constitutive relation. Linear elasticity is almost always a poor model of actual soil behavior, but it does have the advantages of low cost and simple application (7.19). In some cases, simple linear models can be used to determine qualitatively the form of particular slope or embankment deformations. However, this method can lead to significant errors and should be used with caution. Another application of linear elasticity is described by Cole and Burland (7.12), who used finite element analysis to extrapolate from measured field performance. They used field measurements to calibrate their analysis and by trial and error selected the best values of the linear stress-strain relation to match their measurements. These constitutive relations were then used to predict subsequent loads and deformations during further construction involving the same material.

Several other forms of the basic stress-strain relation have been proposed (7.19). Most methods deal with nonlinear elastic forms, such as bilinear (7.20) or hyperbolic (7.49). Multilinear and spline functions (7.15, 7.23) have been used to define the constitutive relation in even greater detail. The theory of plasticity (7.63) offers great potential through the use of yield functions to describe nonlinear effects, together with the coupling of volume changes to stress changes. In the use of elastic theory, volume changes are modeled by a variable Poisson's ratio. Because of the relative incompressibility of undrained saturated soils, Poisson's ratio has a constant value of approximately 0.5. This number may be changed slightly (e.g., to 0.49) to eliminate computational

instability in some programs. When volume changes do occur (for example, during drainage or in partially saturated soils), the variation in Poisson's ratio can be related experimentally to stress changes (7.50) or to the coefficient of earth pressure at rest (K_0). Consolidation effects during loading can also be considered (7.10, 7.41). Some advantages have been shown (7.11) for using the shear modulus (G) and bulk modulus (K) instead of Young's modulus (E) and Poisson's ratio (ν) for formulating elastic finite element problems in soil.

During deformation analysis, it is common to exceed the available strength of soil at specific points long before overall collapse of the slope is imminent. This presents the problem of modeling local yield. Since many soils exhibit strain-softening behavior after yield (Chapter 6), it is appropriate to model this by use of finite element methods. Höeg (7.35) describes one method using an elastic-plastic model, and several other techniques have been proposed (7.17).

Modeling natural deposits of rock and heavily overconsolidated clay requires consideration of joints and similar discontinuities. These features can experience shear translation along the joints, as well as dilatancy and opening of the joints during movement of the slope. The application of the finite element method to rock (7.75) and particularly to dilating jointed rock (7.28) has relevance to slope-stability problems.

Other characteristics of the constitutive behavior of soil and rock have been considered for application to finite element deformation analysis. These include anisotropy (7.14), creep, and tensile cracking (7.23). As greater and greater complexity is added to the modeling parameters, the cost of the finite element analysis increases, often beyond practical limits. However, the major problem limiting the application of deformation analysis to natural slopes and cuts remains the heterogeneous and complex nature of the materials involved.

REFERENCES

7.1   Al-dhahir, Z., Kennard, M. F., and Morgenstern, N. R. Observations on Pore Pressures Beneath the Ask Lagoon Embankments at Fiddler's Ferry Power Station. In In Situ Investigations in Soils and Rocks, British Geotechnical Society, London, 1970, pp. 265-276.

7.2   Barron, R. A. Correspondence. Geotechnique, Vol. 14, No. 4, 1964, pp. 360-361.

7.3   Bishop, A. W. The Use of the Slip Circle in the Stability Analysis of Slopes. Geotechnique, Vol. 5, No. 1, 1955, pp. 7-17.

7.4   Bishop, A. W., and Bjerrum, L. The Relevance of the Triaxial Test to the Solution of Stability Problems. Proc., Research Conference on Shear Strength of Cohesive Soils, Boulder, American Society of Civil Engineers, New York, 1960, pp. 437-501.

7.5   Bishop, A. W., and Morgenstern, N. R. Stability Coefficients for Earth Slopes. Geotechnique, Vol. 10, No. 4, 1960, pp. 129-150.

7.6   Booker, J. R., and Davis, E. H. A Note on a Plasticity Solution to the Stability of Slopes in Homogenous Clays. Geotechnique, Vol. 22, No. 4, 1972, pp. 509-513.

7.7   Brekke, T. L., Noorishad, J., Witherspoon, P. A., and Maini, Y. N. T. Coupled Stress and Flow Analysis of Fractured Dam Foundations and Rock Slopes. Proc., Symposium on Percolation Through Fissured Rock, International Society of Rock Mechanics and International Association of Engineering Geology, Stuttgart, Deutsche Gesellschaft für Erd- und Grundbau, Essen, West Germany, 1972, pp. T4-J, 1-8.

169

7.8 Casagrande, A., and Carrillo, N. Shear Failure of Anisotropic Materials. In Contributions to Soil Mechanics: 1941-1953, Boston Society of Civil Engineers, 1944, pp. 122-135.

7.9 Cedergren, H. R. Seepage, Drainage, and Flow Nets. Wiley, New York, 1967, 489 pp.

7.10 Christian, J., and Boehmer, J. W. Plane Strain Consolidation by Finite Elements. Journal of Soil Mechanics and Foundations Division, American Society of Civil Engineers, New York, Vol. 96, No. SM4, 1970, pp. 1435-1457.

7.11 Clough, R. W., and Woodward, R. J. Analysis of Embankment Stresses and Deformations. Journal of Soil Mechanics and Foundations Division, American Society of Civil Engineers, New York, Vol. 93, No. SM4, 1967, pp. 529-549.

7.12 Cole, K. W., and Burland, J. B. Observations of Retaining Wall Movements Associated With A Large Excavation. Proc., 5th European Conference on Soil Mechanics and Foundation Engineering, Spanish Society for Soil Mechanics and Foundations, Madrid, 1972, pp. 327-342.

7.13 Cruden, D. M., and Krahn, J. A Reexamination of the Geology of the Frank Slide. Canadian Geotechnical Journal, Vol. 10, No. 4, 1973, pp. 581-591.

7.14 D'Appolonia, D. J., and Lambe, T. W. Method for Predicting Initial Settlement. Journal of Soil Mechanics and Foundations Division, American Society of Civil Engineers, New York, Vol. 96, No. SM2, 1970, pp. 523-544.

7.15 Desai, C. S. Nonlinear Analyses Using Splice Functions. Journal of Soil Mechanics and Foundations Division, American Society of Civil Engineers, New York, Vol. 97, No. SM10, 1971, pp. 1461-1480.

7.16 Desai, C. S., ed. Applications of the Finite Element Method in Geotechnical Engineering: A Symposium. U.S. Army Engineer Waterways Experiment Station, Vicksburg, Miss., 1972, 1227 pp.

7.17 Desai, C. S. Theory and Applications of the Finite Element Method in Geotechnical Engineering. In Application of the Finite Element Method in Geotechnical Engineering: A Symposium (Desai, C. S., ed.), U.S. Army Engineer Waterways Experiment Station, Vicksburg, Miss., 1972, pp. 3-90.

7.18 Desai, C. S., and Abel, J. F. Introduction to the Finite Element Method: A Numerical Method for Engineering Analysis. Van Nostrand-Reinhold, New York, 1972, 477 pp.

7.19 Duncan, J. M. Finite Element Analysis of Stresses and Movements in Dams, Excavations, and Slopes. In Application of the Finite Element Method in Geotechnical Engineering: A Symposium (Desai, C. S., ed.), U.S. Army Engineer Waterways Experiment Station, Vicksburg, Miss., 1972, pp. 267-326.

7.20 Dunlop, P., and Duncan, J. M. Development of Failure Around Excavated Slopes. Journal of Soil Mechanics and Foundations Division, American Society of Civil Engineers, New York, Vol. 96, No. SM2, 1970, pp. 471-493.

7.21 Dunlop, P., Duncan, J. M., and Seed, H. B. Finite Element Analyses of Slopes in Soil. Department of Civil Engineering, Univ. of California, Berkeley, Rept. TE-68-3, 1968.

7.22 Eigenbrod, K. D., and Morgenstern, N. R. A Slide in Cretaceous Bedrock at Devon, Alberta. In Geotechnical Practice for Stability in Open Pit Mining (Brawner, C. O., and Milligan, V., eds.), Society of Mining Engineers, American Institute of Mining, Metallurgical and Petroleum Engineers, New York, 1972, pp. 223-238.

7.23 Eisenstein, Z., Krishnayya, A. V., and Morgenstern, N. R. An Analysis of the Cracking at Duncan Dam. Proc., Specialty Conference on Performance of Earth and Earth-Supported Structures, Purdue Univ., Lafayette, Ind., American Society of Civil Engineers, New York, Vol. 1, Part 1, 1972, pp. 765-778.

7.24 Esu, F. Short-Term Stability of Slopes in Unweathered Jointed Clays. Geotechnique, Vol. 16, No. 4, 1966, pp. 321-328.

7.25 Freeze, R. A., and Witherspoon, P. A. Theoretical Analysis of Regional Groundwater Flow: 1—Analytical and Numerical Solutions to the Mathematical Model. Water Resources Research, Vol. 2, No. 4, 1966, pp. 641-656.

7.26 Freeze, R. A., and Witherspoon, P. A. Theoretical Analysis of Regional Groundwater Flow: 3—Quantitative Interpretations. Water Resources Research, Vol. 4, No. 3, 1968, pp. 581-590.

7.27 Gibson, R. E., and Morgenstern, N. R. A Note on the Stability of Cuttings in Normally Consolidated Clays. Geotechnique, Vol. 12, No. 3, 1962, pp. 212-216.

7.28 Goodman, R. E., and Dubois, J. Duplication of Dilatancy in Analysis of Jointed Rocks. Journal of Soil Mechanics and Foundations Division, American Society of Civil Engineers, New York, Vol. 98, No. SM4, 1972, pp. 399-422.

7.29 Hansen, J. B. The Philosophy of Foundation Design: Design Criteria, Safety Factors, and Settlement Limits. In Proc. (Vesic, A. S., ed.), Symposium on Bearing Capacity and Settlement of Foundations, Duke Univ., Durham, N.C., 1965, 1967, pp. 9-14.

7.30 Hansen, J. B., and Gibson, R. E. Undrained Shear Strengths of Anisotropically Consolidated Clays. Geotechnique, Vol. 1, No. 3, 1949, pp. 189-204.

7.31 Harr, M. E. Groundwater and Seepage. McGraw-Hill, New York, 1962, 315 pp.

7.32 Hendron, A. J., Jr., Cording, E. J., and Aiyer, A. K. Analytical and Graphical Methods for the Analysis of Slopes in Rock Masses. U.S. Army Engineer Nuclear Cratering Group, Livermore, Calif., Technical Rept. 36, 1971.

7.33 Herbert, R., and Rushton, K. R. Ground-Water Flow Studies by Resistance Networks. Geotechnique, Vol. 16, No. 1, 1966, pp. 53-75.

7.34 Heuze, F. E., and Goodman, R. E. Three-Dimensional Approach for Design of Cuts in Jointed Rock. In Stability of Rock Slopes (Cording, E. J., ed.), Proc., 13th Symposium on Rock Mechanics, Univ. of Illinois at Urbana-Champaign, American Society of Civil Engineers, New York, 1972, pp. 397-404.

7.35 Höeg, K. Finite Element Analysis of Strain-Softening Clay. Journal of Soil Mechanics and Foundations Division, American Society of Civil Engineers, New York, Vol. 98, No. SM1, 1972, pp. 43-58.

7.36 Hoek, E. Methods for the Rapid Assessment of the Stability of Three-Dimensional Rock Slopes. Quarterly Journal of Engineering Geology, Vol. 6, No. 3, 1973, pp. 243-255.

7.37 Hoek, E., and Bray, J. W. Rock Slope Engineering. Institution of Mining and Metallurgy, London, 1974, 309 pp.

7.38 Hunter, J. H., and Schuster, R. L. Stability of Simple Cuttings in Normally Consolidated Clays. Geotechnique, Vol. 18, No. 3, 1968, pp. 372-378.

7.39 Hutchinson, J. N. A Landslide on a Thin Layer of Quick Clay at Furre, Central Norway. Geotechnique, Vol. 11, No. 2, 1961, pp. 69-94.

7.40 Hutchinson, J. N. A Reconsideration of the Coastal Landslides at Folkestone Warren, Kent. Geotechnique, Vol. 9, No. 1, 1969, pp. 6-38.

7.41 Hwang, C. T., Morgenstern, N. R., and Murray, D. W. Application of the Finite Element Method to Consolidation Problems. In Application of the Finite Element Method in Geotechnical Engineering: A Symposium (Desai, C. S., ed.), U.S. Army Engineer Waterways Experiment Station, Vicksburg, Miss., 1972, pp. 739-765.

7.42 Janbu, N. Slope Stability Computations. In Embankment-Dam Engineering (Hirschfeld, R. C., and Poulos, S. J., eds.), Wiley, New York, 1973, pp. 47-86.

7.43 Janbu, N., Bjerrum, L., and Kjaernsli, B. Soil Mechanics Applied to Some Engineering Problems. Norwegian Geotechnical Institute, Oslo, Publ. 16, 1956, pp. 5-26.

7.44 John, K. W. Engineering Analysis of Three-Dimensional Stability Problems Utilizing the Reference Hemisphere. Proc., 2nd Congress, International Society of Rock Mechanics, Belgrade, Vol. 3, 1970, pp. 385-391.

7.45 Kankare, E. Failures at Kimola Floating Canal in Southern Finland. Proc., 7th International Conference on Soil Mechanics and Foundation Engineering, Mexico City, Vol. 2, 1969, pp. 609-616.

7.46 Kealy, C. D., and Busch, R. A. Determining Seepage Characteristics of Mill-Tailings Dams by the Finite Element Method. U.S. Bureau of Mines, Rept. RI 7477, 1971, 113 pp.

7.47 Kenney, T. C., and Chan, H. T. Field Investigation of Permeability Ratio of New Liskeard Varved Soil. Canadian Geotechnical Journal, Vol. 10, No. 3, 1973, pp. 473-488.

7.48 Kenney, T. C., and Uddin, S. Critical Period for Stability of an Excavated Slope in Clay Soil. Canadian Geotechnical Journal, Vol. 11, No. 4, 1974, pp. 620-623.

7.49 Kondner, R. L., and Zelasko, J. S. A Hyperbolic Stress-Strain Formulation for Sands. Proc., 2nd Pan-American Conference on Soil Mechanics and Foundations Engineering, Brazil, American Society of Civil Engineers, New York, Vol. 1, 1963, pp. 289-324.

7.50 Kulhawy, F. H., and Duncan, J. M. Stresses and Movements in Oroville Dam. Journal of Soil Mechanics and Foundations Division, American Society of Civil Engineers, New York, Vol. 98, No. SM7, 1972, pp. 653-655.

7.51 Lefebvre, G., Duncan, J. M., and Wilson, E. L. Three-Dimensional Finite Element Analysis of Dams. Journal of Soil Mechanics and Foundations Division, American Society of Civil Engineers, New York, Vol. 99, No. SM7, 1973, pp. 495-507.

7.52 Little, A. L., and Price, V. E. The Use of an Electronic Computer for Slope Stability Analysis. Geotechnique, Vol. 8, No. 3, 1958, pp. 113-120.

7.53 Lo, K. Y. Stability of Slopes in Anisotropic Soils. Journal of Soil Mechanics and Foundations Division, American Society of Civil Engineers, New York, Vol. 91, No. SM4, 1965, pp. 85-106.

7.54 Louis, C. Strömungsvorgänge in klüftigen Medien und ihre Wirkung auf die Standsicherheit von Bauwerken und Boschungen im Fels. Institut Bodenmechanik und Felsmechanik, Universitat Fridericiana Karlsruhe, Vol. 30, 1967, 121 pp.

7.55 Morgenstern, N. R. The Influence of Groundwater on Stability. In Stability in Open Pit Mining (Brawner, C. O., and Milligan, V., eds.), Society of Mining Engineers, American Institute of Mining, Metallurgical and Petroleum Engineers, New York, 1971, pp. 65-81.

7.56 Morgenstern, N. R., and Guther, H. Seepage Into an Excavation in a Medium Possessing Stress-Dependent Permeability. Proc., Symposium on Percolation Through Fissured Rock, International Society of Rock Mechanics and International Association of Engineering Geology, Stuttgart, Deutsche Gesellschaft für Erd- und Grundbau, Essen, West Germany, 1972, pp. T2-C, 1-15.

7.57 Morgenstern, N. R., and Price, V. E. The Analysis of the Stability of General Slip Surfaces. Geotechnique, Vol. 15, No. 1, 1965, pp. 79-93.

7.58 Morgenstern, N. R., and Price, V. E. A Numerical Method for Solving the Equations of Stability of General Slip Surfaces. Computer Journal, Vol. 9, No. 4, 1967, pp. 388-393.

7.59 Müller, L. The Rock Slide in the Vaiont Valley. Felsmechanik und Ingenieur Geologie, Vol. 2, No. 3-4, 1964, pp. 148-212.

7.60 Nakase, A. Stability of Low Embankment on Cohesive Soil Stratum. Soils and Foundations, Vol. 10, No. 4, 1970, pp. 39-64.

7.61 Patton, F. D., and Deere, D. U. Geologic Factors Controlling Slope Stability in Open Pit Mines. In Stability in Open Pit Mining (Brawner, C. O., and Milligan, V., eds.), Society of Mining Engineers, American Institute of Mining, Metallurgical and Petroleum Engineers, New York, 1971, pp. 23-48.

7.62 Pilot, G., and Moreau, M. La Stabilité des Remblais Sur Sols Mous. Eyrolles, Paris, 1973.

7.63 Prevost, J-H., and Höeg, K. Effective Stress-Strain-Strength Model for Soils. Journal of Geotechnical Engineering Division, American Society of Civil Engineers, New York, Vol. 101, No. GT3, 1975, pp. 259-278.

7.64 Sevaldson, R. A. The Slide in Lodalen, October 6, 1954. Geotechnique, Vol. 6, No. 4, 1956, pp. 167-182.

7.65 Sharp, J. C., and Maini, Y. N. T. Fundamental Considerations on the Hydraulic Characteristics of Joints in Rock. Proc., Symposium on Percolation Through Fissured Rock, Stuttgart, International Society of Rock Mechanics and International Association of Engineering Geology, Deutsche Gesellschaft für Erd- und Grundbau, Essen, West Germany, 1972, pp. T1-F, 1-15.

7.66 Skempton, A. W., and Hutchinson, J. N. Stability of Natural Slopes and Embankment Foundations. Proc., 7th International Conference on Soil Mechanics and Foundation Engineering, Mexico, State-of-the-Art Volume, 1969, pp. 291-340.

7.67 Spencer, E. A Method of Analysis of the Stability of Embankments Assuming Parallel Inter-Slice Forces. Geotechnique, Vol. 17, No. 1, 1967, pp. 11-26.

7.68 Taylor, D. W., Stability of Earth Slopes. Journal of Boston Society of Civil Engineers, Vol. 24, No. 3, 1937, pp. 197-246.

7.69 Taylor, D. W. Fundamentals of Soil Mechanics. Wiley, New York, 1948, 700 pp.

7.70 Taylor, R. L., and Brown, C. B. Darcy Flow Solutions With a Free Surface. Journal of Hydraulics Division, American Society of Civil Engineers, New York, Vol. 93, No. HY2, 1967, pp. 25-33.

7.71 Terzaghi, K., and Peck, R. B. Soil Mechanics in Engineering Practice. Wiley, New York, 1967, 729 pp.

7.72 Toth, J. A Theoretical Analysis of Groundwater Flow in Small Drainage Basins. Proc., 3rd Hydrology Symposium, National Research Council of Canada, Ottawa, 1962, pp. 75-96.

7.73 Whitman, R. V., and Bailey, W. A. Use of Computers for Slope Stability Analysis. Journal of Soil Mechanics and Foundations Division, American Society of Civil Engineers, New York, Vol. 93, No. SM4, 1967, pp. 475-498.

7.74 Wittke, W., Rissler, P., and Semprich, S. Three-Dimensional Laminar and Turbulent Flow Through Fissured Rock According to Discontinuous and Continuous Models. Proc., Symposium on Percolation Through Fissured Rock, International Society of Rock Mechanics and International Association of Engineering Geology, Stuttgart, Duetsche Gesellschaft für Erd- und Grundbau, Essen, West Germany, 1972, pp. T1-H, 1-18 (in German).

7.75 Zienkiewicz, O. C., Best, B., Dullage, C., and Stagg, K. G. Analysis of Nonlinear Problems in Rock Mechanics With Particular Reference to Jointed Rock Systems. Proc., 2nd Congress, International Society of Rock Mechanics, Belgrade, Vol. 3, 1970, pp. 501-509.

Chapter 8

Design and Construction of Soil Slopes

David S. Gedney and William G. Weber, Jr.

The design of stable slopes in soil has been extensively studied by engineers and geologists. In recent years, substantial advancements have been made in understanding the engineering characteristics of soils as they relate to stability. Chapters 6 and 7 describe the state of the art regarding the determination of pertinent soil parameters and the recommended approaches to engineering analysis. These techniques allow the design and construction of safe and economic slopes under varying conditions. This chapter applies the basic principles established in Chapters 6 and 7 to procedures for the design of stable slopes. The procedures can also be applied to preconstructed slopes and to correction of existing landslides.

PHILOSOPHY OF DESIGN

There are several basic considerations in the design of stable slopes. First, because of the nature of soils and the geologic environment in which they are found, each slope design is different. Second, the basic mechanics applied to estimate the stability of a cut slope in soil are the same as those used to estimate the stability of a fill slope. Third, finding the correct method of stability analysis solves only part of the design problem. Designing a stable slope includes field investigation, laboratory investigation, and construction control. The details involved in this work cannot be standardized because maximum flexibility is needed as each problem is assessed. Judgment, experience, and intuition, coupled with the best data-gathering and analytical techniques, all contribute to the solution.

SAFETY FACTOR

The specific analytical techniques used to predict the stability of slopes are explained in Chapter 7. In all cases, the geotechnical engineer determines the safety factor, which is defined several ways but most commonly as

1. The ratio of resisting forces to driving forces along a potential failure surface;
2. The ratio of resisting moments to driving moments about a point;
3. The ratio of available shear strength to the average shear stress in the soil along a potential failure surface; and
4. The factor by which the shear strength parameters may be reduced in order to bring the slope into a state of limiting equilibrium along a given slip surface.

The last definition is used in Chapter 7, and, unless otherwise noted, effective stress parameters are implicit.

Ideally, failure is represented by factor of safety values less than one, and stability is represented by values greater than one. The geotechnical engineer must be aware that the safety factor for a given slope depends heavily on the quality of the data used in the analysis. In addition, the various methods used to compute safety factors give wide ranges of values, except when the ratio equals unity.

The problem of determining a meaningful safety factor is complicated by factors such as interpretation of field and laboratory data, uncertainty of construction control, and the designer's incomplete information about the design problem. In any case, using the best information attainable and the procedures outlined in Chapters 6 and 7 allows the engineer to compute a minimum factor of safety against failure as a basis for comparing design alternatives.

After consideration of the factors that influence design and the consequences of failure, the reasonableness of reducing the safety factor can be established. In highway engineering, slope designs generally require safety factors in the range of 1.25 to 1.50. Higher factors are required if there is a high risk for loss of human life or uncertainty regarding the pertinent design parameters. Likewise, lower safety factors can be used if the engineer is fairly confident of the accuracy of input data and if good construction control can be executed.

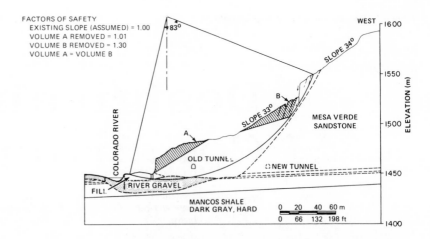

Figure 8.1. Stabilization of the Cameo slide above a railroad in the Colorado River Valley by partial removal of the head (8.1, 8.22). Stability analysis determined that removal of volume B was more effective than removal of volume A.

FACTORS OF SAFETY
EXISTING SLOPE (ASSUMED) = 1.00
VOLUME A REMOVED = 1.01
VOLUME B REMOVED = 1.30
VOLUME A = VOLUME B

DESIGN PROCEDURES

The slope-stability design procedures outlined in Chapter 7 clearly involve a relation between available shear strength and applied shear stress within a soil mass. The analytical techniques allow comparison of various design alternatives, including effects of those alternatives on the stability of the slope and the economy of the solutions. In addition, Chapter 7 discusses the various shapes of potential failure surfaces, including a circular arc, a planar surface, and the Morgenstern-Price variations (8.18).

The preliminary design process may begin by considering various published stability charts based on simplified assumptions. Such a study may be adequate in some cases to decide whether a standard slope angle can be used. In all cases, the design process must include consideration of the full life span of the slope being studied, because soil strength and groundwater conditions usually change with time. At the minimum, the analysis should study conditions expected immediately after construction (end-of-construction case) and at some longer time after construction.

As indicated in Chapter 7, there is little difference among the results obtained from various methods of stability analysis performed immediately after construction. Since design problems in cohesionless soils are relatively minor, except for instances of dynamic loading, reasonable assumptions regarding shear strength may be used with appropriate safety factors. In cohesive soils, the total stress analysis with appropriate laboratory-determined strengths can be used for simplicity.

One should thoroughly study the background presented in Chapters 6 and 7 before proceeding with any of the design procedures outlined in this chapter. These procedures for stable slope design are separated into three broad categories:

1. Avoid or eliminate the problem;
2. Reduce the forces tending to cause movement; and
3. Increase the forces resisting the movement.

A summary of these procedures is given in Table 8.1.

Avoid Problem

For most highway design studies, a geological reconnais-

sance is an important preliminary part of the project development. During reconnaissance, potential stability problems, such as poor surface drainage, seepage zones on existing natural slopes, hillside creep, and ancient landslides, should be carefully noted. Early recognition of known troublesome areas encourages alternative studies for future highway location. If relocation is not possible, adjustments to the line and grade of the highway should be considered.

The most difficult landforms to detect, and the most costly to deal with in construction, are the geologically ancient landslides. Quite often, natural weathering processes or human changes to the environment all but obscure these landforms; however, a field examination by a trained geologist or geotechnical engineer and aerial photographs (Chapter 3) will reveal certain physical incongruities, such as hummocky terrain, blocked regional or local drainage patterns, ancient slide scarps, and vegetation differences.

Since old landslides and talus slopes continue to move downslope until driving and resisting forces are balanced, these slopes may have widely varying abilities to resist new loadings, either internal or external. For instance, such slopes may have perceptible movements during periods of heavy rainfall (high seepage force or increased elevation of groundwater or both). In any stability analysis in which the factor of safety against movement is at or near unity, the influence of even a slight increase in the seepage force or a slight reduction in the resisting forces due to raised groundwater levels is significant. Thus, the decision to construct transportation facilities through or over ancient landslides must be carefully studied and appropriate consideration given to remedial treatment and long-term stability.

Removal of Materials

If relocation or realignment of a proposed roadway is not practical, either complete or partial removal of the unstable materials should be among the alternative design considerations. Figure 8.1 shows an example of one such study. Economics, as well as the relative risk to slope stability, will quite naturally play an important role in the final course of action selected.

The removal of potentially unstable materials can vary from simple stripping of a near-surface layer a few meters thick before embankment construction to a more complicated and costly operation such as that encountered in a sidehill cut along the Willamette River in West Linn, Oregon,

Table 8.1. Summary of slope design procedures.

Category	Procedure	Best Application	Limitation	Remarks
Avoid problem	Relocate highway	As an alternative anywhere	Has none if studied during planning phase; has large cost if location is selected and design is complete; also has large cost if reconstruction is required	Detailed studies of proposed relocation should ensure improved conditions
	Completely or partially remove unstable materials	Where small volumes of excavation are involved and where poor soils are encountered at shallow depths	May be costly to control excavation; may not be best alternative for large slides; may not be feasible because of right-of-way requirements	Analytical studies must be performed; depth of excavation must be sufficient to ensure firm support
	Bridge	At sidehill locations with shallow-depth soil movements	May be costly and not provide adequate support capacity for lateral thrust	Analysis must be performed for anticipated loadings as well as structural capability to restrain landslide mass
Reduce driving forces	Change line or grade	During preliminary design phase of project	Will affect sections of roadway adjacent to slide area	
	Drain surface	In any design scheme; must also be part of any remedial design	Will only correct surface infiltration or seepage due to surface infiltration	Slope vegetation should be considered in all cases
	Drain subsurface	On any slope where lowering of groundwater table will effect or aid slope stability	Cannot be used effectively when sliding mass is impervious	Stability analysis should include consideration of seepage forces
	Reduce weight	At any existing or potential slide	Requires lightweight materials that are costly and may be unavailable; may have excavation waste that creates problems; requires consideration of availability of right-of-way	Stability analysis must be performed to ensure proper use and placement area of lightweight materials
Increase resisting forces	Drain subsurface	At any slide where water table is above shear plane	Requires experienced personnel to install and ensure effective operation	
	Use buttress and counterweight fills	At an existing slide, in combination with other methods	May not be effective on deep-seated slides; must be founded on a firm base	
	Install piles	To prevent movement or strain before excavation	Will not stand large strains; must penetrate well below sliding surface	Stability analysis is required to determine soil-pile force system for safe design
	Install anchors	Where rights-of-way adjacent to highway are limited	Involves depth control based on ability of foundation soils to resist shear forces from anchor tension	Study must be made of in situ soil shear strength; economics of method is function of anchor depth and frequency
	Treat chemically	Where sliding surface is well defined and soil reacts positively to treatment	May be reversible action; has not had long-term effectiveness evaluated	Laboratory study of soil-chemical treatment must precede field installation
	Use electroosmosis	To relieve excess pore pressures at desirable construction rate	Requires constant direct current power supply and maintenance	
	Treat thermally	To reduce sensitivity of clay soils to action of water	Requires expensive and carefully designed system to artificially dry out subsoils	Methods are experimental and costly

where a section of I-205 required extensive excavation to depths as great as 70 m (230 ft).

In the latter case, analytical studies predicted the need for flatter than the normal 2:1 slope because of weakened flat-lying deposits of clay shales just above the base of the proposed roadway ditch line. Right-of-way considerations for flatter slopes included an emergency water supply reservoir for the city of West Linn immediately adjacent to the present highway property lines. Various alternative design schemes for stability were studied, including grade and alignment changes, structural support walls, and complete relocation; all of these alternatives proved to be much

more costly than purchasing additional highway right-of-way and replacing the municipal water supply system. In addition, adjacent projects were known to be deficient in borrow material for required embankment construction.

The final design used to complete the project included the excavation of a wide bench zone at or near the roadway level and the use of flat slope ratios to ensure a greater than required safety factor against potential failure. This example serves to underscore the need for accurate stability studies, not only to compare various design alternatives but to allow the design engineer to properly select the critical locations within a slope in need of treatment. Lack of such analysis could have substantially increased construction costs on the West Linn project.

Bridging

In some instances, removal of especially steep and long narrow unstable slopes is too costly. One alternative design is to span the unstable area with a land bridge or a structure whose support is founded on piles placed well below the unstable foundation materials (8.1). Stability studies must ascertain that the bridge is indeed founded at sufficient depth below the unstable materials and not just penetrating into a more stable stratum. If supports must penetrate through the moving soil, as shown in Figure 8.2, the foundation piling must be designed to withstand the predicted lateral forces. Bridging may also include limited excavation and the use of surface and subsurface drainage.

Reduce Driving Forces

Since the stability of soil slopes is a limiting equilibrium problem in which the external forces acting on a soil mass are at least balanced, the design of stable slopes must address ways to ensure proper safety from the forces tending to cause movements. Since the driving forces are essentially gravitational because of the weight of the soil and water, the simplest approach to reducing such forces is to reduce the mass that is involved. Flattened slopes, benched slopes, reduced cut depths, internal soil drainage, and lightweight

Figure 8.2. Landslide avoidance by bridging near Santa Cruz, California (8.24).

fill all represent feasible treatments. The reduction of driving forces can be divided into three main categories:

1. Change of line or grade or both,
2. Drainage, and
3. Reduction of weight.

The stability of embankment slopes and natural slopes cannot necessarily be approached in the same way. Except in certain unique instances, the stability of embankment slopes increases with time because of consolidation and strength increases in the slope-forming materials. One noticeable exception could be embankments composed of degradable compacted shale, which will deteriorate with time and result in subsequent settlement and distortion or failure of the fills. In cut slopes, the long-term stability may be far less than that available at the time of construction. The ability of a cut slope to withstand the effects of time and stress change is discussed in Chapter 6.

Talus slopes often have marginal stability and deserve particular attention. Talus can be defined as rock fragments that have any size or shape and have been heterogeneously deposited by nature at the base of steep slopes. Runoff from normal rainfall may cause a sufficient increase in seepage forces to initiate movement within talus slopes. Recognition of talus slope forms is important in the predesign process; such slopes should be avoided during construction unless other alternatives are not available. If talus slopes must be disturbed by construction activity, careful analysis should consider the benefits of internal drainage to control potential slope movements that may be triggered by the buildup of large internal water pressures.

Change of Line or Grade

Early in the design stage, cut and fill slopes should be evaluated for potential stability. If conditions warrant, adjustments to the line and grade can be effected to minimize or completely eliminate the stability problem. This approach can also be applied to landslides during and after construction. The economics of various alternative solutions should ensure the feasibility of this approach. An example of a grade revision to prevent a cut slope movement is shown in Figure 8.3.

Line or grade changes are usually associated with a reduction of driving forces. Movement of the roadway alignment

Figure 8.3. Grade change effected during construction to preclude failure at cut slope.

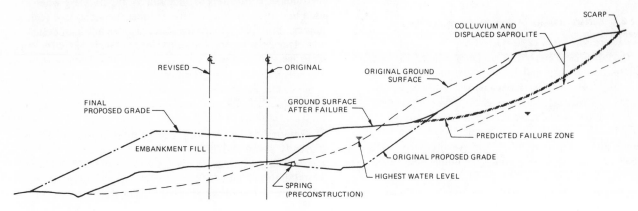

175

away from the toe of a potential or existing slide area will prevent having to remove the toe support. When it is necessary to move the alignment away from an existing slide as a corrective measure, a buttress fill is usually placed to support the sliding mass. If a shift in alignment is not possible, the grade may have to be raised over the buttress fill. In this case, additional costs will accrue, since a transition zone on each side of the grade change will be required.

Changes to effect reduction in the driving forces during construction operations are not only difficult but expensive. To flatten construction slopes often requires additional right-of-way and could involve alignment shifts that affect the design on either side of the troubled area. The cost-effectiveness of geotechnical studies is greatest during the preliminary design stages of any project.

Surface Drainage

Of all possible design schemes considered for the correction of existing or potential landslides, proper drainage of water is probably the most important. Drainage will both reduce the weight of the mass tending to slide and increase the strength of the slope-forming material.

Adequate surface drainage is necessary in new cuts, as well as in completed slopes where movement has occurred. The design of cut slopes must always take into consideration the natural drainage patterns of the area and the effect that the constructed slope will have on these drainage patterns. Two items that should be evaluated are (a) surface water that will flow across the face of the cut slope and (b) surface water that will seep into the soil at the head of the cut. These conditions produce erosion and increase the tendency for potential surface slumps and localized failures on the slope face. As shown in Figure 8.4, diversion ditches and interceptor drains are widely used as erosion control measures in situations in which large volumes of runoff are anticipated. When trenches with a definite grade are constructed, the surface runoff and seepage are intercepted.

Good surface drainage is strongly recommended as part of the treatment for any slide or potential slide (8.8). Every effort should be made to ensure that surface waters are carried away from a slope. Such considerations become important when a failure has already occurred. Unless sealed, cracks behind the scarp face of a slide can carry large volumes of surface waters into the failure zone and result in serious consequences. Even the obvious activity of reshaping the surface of a landslide mass can be extremely

Figure 8.4. Surface drainage of slope by diversion ditch and interceptor drain.

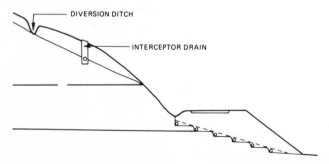

Figure 8.5. Slope protected by pneumatically applied mortar.

Figure 8.6. Horizontal and vertical drains to lower groundwater in natural slopes.

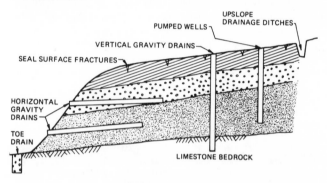

beneficial, in that unnoticed cracks are sealed and water-collecting surface depressions are eliminated.

Slope treatment per se may involve a number of alternatives, all designed to promote rapid runoff and improve slope stability. Some of these measures are (a) seeding or sodding and (b) using gunite, riprap, thin masonry or concrete slope paving, and rock fills. Gunite and thin masonry or concrete slope paving have been used successfully to protect weak shales or claystones from rapid weathering (Figure 8.5). The use of asphalt paving to prevent infiltration of surface water is also common in some areas. These methods of controlling surface runoff are effective when used in conjunction with various subsurface drainage techniques. Surface drainage measures require minimal design and offer positive protection to slopes along transportation facilities.

Subsurface Drainage

The removal of water within a slope by subsurface drainage is usually costly and difficult. Methods frequently used to accomplish subsurface drainage are the installation of horizontal drains (Figure 8.6), vertical drainage wells (Figures 8.6 and 8.7), and drainage tunnels. The drainage-related

Figure 8.7. Vertical wells to lower groundwater in roadway slope.

Figure 8.8. Corrective measures for Castaic-Alamos Creek slides, California (8.10).

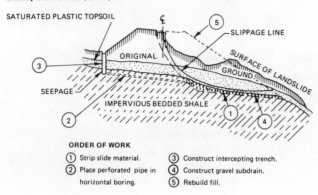

ORDER OF WORK
① Strip slide material.
② Place perforated pipe in horizontal boring.
③ Construct intercepting trench.
④ Construct gravel subdrain.
⑤ Rebuild fill.

expense is generally less when these measures are incorporated into the preliminary design process than when they are included as remedial measures during or following construction. Occasionally, attempts are made to intercept subsurface flows above the sliding mass; however, the expense usually precludes treatment by this procedure for all but special cases. Since seepage forces act to increase the driving force on a landslide, the control of subsurface water is of major importance to the geotechnical engineer. If the preliminary investigation reveals the presence of groundwater, if design studies predict slide movement, and if positive subsurface drainage can preclude failure, a suitable design should be prepared for cost comparison with other alternatives.

Subsurface drainage is equally important in cut areas and under proposed embankments. The effectiveness and frequency of use of the various types of drainage treatment vary according to the geology and the climatic conditions. It is generally agreed, however, that groundwater constitutes the most important single contributory cause for the majority of landslides; thus, in many areas of the country the most generally used successful methods for both prevention and correction of landslides consist entirely or partially of groundwater control (8.8, 8.29). Figure 8.8 shows the use of both surface and subsurface drainage to satisfy

stability requirements for finished slopes on a highway project. Although most types of subsurface-drainage treatment are applicable to the prevention and correction of landslides in both embankment and excavation areas, the differences in methods are considered of sufficient importance to justify separate discussion of the subsurface-drainage treatments applied to these two general types of landslides.

Embankment Areas

Landslides may occur when the imposed embankment load results in shear stresses that exceed the shear strength of the foundation soil or when the construction of the embankment interferes with the natural movement of groundwater. Therefore, two factors must be considered in the investigation of possible landslides: (a) weak zones in the foundation soil that may be overstressed by the proposed embankment load and (b) subsurface water that may result in the development of hydrostatic pressures so as to induce slide movement by a significant reduction in the shear strength of the soil. Because there often is no apparent surface indication of unstable slope conditions, a careful exploration must be made if these conditions are to be predicted before construction. Some of the methods of preventing landslides related to drainage are discussed below (8.4).

If a surface layer of weak soil is relatively shallow and underlain by stable rock or soil, the most economical treatment is usually to strip the unsuitable material, as shown in Figure 8.9. If seepage is evident after stripping or if there is a possibility that it may develop during wet cycles, a layer of pervious material should be placed before the embankment is constructed. This layer may consist of clean pit-run gravel, free-draining sand, or other suitable local materials. If springs or concentrated flows are encountered, drain pipes may also be required.

Where subsurface water or soil of questionable strength is found at such great depths that stripping is uneconomical, stabilization trenches have been used successfully to prevent landslides. Stabilization trenches (Figure 8.10) are usually excavated with the steepest side slopes that will be stable for the construction period.

The trench should extend below any water-bearing layers and into firm material. A layer of pervious material is used as a lining within the excavation, and an underdrain pipe is used as a collector. The trench is backfilled, and the embankment is constructed. If the unstable area is in a natural depression of limited areal extent, one trench normal to the centerline of the road may be sufficient. In

Figure 8.9. Stripping of unstable surface material as a slide-prevention measure on Redwood Highway, Humboldt County, California (8.24). Filter material ensures drainage at base of new embankment fill.

177

Figure 8.10. Stabilization trench with pervious material and perforated pipe for subsurface drainage (8.24).

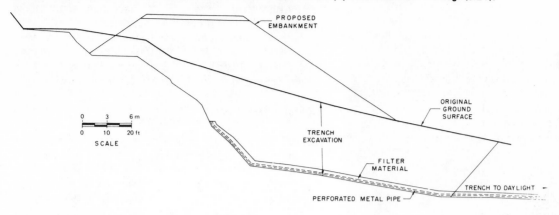

Figure 8.11. Drainage tunnels to prevent landslides near Crockett, California (8.24).

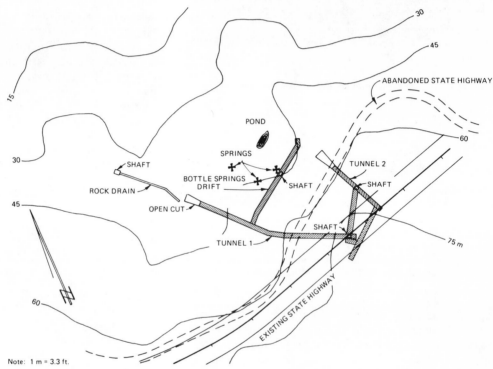

the case of large areas, an extensive system of stabilization trenches may be necessary, frequently arranged in a herringbone pattern. In addition to providing subdrainage, the trenches add considerable structural strength to the foundation.

Stabilization trenches facilitate drainage and provide increased resistance to possible sliding due to the effect of "keying" the compacted backfill of the trench section into the firm material beneath the trench. This procedure has generally been effective in preventing embankment landslides, but a few failures have occurred because the trenches were not carried down to firm material or they were too widely spaced.

Stabilization trench design requires a thorough subsurface investigation program, which must adequately define the subsurface soil layers and locate all water levels in the zone affected by the proposed embankment. One method of designing stabilization trenches is as follows: A line on a 1:1 slope is projected from the outside edge of the top of

the embankment to a point of intersection with the surface of competent material; this point locates the outside toe of the trench. Some deviation from the above concept is tolerable, and may even be required, to provide a fairly uniform trench alignment and grade.

Where the depth to subsurface water is so great that the cost of stripping or placing drainage trenches is prohibitive, drainage tunnels are sometimes used (8.23). Although originally and more commonly used as a correctional treatment, drainage tunnels are sometimes constructed as a preventive measure. The use of drainage tunnels was fairly common at one time by both railroad companies and some highway departments, but at present this method is used rather infrequently largely because of the high construction cost.

An elaborate installation of drainage tunnels, together with an ingenious hot-air furnace for drying the soil, was used to control a large slide near Santa Monica, California (8.15). The use of drainage tunnels in Oregon has also been reported (8.23). These tunnels, usually about 1 m (3 ft) wide by 2 m (6 ft) high (in cross section), must be excavated manually; since skilled tunnel workers are not normally employed on such construction projects, other methods of treatment that permit the use of construction equipment are likely to cost less than the tunnels. Figure 8.11 shows an installation of drainage tunnels on a highway project.

Horizontal drains have supplanted drainage tunnels in most cases. Like drainage tunnels, they were first installed as corrective treatment. Although they are still used for this purpose, horizontal drain installations are now commonly used as a preventive measure for slope instability. A horizontal drain is a small-diameter well drilled into a slope on approximately a 5 to 10 percent grade and fitted with a perforated pipe. Pipes should be provided to carry the collected water to a safe point of disposal to prevent surface erosion.

Both vertical and horizontal drains were used in a slide that occurred in 1968 during construction on I-580 at Altamont Pass in California. Figure 8.12 shows such an installation. The slide extended along 310 m (1015 ft) of roadway with about 30 m (100 ft) of embankment. Remedial measures were (a) installation of a line of vertical drainage wells along the edge of the eastbound lanes; (b) construction of a berm between the eastbound and westbound lanes and a berm adjacent to the eastbound lanes; (c) installation of horizontal drains in five general areas to control groundwater, relieve excess hydrostatic pressure, and intercept and drain the vertical wells; and (d) completion of the construction of the embankment at a controlled rate of loading. The vertical wells were about 1 m (3 ft) in diameter, 12 m (40 ft) deep, and belled at the bottom so that they interconnected to form a somewhat continuous curtain. The drain had a 20-cm (8-in) perforated pipe in the center for the full depth of the vertical drain and was backfilled with pervious material. The horizontal drains were then drilled to intersect and drain the belled portion of the vertical well. The 20-cm perforated pipe was used to observe the water tables and to monitor

the success of the system. Inspection of the system during September 1973 indicated that all water tables were successfully maintained at levels near the bottoms of the vertical drains.

A similar combination drainage system was used on a landslide on I-80 near Pinole, California. This roadway had been open to traffic for several years when a 23-m (75-ft) embankment failed abruptly and closed the freeway in both directions. A drainage gallery formed by a line of vertical wells with overlapping belled bases was placed on each side of the embankment, as shown in Figure 8.13. The lower line of vertical wells was drained with a 30-cm (12-in) pipe, and the uphill line of vertical wells was drained by means of horizontal drains. Berms were used to support the material placed in the failed area. A field and laboratory investigation of the Pinole slide included borings, installation of inclinometers, casings, and laboratory triaxial tests. The existence of water pressure in the layered subsoils was evidenced by a rise of water in the borings of 3 to 4.5 m (10 to 15 ft) when pervious strata were encountered.

Based on the observed excessive seepage at the upstream toe of the fill and the water level data, engineers from the California Division of Highways concluded that hydrostatic pressures had indeed triggered this failure. Two vertical wells were immediately installed upstream of the failed embankment and pumped to a depth of 10 m (33 ft); they produced water at the rate of 5400 L/d (1425 gal/d). Twelve horizontal drains were then installed, varying from 170 to 250 m (560 to 820 ft) in length, and these produced a total flow of 38 000 to 46 000 L/d (10 000 to 12 000 gal/d). In a 6-week period this subdrainage system lowered the groundwater 2 m (6.5 ft) at the upstream toe, 0.7 m (2.3 ft) beneath the sliding mass, and 0.3 m (1 ft) at the downstream toe. The triaxial tests indicated that the impervious soils forming the mass of the foundation material had cohesion values ranging from 25 to 145 kPa (3.5 to 20 lbf/in^2) with a negligible friction angle. The minimum factor of safety was calculated to be 1.01 when the failure oc-

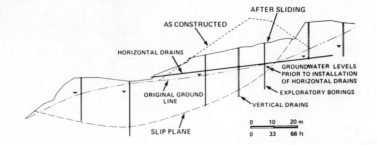

Figure 8.12. Horizontal and vertical drains to prevent slides (8.24).

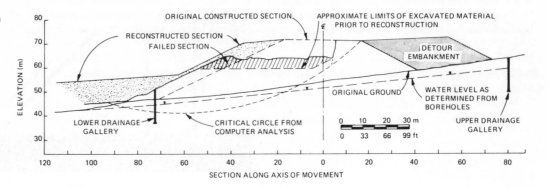

Figure 8.13. Cross section along axis of movement of landslide on I-80 near Pinole, California.

179

curred. The location of the critical circle was confirmed by movements observed from slope indicator readings. A factor of safety of 1.4 was obtained for the conditions after treatment and reconstruction (8.31).

Figure 8.14 shows a system of horizontal drains that was installed as a slide-correction measure; similar installations are frequently used as corrective treatments at other locations.

Vertical drain wells have also been installed under embankments to accelerate the consolidation of weak compressible foundation soils. Discussion of the various uses of vertical drains for this purpose is beyond the scope of this chapter, but many excellent references to vertical sand drain design and construction practices are available in the soil mechanics literature (e.g., 8.7, 8.16).

The continuous siphon is an excellent method devised by the Washington State Highway Commission for providing a drainage outlet for drainage wells or sumps (Figure 8.15). This siphon arrangement can be used to drain trenches, wells, or sumps and is less costly than tunnels, drilled-in pipes, or similar conventional outlet systems. In addition, it permits the installation of subdrainage systems

in areas that do not have readily accessible outlets. A continuous siphon method has the usual limitation of depth, but is useful where applicable.

Excavation Areas

All of the subsurface drainage methods discussed in connection with prevention of landslides in embankments can also be applied to prevention of landslides in excavation areas. Drainage is sometimes installed to intercept subsurface water above the limits of the excavation, but there is seldom any assurance that such interceptor trenches will effectively cut off all groundwater that might contribute to slope failure. If deep trenches are required, the cost is frequently prohibitive, considering the probable effectiveness of the drainage trenches.

Deep trench drains (which, when finally extended deep enough, did work effectively to halt a large slope movement) were used during the construction of a portion of I-81 near Hollins, Virginia (Figure 8.16); this section of highway required a small cut about 10 m (33 ft) deep with 2:1 slopes. The removed material consisted primarily of

Figure 8.14. Horizontal drains used to control large land movements (8.1, 8.14).

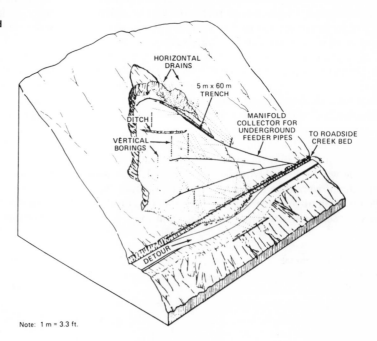

Note: 1 m = 3.3 ft.

Figure 8.15. Washington siphon used by Washington Department of Transportation to lower water level and stabilize landslides (8.24).

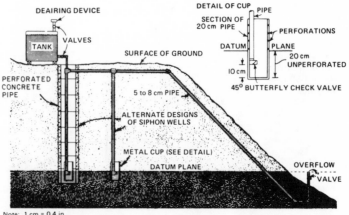

Note: 1 cm = 0.4 in.

colluvium and residual soil weathered from a deeper shale bedrock. During construction a small landslide occurred, and the first attempts toward stabilization called for reducing the driving forces by flattening the slope to 3:1. Although stability analyses based on the assumption that the water table was at the ground surface predicted a safe condition, the regraded slope remained stable only until the following spring, when a second much larger slide occurred. It was obvious that a close relation existed between rainfall intensity and slope movement. Also, calculations in stability studies indicated that the average soil shear strength was much lower than originally assumed.

After consideration of many alternatives, including complete relocation of the roadway, construction of a drilled-pile restraint structure, and complete removal of the sliding mass, a remedial scheme was designed to further unload the slope and to ensure interception of subsurface waters. The design included (a) installing a trench drain around the slide scarp and up the face of the slide more or less at right angles to the roadway; (b) flattening the slope to 4:1; and (c) cutting an intermediate bench at approximately the midheight of the slope. Unfortunately, the trench around the scarp area did not totally halt the heavy subsurface water flow, and the movement continued during periods of heavy rainfall. Some 9 years of costly maintenance followed until the headward progression of the slide necessitated large increases in rights-of-way. The final remedial scheme consisted of (a) using a large rock buttress to restrain the slope above the existing scarp face and to effect deeper drainage interception; (b) placing large granular drainage trenches in two channels down the slope within the flow debris; and (c) regrading the final slope to attain full surface drainage and allow grass establishment. The final slope has remained stable for the past 3 years. The total cost for remedial construction was more than $1 million. However, this proved to be at least $1 million less than the closest alternative, which was to completely relocate the highway away from the slide area (*8.13*).

The most widely used method of subsurface drainage for cut slopes is probably the use of horizontal drains, which are described in the previous section. In excavation areas the drains are installed as the cut is excavated (Figure 8.17), often from one or more benches in the cut slope. Numerous cut slopes drained by this method have remained stable in spite of unfavorable soil formations and the presence of large amounts of subsurface water. If the treatment is delayed until after a landslide has developed, the cost of correcting the slide by subsurface drainage will be much greater and have much less chance of success.

Reduction of Weight

Another technique for reducing the driving forces is referred to as selective "unloading" of a slide. Unloading refers to removal or excavation of a sufficient quantity of slope-forming material at the head of the slide to ensure stability of the mass. This approach is ineffective for infinite slopes or for flow types of earth movement, as discussed in Chapter 6. The required quantity of material to be removed must be carefully predicted by stability analyses using high-quality laboratory and field data. In addition, economics and material usage may dictate whether unloading proce-

Figure 8.16. Aerial view of Hollins slide on I-81 near Roanoke, Virginia.

dures are reasonable on any project. The design of removal procedures must always consider the stability of the slope behind the area to be removed. In some instances, either through project needs for borrow materials or through consideration of the size of total volumes of slide materials, simply to remove the total slide mass is feasible. This procedure is usually limited to slides in which material volumes are relatively small, and it is an effective means of reducing problems when used during the design stage. In addition, the use of variable or flattened slopes at the top of a cut will often aid in unloading a potential slide area.

Slope flattening was used effectively on a 98-m (320-ft) cut for a southern California freeway (Figure 8.18). A failure took place during construction on a 1:1 benched cut slope composed predominantly of sandstone and interbedded shales. After considerable study and analysis, the slope was modified to 3:1, and the final roadway grade was raised some 18 m (60 ft) above the original design elevation. Moreover, to provide additional stability, earth buttresses were placed from roadway levels to a height of 21 m (69 ft) along the final slopes.

In the past, benching has been used by some engineers to reduce the driving force on a potential or existing slide. However, both field experience and stability analyses indicate that this objective is not always achieved (*8.25*). Hence, a careful study and review of alternatives is recommended when benching is proposed. The use of benching to reduce the driving forces is not generally recommended, but

Figure 8.17. Combined benching and drainage for slope stabilization at Dyerville cut on US-101 in California (8.29).

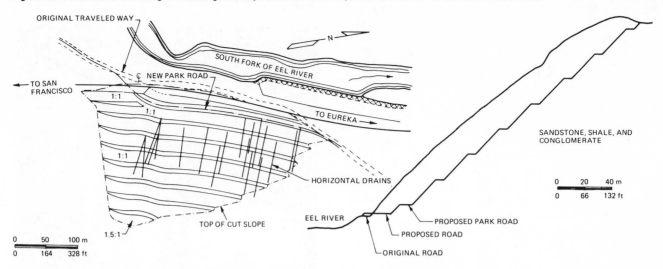

Figure 8.18. Slope flattening and grade change at Mulholland cut on San Diego Freeway (8.29).

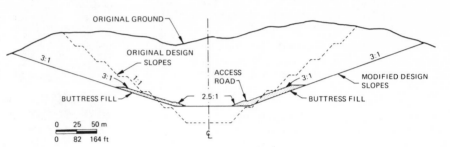

Figure 8.19. Use of lightweight fill and gravel counterweight.

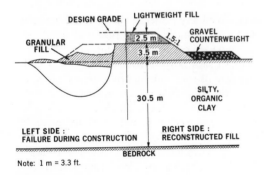

Note: 1 m = 3.3 ft.

Figure 8.20. Use of styrofoam layer as lightweight fill to reduce possibility of potential slope failure in an embankment (8.9).

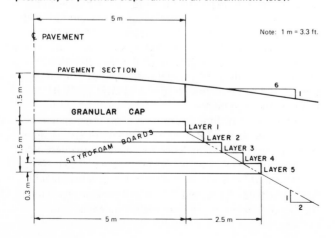

benches do serve a useful purpose in (a) controlling surface runoff if the bench is properly designed and has paved ditches and (b) providing work areas for the placing of horizontal drains.

In embankment construction, lightweight materials such as slag, encapsulated sawdust, expanded shale, cinders, and seashells, have been used successfully to reduce the driving force (Figure 8.19). Polystyrene foam has also been used recently as a lightweight fill material to reduce the stresses in a fill foundation. In an example in Michigan (8.9), a 1.5-m (5-ft) lift thickness of styrofoam in 0.3 by 0.6 by 2.4-m (1 by 2 by 8-ft) boards was placed with staggered joint patterns (Figure 8.20). The foam backfill was covered by 1-mm (4-mil) polyethylene sheeting to protect the foam from possible spills of petroleum-based liquids that might seep through a 1.5-m-thick pavement and granular fill cap. The roadway pavement structure was then placed on the granular cap material (Figure 8.20). Since the polystyrene foam had a density of 768 kg/m^3 (48 lb/ft^3) compared to 1928 kg/m^3 (120 lb/ft^3) for the normal sand backfill, a rather significant weight reduction was realized with the 1.5-m substitution.

In areas where wood product waste is available at reasonable cost, highway departments have used sawdust or wood chips as lightweight fill. Nelson and Allen (8.19) report such a case in which a landslide was stopped by removing earth from the landslide head and replacing it with encapsulated sawdust and wood fiber (Figure 8.21). Since exposed wood products above the groundwater table are known to decay with time, asphalt encapsulation commonly is applied as a retardant to the decay process.

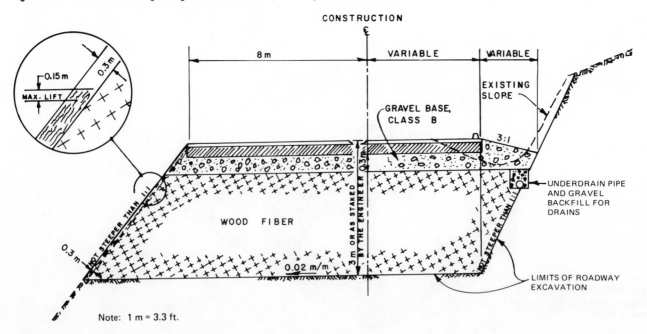

Note: 1 m = 3.3 ft.

Seashells have been used as lightweight fill wherever the shells can be obtained in sufficient quantities. A layer about 1.5 to 2 m (5 to 7 ft) thick over surface swamp deposits forms a foundation that can support construction hauling equipment and effectively reduce the foundation stresses caused by the fill.

Expanded shale aggregates have found excellent, but somewhat expensive, use in embankment construction where fill slope movements suggest potential long-term instability. In one northeastern state, these lighter heat-expanded materials have been used extensively instead of normal fill soils to stabilize high fills on which bridge abutments are constructed. In most instances, the average weight of the shale material is about two-thirds that of a normal earth fill. However, because of the high cost of this material, other alternatives for reducing driving forces will probably provide better design alternatives.

Increase Resisting Forces

The third general method for stabilizing earth slopes is to increase the forces resisting the mass movement. As explained in Chapter 7, two approaches to improving stability are (a) to offset or counter the driving forces by an externally applied force system and (b) to increase the internal strength of the soil mass so that the slope remains stable without external assistance. Both techniques should be considered during design studies to ensure the best engineering and most economical solution. Similar techniques are used to correct landslides that occur during or after construction. The basic principles of soil shear strength (Chapter 6) and the importance of groundwater, excess hydrostatic pressure, and seepage pressure on soil strength should be reviewed.

A multitude of methods is available to the geotechnical engineer to increase the resisting forces on a potential or existing landslide. Although the techniques may vary

widely, they may be reduced to two basic principles: (a) application of a resisting force at the toe of the slide and (b) increase in the strength of the material in the failure zone. Three systems presented (buttress or counterweight fills, pile systems, and anchor systems) basically apply a resisting force at the toe of the sliding mass; the remaining systems (subsurface drainage, chemical treatment, electro-osmosis, and thermal treatment) are essentially methods for increasing the strength of the material in the failure zone.

Apply External Force

Buttress or Counterweight Fills

Buttress (Figure 8.22) or counterweight fill design for slope stability involves one basic principle: to provide sufficient dead weight or artificially reinforced restraint near the toe of the unstable mass to prevent movement. Stability analyses based on the unretained slope geometry and available soil shear strengths predict the forces tending to cause movement and those that exist within the soil mass to resist the movement. A buttress design provides an additional resistance component near the toe of the slope to ensure an adequate safety factor against failure.

The ability of any restraining structure to perform as a designed stabilizing mass is a function of resistance of the structure to (a) overturning, (b) sliding at or below its base, and (c) shearing internally. An overturning analysis is performed by treating the buttress as a gravity structure and resolving the force system to ensure the proper location of the resultant. Potential sliding at or below the base requires a similar analysis, and care must be taken in both the design and the construction phases to ensure adequate depth for founding the buttress and prescribed quality for the layer on which the buttress is placed. Internal shear requires that the designer check the cross-sectional area at

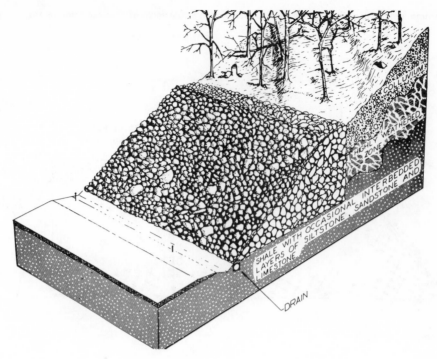

Figure 8.22. Rock buttress used to control unstable slope.

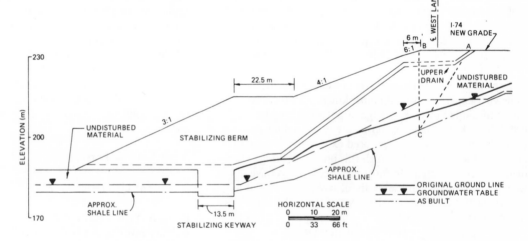

Figure 8.23. Stabilization berm used to correct landslide in shale on I-74 in Indiana.

various elevations within the buttress or counterweight fill to ensure that the resisting structure does not fail by shear within itself.

Several important highway sections have been constructed with or treated remedially by a buttress type of restraining structure. The construction of I-74 in southeastern Indiana in the 1960s included the placement of many kilometers of embankment. The borrow material used in the embankments was predominantly local shale materials that were interbedded with limestone and sandstone. Unfortunately, shales deteriorate with time when exposed to the environment, and cut slopes in fresh shale steeper than 1:1 will deteriorate and slough on the surface until a final stable slope of about 2:1 is attained. These same shale materials, when placed and manipulated into an embankment by the use of accepted construction techniques, will similarly degrade with time. Ultimately, the embankment slopes may slough and eventually fail. The first indications of the degradation process are localized

depressions of the roadway; these gradually spread laterally to include large areas of the pavement surface. These depressions occur as the embankment volume decreases because voids occur between rock blocks and become filled with soil and degraded rock fragments.

Several fill slopes that did shear along I-74 have been thoroughly investigated and analyzed, and alternative remedial treatments have been evaluated. On one slide, known locally as the Chicken slide, careful studies of the in situ shear strength of the shales (ϕ = 14° to 16°) versus the original strengths used in the preconstruction studies showed an approximate reduction of one-half from the as-built condition (*8.12*). The alternatives considered by the Indiana State Highway Commission were (a) relocate, (b) remove and replace, (c) buttress with earth and rock counterweight, and (d) buttress with reinforced earth wall.

Each alternative was thoroughly studied, and appropriate cost figures were determined. By far the least expensive and least disruptive to traffic were the two buttress

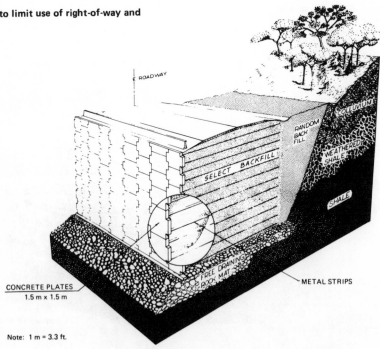

CONCRETE PLATES
1.5 m x 1.5 m

Note: 1 m = 3.3 ft.

alternatives. Since cost estimates for both solutions were close to $1 million, the highway commission advertised for bids. Based on contractor bid prices ($700 000 versus $1 000 000 for reinforced earth), the earth and rock counterweight design was finally selected. A cross section of the design, as finally constructed, is shown in Figure 8.23.

Reinforced earth, as the name implies, is a construction material that involves the designed use of backfill soil and thin metal strips to form a mass that is capable of supporting or restraining large imposed loads (Figure 8.24). The face of a reinforced earth wall is usually vertical, and the backfill material is confined behind either metal or unreinforced concrete facings. Reinforced earth is finding increased use in highway construction, particularly when it is used as a buttress type of retaining structure. As a buttress, reinforced earth acts as a gravity structure placed on a stable foundation, and it must be designed to resist the slope driving forces, i.e., overturning, shearing internally, and sliding at or below the base.

The Tennessee Department of Transportation selected a reinforced earth buttress wall to correct a large landslide on a section of I-40 near Rockwood (8.26). Alternative designs were prepared for a rock buttress and the reinforced earth structure. Cost estimates ($505 000 for reinforced earth versus $930 000 for a rock buttress), ease of construction, and time for construction were the principal reasons for selecting the reinforced earth wall. The slope-forming materials were essentially a thick surface deposit of colluvium underlain by residual clays and clay shales (Figure 8.25). The groundwater table was seasonally variable, but was generally found to be above the colluvium-residuum interface. This particular landslide occurred within an embankment placed as a sidehill fill directly on a colluvium-filled drainage ravine. Because of blocked subsurface drainage and weakened foundation soils, the fill failed some 4 years after construction. Instrumentation, including slope inclinometer casings, placed the failure surface at the con-

tact zone between the colluvium-residuum materials.

Final design plans called for careful excavation of the failed portion of the fill to a firm unweathered shale base, installation of a highly permeable drainage course approximately 10 m (33 ft) wide below the wall area, placement of the reinforced earth wall, and final backfill operations behind the reinforced earth mass (Figure 8.26). The wall was designed for a minimum safety factor of 1.5 against failure. The 253-m long by 10-m high (830-ft by 3.3-ft) wall was completed in approximately 60 d (Figure 8.27).

Other types of buttress or restraining structures commonly used include timber bulkheads; timber, metal, and concrete cribbing; rubble and masonry retaining walls; reinforced concrete retaining walls; and various forms of anchor walls (Figures 8.28, 8.29, 8.30, and 8.31).

Pile Systems

In many urban locations, flattened slopes or counterweight fills are not feasible solutions to cut slope stability problems. Right-of-way limitations and the presence of existing private and commercial structures require much closer attention to the relative risk acceptable in a proposed stability solution. One positive approach is the use of large-diameter piles placed as a preexcavation restraining system. In this system, the forces tending to cause movement are carefully predicted, and the additional restraint necessary to offset soil movement is provided by a closely spaced vertical pile wall. The cast-in-place piles may be designed and placed as cantilevers or tied back with an anchor system. Either alternative requires the pile cross section to resist the full earth thrust imposed by the soil as excavation progresses.

Perhaps the best known application of this design is on a section of I-5 near Seattle, Washington (8.21). Cuts in a major interchange area in heavily overconsolidated marine clays were designed with slope ratios based on laboratory-determined undrained shear strength parameters from re-

185

Figure 8.25. Fill failure on I-40 near Rockwood, Tennessee (*8.26*).

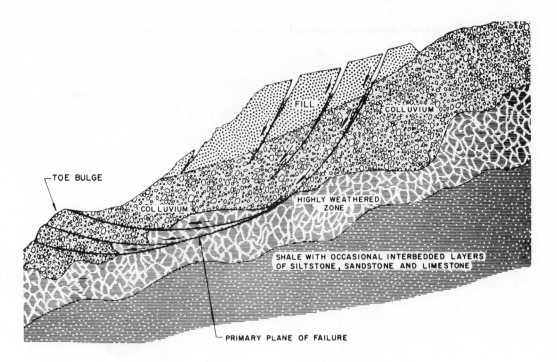

TOE BULGE

FILL

COLLUVIUM

COLLUVIUM

HIGHLY WEATHERED ZONE

SHALE WITH OCCASIONAL INTERBEDDED LAYERS OF SILTSTONE, SANDSTONE AND LIMESTONE

PRIMARY PLANE OF FAILURE

Figure 8.26. Cross section of reinforced earth wall to correct fill failure on I-40 near Rockwood, Tennessee (*8.26*).

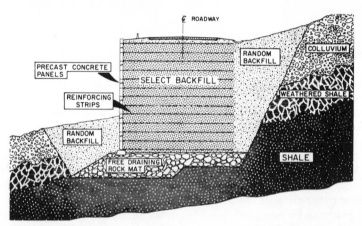

℄ ROADWAY

PRECAST CONCRETE PANELS

REINFORCING STRIPS

SELECT BACKFILL

RANDOM BACKFILL

COLLUVIUM

WEATHERED SHALE

RANDOM BACKFILL

FREE DRAINING ROCK MAT

SHALE

Figure 8.27. Aerial view of final correction to fill failure using reinforced earth, I-40 near Rockwood, Tennessee (*8.26*).

Figure 8.28. Concrete crib wall and gravel backfill installed to prevent movement in Arcata, California (*8.24*).

Figure 8.29. Uses of retaining walls for slope stabilization (*8.27*).

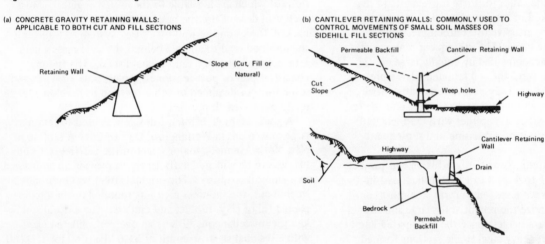

(a) CONCRETE GRAVITY RETAINING WALLS: APPLICABLE TO BOTH CUT AND FILL SECTIONS

Retaining Wall

Slope (Cut, Fill or Natural)

(b) CANTILEVER RETAINING WALLS: COMMONLY USED TO CONTROL MOVEMENTS OF SMALL SOIL MASSES OR SIDEHILL FILL SECTIONS

Permeable Backfill

Cantilever Retaining Wall

Cut Slope

Weep holes

Highway

Cantilever Retaining Wall

Highway

Soil

Drain

Bedrock

Permeable Backfill

Figure 8.30. Gabion-wall retaining structure.

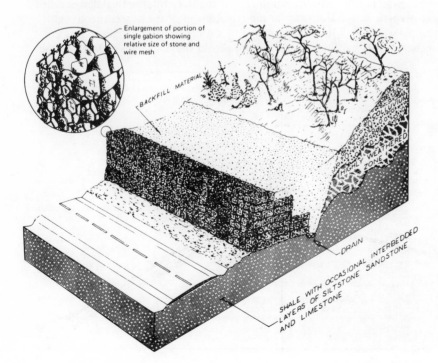

Enlargement of portion of single gabion showing relative size of stone and wire mesh

BACKFILL MATERIAL

HIGHLY WEATHERED ZONE

COLLUVIUM

DRAIN

SHALE WITH OCCASIONAL INTERBEDDED LAYERS OF SILTSTONE SANDSTONE AND LIMESTONE

Figure 8.31. Design of strut to correct cut slope failure on I-94 in Minneapolis-St. Paul (*8.28*).

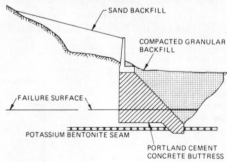

SAND BACKFILL

COMPACTED GRANULAR BACKFILL

FAILURE SURFACE

POTASSIUM BENTONITE SEAM

PORTLAND CEMENT CONCRETE BUTTRESS

187

Figure 8.32. Cylinder pile wall system proposed to stabilize deep-seated slope failure on I-94 in Minneapolis-St. Paul (8.28).

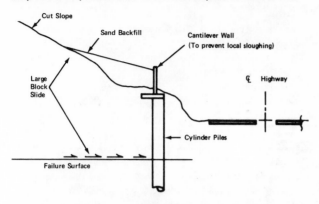

covered soil samples. Some weeks after the cuts had been opened, movements in the form of bulges and sloughing began on several slope faces. Detailed analysis by geotechnical engineers laid the cause for these movements to the release of large "locked-in" soil stresses. The stresses were caused geologically by massive overburden pressures previously applied to the clays. The subsequent removal of these loads was not accompanied by equally large elastic rebounds by the soil; thus, the soil retained a large prestress history. During roadway excavations, substantial cuts were made through these prestressed soils, and the removal of the lateral support, together with the increased moisture content, permitted expansion and subsequent loss of soil strength.

Further slope flattening was not feasible on this project, and large-diameter (1 to 4 m, 3 to 13 ft) drilled, cast-in-place concrete shafts were spaced to form an almost continuous wall to minimize the potential for large soil strains as the excavations were made. Since the anticipated lateral soil forces were large, heavy steel H-pile sections formed the cores of the shafts, and high-strength concrete was placed around those sections. Only minor lateral expansion occurred during the excavation process, and the shafts were designed by using the fully mobilized shear strength of the clay. Drilled shafts were also used in remedial work for the slope of the Potrero Hill cut in San Francisco (8.30)

and in the I-94 cuts (Figure 8.32) in and around Minneapolis-St. Paul (8.33).

Recorded attempts to use driven steel piles or wood piles of nominal diameter to retard or prevent landslides have seldom been successful. For most earth or rock movements, such piles are incapable of providing adequate shearing resistance. In addition, when they are used in even small earth slides, movement of soil between and around the piles must be considered. Quite often, a major earth movement develops a rupture surface in the soil below the pile tips. All such pile schemes should be carefully designed by using realistic soil parameters (8.2). The forces involved in even the smallest landslide are large, and for the piles to be effective they must have sufficient cross section and depth to prevent movement.

Anchor Systems

One type of anchor system is tied-back walls, many variations of which are available to the design engineer. Most such wall designs use the basic principle of carrying the backfill forces on the wall by a "tie" system to transfer the imposed load to an area behind the slide mass where satisfactory resistance can be established. The ties may consist of pre- or post-tensioned cables, rods, or wires and some form of deadmen or other method to develop adequate passive earth pressure.

A good example of such a design was used to retain a large movement in Washington, D.C. (8.20). A section of New York Avenue, a major street in the District of Columbia, was to be widened sufficiently to provide an additional two lanes of roadway. The original street was built in a sidehill cut; the shoulder area was founded on an uncompacted fill of clay, rubble, and cinders. The natural soil below the miscellaneous fill was an overconsolidated clay with a residual shear strength of 31.6 kPa (660 lbf/ft²) (60 percent of laboratory-measured peak value) and a sensitivity of about 4. Since the new construction required an additional width for the roadway fill, the final design called for removal and replacement of the miscellaneous existing embankment as a first order of work. Right-of-way considerations did not permit the 4:1 slopes dictated by the

Figure 8.33. Section of tied-back wall to correct slide condition on New York Avenue in Washington, D.C. (8.20).

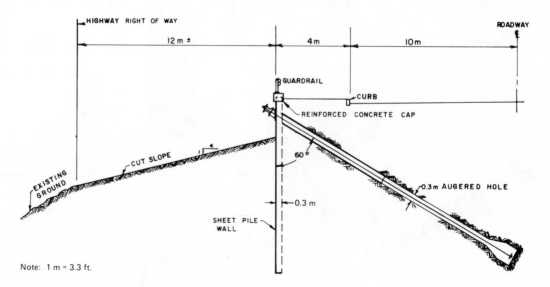

Note: 1 m = 3.3 ft.

stability studies of the widened embankment section.

A comparison of the various design alternatives indicated the economic use of a tied-back sheet pile wall (Figure 8.33). The depth of embedment for the tie-back system varied from 12 to 21 m (40 to 69 ft) to ensure adequate anchorage in the stiff clays. A system of belled concrete anchors into which 3.2-cm (1.25-in) diameter high-strength steel rods were cast acted as the principal support against pull-out. The rods were pretensioned against the sheet-pile wall system to the full design loads following installation. A maximum allowable long-term shear stress of 23.9 kPa (500 lbf/ft²) for the clay was used in analysis; this stress provided a safety factor of about 2. Instruments left in place to facilitate the long-term recording of actual stress levels indicate that the tensile stresses in the steel were well within predicted values.

Increase Internal Strength

Subsurface Drainage

One of the most effective treatments of landslides is to increase the forces available to resist motion by increasing the shear strength of the soil through subsurface drainage (increase in the effective normal stress on the failure surface). This treatment is discussed earlier in this chapter.

Chemical Treatment

Various schemes have been tried by researchers and practitioners to treat unstable soil slopes with injected chemicals. One interesting application under research in California is a patented ion-exchange technique described by Smith and Forsyth (8.30). The ion-exchange technique consists of treating the clay minerals along the plane of potential movement with a concentrated chemical solution. The actual chemicals used in the ion exchange are determined by the clay mineralogy of the soil to be treated and by the prevailing groundwater conditions in the slide mass.

To chemically change a clay soil by ion exchange, some cations in the clay minerals are replaced with different cations that are introduced by chemical solution. In a saturated clay, the rate of migration through the soil structure appears to be much greater for cations than for water. Cation replacement can result in as much as a 200 to 300 percent increase in soil shear strength. Since the initial strength in the shear zone is low for most clays in which this method can be used, this relatively minor increase in strength may be sufficient to stabilize a landslide. Although the ion-exchange technique was successfully used in northern California to stabilize a slide (8.17), it may have little chance for success on other landslides.

Other chemical treatments used with varying degrees of success are lime or lime-soil mixtures, cement grout, and potassium injections. Perhaps the chemical treatment most used in attempts to increase soil shear strength is lime. High-pressure injections of lime slurry have been used in several states with limited success. One successful treatment was reported by Handy and Williams (8.11); approximately 45 000 kg (100 000 lb) of quicklime were placed in predrilled 0.2-m (0.5-ft) diameter holes on 1.5-m (5-ft) centers throughout an extensive slide area. The lime mi-

grated a distance of 0.3 m (1 ft) from the drilled holes in 1 year. Slide movements subsequently ceased, and the area has remained stable to date.

An interesting application of cement grout occurred on a section of I-40 along the Pigeon River in North Carolina. A 90-m (300-ft) benched cut slope for the roadway began moving forward, threatening the road and a large water supply reservoir on the downhill side of the roadway. Subsequent investigations and analyses showed the roadway foundation area to consist of broken rubble and talus debris to great depths. Instruments placed movement along a definite plane where rock voids were large. Large volumes of cement grout were injected into the voids of this layer and surrounding areas in an attempt to increase the shearing resistance of the slope foundation. Although great volumes of grout were required at considerable expense, the slope did become stable, and the water supply for a major city was not lost.

Electroosmosis

One method that effectively increases soil shear strength in situ is electroosmosis (8.5, 8.6). This technique, although extremely expensive, causes migration of pore water between previously placed electrodes; the loss of pore water, in turn, causes consolidation of the soil and a subsequent increase in shear strength. Casagrande, Loughney, and Matich (8.6) describe a highway project that required an excavation approximately 4.5 m (15 ft) in depth and some 24 by 12 m (80 by 40 ft) in area to install a bridge foundation support system. The side slopes were predominantly a saturated, somewhat uniform silt material placed on a 2.5:1 slope. During excavation a slide occurred. After consideration of alternatives such as freezing, chemical injection, slope flattening, and restraining walls, the designers selected electroosmosis as the most economical solution. Some 3 months were required to lower the groundwater sufficiently to proceed with construction; however, the final slope excavation was steepened to 1:1 and the project was successfully completed. Long-term solutions using electroosmosis must give full consideration to the need for a constant supply of direct current and the need for maintenance personnel to periodically check the system for replacement of field-installed electrodes.

One interesting variation of the electroosmotic effect is that suggested by Veder (8.32). Where landslides occur at the contact zone between soil layers, Veder reports that differences in water content between the layers means a difference in electropotential. This difference in potential creates a gradient that forces water to move through the soil toward the region of lower potential. Veder suggests that the insertion of metallic conductors into the soil to create short-circuit electrodes will halt soil-water movement. Thus, the imposed short-circuit effectively acts in the reverse of electroosmosis where an external source of direct current is used to cause soil water migration. Veder reports several cases in which this procedure has successfully stabilized landslides.

Thermal Treatment

For several years the Russians have experimented and reported on the success of thermal treatment of plastic and loessial soils. The high temperature treatments cause a permanent drying of the embankment or cut slope. Hill (*8.15*) discusses the use of thermal treatment in the United States. Beles and Stanculescu (*8.3*) describe an interesting use of thermal methods to reduce the in situ water contents of heavy clay soils in Romania. Applications to highway landslides and to unstable railroad support fills are cited.

Combination of Treatments

In most applications, a combination of the various methods outlined above is used. A buttress fill may be combined with a subdrainage system to provide a resisting force that allows the drainage system to become effective, thus tending to increase the stability of the slope with time. Vertical wells may be pumped to stabilize a cut during construction; however, horizontal drains usually are installed as a long-term solution, and pumping the vertical wells is then discontinued. This procedure was used in California to stabilize the Pinole slide (Figure 8.13); two vertical wells were pumped immediately to relieve the water pressure while the horizontal drains were installed. Then, berms were placed as the embankment was reconstructed to ensure the integrity of the reconstruction while the horizontal drains continued to reduce the water pressure. The long-term stability of this treatment requires that the horizontal drains function properly for the life of the structure. On another project in California, several different approaches were also used to stabilize a slow moving landslide by chemical injection to effect a strength increase within the predetermined slide zone. Then, horizontal drains were installed a year later to effect positive drainage by gravity flow deep within the slide and thereby ensure long-term stability to the area.

Each case in which combinations of various methods have been used both in design and construction represents a study of carefully considered and applied engineering principles to reach a reasonably economic solution. Perhaps the relation between design and actual construction is somewhat unique in geotechnical engineering, because a failure of the slope may result if various combinations of soil strength, groundwater levels, and slope geometrics that occur during construction are not fully considered in design.

TOE EROSION

Toe erosion, as used in this chapter, is the removal of material from the base or toe of a slide or natural slope by natural forces. Although wind erosion can be appreciable, the most common type of toe erosion encountered by a geotechnical engineer is that caused by moving water in rivers, streams, or oceans eroding slope-forming materials. The general solution for this type of problem is to protect the toe of a slope by either a riprap surface layer or a free-draining durable rock layer placed at the base of the slope to an elevation of about 1 m (3.3 ft) above the expected mean high water level.

The erosion of natural or human-made slopes by rivers or streams is a major cause of land instability. Geologic studies refer to the erosion of stream valleys, cliff formations on oceanfronts, and loss of land due to moving waters. Engineers are faced with these problems in design and con-

struction of transportation facilities. Careful attention must be given to the protection of earth slopes in any channel design. Protection may be in the form of (a) riprap or other suitable material or (b) a lining of reinforced concrete with designed hydraulic features to ensure dissipation of the destructive forces of the anticipated flow. One should never assume that a slope adjacent to natural watercourses is adequate until thorough hydraulic studies are made and corresponding protection of the slope provided for the anticipated long-term effects of the water.

Various buttress systems have been used successfully in situations in which the general lack of space precludes other treatments, particularly where a facility follows a river or ocean face. Pile systems have had little success where the ocean is eroding away the toe of a cliff. Various surface and subsurface drainage systems have been used in combination with buttress systems, and, if properly designed and constructed, these will be successful. In general, the solution to the toe erosion problem is to install a system that prevents further loss of support for the slope and to use other means for increasing the resisting forces. Thus, a combination of several methods will generally be required.

REFERENCES

8.1   Baker, R. F., and Marshall, H. C. Control and Correction. In Landslides and Engineering Practice (Eckel, E. B., ed.), Highway Research Board, Special Rept. 29, 1958, pp. 150-188.

8.2   Baker, R. F., and Yoder, E. J. Stability Analyses and Design of Control Methods. In Landslides and Engineering Practice (Eckel, E. B., ed.), Highway Research Board, Special Rept. 29, 1958, pp. 189-216.

8.3   Beles, A. A., and Stanculescu, I. I. Thermal Treatment as a Means of Improving the Stability of Earth Masses. Geotechnique, Vol. 8, No. 4, 1958, pp. 158-165.

8.4   California Department of Transportation. Foundation Exploration, Testing, and Analysis Procedures. In Materials Manual, California Department of Transportation, Sacramento, Vol. 6, 1973.

8.5   Casagrande, L. Electro-Osmosis. Proc., 2nd International Conference on Soil Mechanics and Foundation Engineering, Rotterdam, Vol. 1, 1948, pp. 218-223.

8.6   Casagrande, L., Loughney, R. W., and Matich, M. A. J. Electro-Osmotic Stabilization of a High Slope in Loose Saturated Silt. Proc., 5th International Conference on Soil Mechanics and Foundation Engineering, Paris, Vol. 2, 1961, pp. 555-561.

8.7   Casagrande, L., and Poulos, S. On the Effectiveness of Sand Drains. Canadian Geotechnical Journal, Vol. 6, No. 3, 1969, pp. 287-326.

8.8   Cedergren, H. R. Seepage, Drainage, and Flow Nets. Wiley, New York, 1967, 489 pp.

8.9   Coleman, T. A. Polystyrene Foam Is Competitive, Lightweight Fill. Civil Engineering, Vol. 44, No. 2, 1974, pp. 68-69.

8.10   Dennis, T. H., and Allan, R. J. Slide Problem: Storms Do Costly Damage on State Highways Yearly. California Highways and Public Works, Vol. 20, July 1941, pp. 1-3, 10, 23.

8.11   Handy, R. L., and Williams, W. W. Chemical Stabilization of an Active Landslide. Civil Engineering, Vol. 37, No. 8, 1967, pp. 62-65.

8.12   Haugen, J. J., and DiMillio, A. F. A History of Recent Shale Problems in Indiana. Highway Focus, Vol. 6, No. 3, July 1974, pp. 15-21.

8.13   Herbold, K. D. Cut Slope Failure in Residual Soil. Proc., 23rd Highway Geology Symposium, 1972, pp. 33-49.

8.14   Herlinger, E. W., and Stafford, G. Orinda Slide. California Highways and Public Works, Vol. 31, Jan.-Feb. 1952, pp. 45, 50-52, 59-60.

8.15 Hill, R. A. Clay Stratum Dried Out to Prevent Landslips. Civil Engineering, Vol. 4, No. 8, 1934, pp. 403-407.

8.16 Johnson, S. J. Foundation Precompression With Vertical Sand Drains. Journal of Soil Mechanics and Foundations Division, American Society of Civil Engineers, New York, Vol. 96, No. SM1, 1970, pp. 145-175.

8.17 Mearns, R., Carney, R., and Forsyth, R. A. Evaluation of the Ion Exchange Landslide Correction Technique. Materials and Research Department, California Division of Highways, Rept. CA-HY-MR-2116-1-72-39, 1973, 26 pp.

8.18 Morgenstern, N. R., and Price, V. E. The Analysis of the Stability of General Slip Surfaces. Geotechnique, Vol. 15, No. 1, 1965, pp. 79-93.

8.19 Nelson, D. S., and Allen, W. L., Jr. Sawdust as Lightweight Fill Material. Highway Focus, Vol. 6, No. 3, July 1974, pp. 53-66.

8.20 O'Colman, E., and Trigo Ramirez, J. Design and Construction of Tied-Back Sheet Pile Wall. Highway Focus, Vol. 2, No. 5, Dec. 1970, pp. 63-71.

8.21 Palladino, D. J., and Peck, R. B. Slope Failures in an Overconsolidated Clay, Seattle, Washington. Geotechnique, Vol. 22, No. 4, 1972, pp. 563-595.

8.22 Peck, R. B., and Ireland, H. O. Investigation of Stability Problems. American Railway Engineering Association, Chicago, Bulletin 507, 1953, pp. 1116-1128.

8.23 Roads and Streets. Curing Slides With Drainage Tunnels. Roads and Streets, Vol. 90, No. 4, 1947, pp. 72, 74, 76.

8.24 Root, A. W. Prevention of Landslides. In Landslides and Engineering Practice (Eckel, E. B., ed.), Highway Research Board, Special Rept. 29, 1958, pp. 113-149.

8.25 Royster, D. L. To Bench, or Not to Bench. Proc., 48th Tennessee Highway Conference, Univ. of Tennessee, Knoxville, Tennessee Department of Highways, Nashville, 1966, pp. 52-73.

8.26 Royster, D. L. Construction of a Reinforced Earth Fill Along I-40 in Tennessee. Proc., 25th Highway Geology Symposium, 1974, pp. 76-93.

8.27 Schweizer, R. J., and Wright, S. G. A Survey and Evaluation of Remedial Measures for Earth Slope Stabilization. Center for Highway Research, Univ. of Texas at Austin, Research Rept. 161-2F, 1974, 123 pp.

8.28 Shannon and Wilson, Inc. Slope Stability Investigation, Vicinity of Prospect Park, Minneapolis-St. Paul. Shannon and Wilson, Seattle, 1968.

8.29 Smith, T. W., and Cedergren, H. R. Cut Slope Design in Field Testing of Soils and Landslides. American Society for Testing and Materials, Philadelphia, Special Technical Publ. 322, 1962, pp. 135-158.

8.30 Smith, T. W., and Forsyth, R. A. Potrero Hill Slide and Correction. Journal of Soil Mechanics and Foundations Division, American Society of Civil Engineers, New York, Vol. 97, No. SM3, 1971, pp. 541-564.

8.31 Smith, T. W., Prysock, R. H., and Campbell, J. Pinole Slide, I-80, California. Highway Focus, Vol. 2, No. 5, Dec. 1970, pp. 51-61.

8.32 Veder, C. Phenomena of the Contact of Soil Mechanics. Proc., 1st International Symposium on Landslide Control, Kyoto and Tokyo, Japan Society of Landslide, 1972, pp. 143-162 (in English and Japanese).

8.33 Wilson, S. D. Landslide Instrumentation for the Minneapolis Freeway. Transportation Research Board, Transportation Research Record 482, 1974, pp. 30-42.

PHOTOGRAPH AND DRAWING CREDITS

Chapter 9

Engineering of Rock Slopes

Douglas R. Piteau and F. Lionel Peckover

In transportation corridors, the objective of rock-slope engineering is to maintain slopes for maximum safety and efficiency. Minimizing rock excavation and predicting the safety and ultimate behavior of rock slopes, whether for highway, railway, spillway, quarry, dam site, or opencut mine, are common objectives of civil, geology, and mining engineers. The rational design of rock slopes is particularly important if slopes are steep, if safety is important, and if slope design significantly affects project costs.

The present empirically based, cut-and-try methods and techniques used to design rock slopes are inadequate. New principles and improved technological capability are needed, particularly in mining operations, in which the trend is toward open-pit mines, and in transportation facility construction, which may be increasingly required in steep terrain.

That rock slopes must be treated differently from soil slopes cannot be emphasized enough. In the analysis of rock slopes, one must recognize the differences in the basic characteristics and behavior of soil and rock. Unlike a soil mass, which is a relatively homogeneous and continuous medium composed of uncemented particles, a rock mass is a heterogeneous and discontinuous medium composed essentially of partitioned solid blocks that are separated by discontinuities. The geometry of the spaces between the solid materials and the interlocking properties of the components of soil and rock are totally different. Failure in soil tends to occur within the soil mass, and the direction of the surface of failure tends not to depend on variations of soil properties. The surface of failure in hard rock masses, however, tends to follow the preexisting discontinuities and not to occur through the intact rock to any great extent unless the rock is quite soft. The shear strength of a rock mass is determined largely by the presence of the discontinuities, and the result is that the rock mass is anisotropic in its strength and deformational properties.

The design of rock slopes involves both engineering and geology and, in addition, a combination of the knowledge of precedent with the art of estimation and judgment. The engineer-geologist must obtain quantitative information on those factors that are necessary for making calculations of the probable stability of the slope. Those factors include structural geology, local topography, drainage, hydrogeology, tectonic history, and other environmental features that may add to or detract from the stability of the slope.

Whether a slope will be stable or unstable will depend on how the forces that tend to resist failure compare with those that tend to cause failure. This concept defines the factor of safety for the slope as the ratio of the sum of the resisting forces that act to prevent failure to the sum of the driving forces that tend to cause failure. To a considerable extent, the problem of rock-slope design and related aspects is one of applied mechanics, and the necessary margin of safety to be provided in any particular case is a question of judgment.

The terms of reference and related slope-design problems of an open-pit excavation, for example, may be entirely different from those of highway and railway cuts. For that matter, slope-design problems and requirements for highways and for railways can also differ markedly. Highways can usually have a greater degree of slope instability on their rights-of-way than railways. Unlike automobiles, trains cannot steer or brake readily and can be derailed by rocks no larger than 30 cm (1 ft). Therefore, a remedial measure that may be used on a highway slope may be entirely unacceptable on a railway slope. No particular attempt has been made here, however, to separate railway and highway rock-slope engineering techniques.

Rock-slope engineering is concerned not with large landslides but with rock falls of individual blocks, translation of small rock masses, and occasional slides of accumulated debris from gullies, talus slopes, and postglacial slide areas. This chapter discusses the aspects of rock-slope engineering relevant to designing cut slopes and maintaining the long-term efficiency and safety of existing slopes. Discussed are the significant factors in rock-slope design, the procedures

Figure 9.1. Well-developed, steeply dipping set of joints along highway at Slocan Lake, British Columbia. Joints control slope stability almost entirely.

Figure 9.2. Well-developed discontinuities dipping toward highway at Porteau Bluffs, British Columbia. Two significant slides have occurred in this area.

for the analysis of rock-slope stability, the general planning involved in extensive rock-slope engineering along transportation corridors, and the stabilization, protection, and warning measures that can be used to remedy rock-slope problems.

SIGNIFICANT FACTORS IN DESIGN OF ROCK SLOPES

This section describes the basic factors that are significant to the stability of rock slopes. Additional information is given by Stacey (9.105), Piteau (9.87), Hoek and Bray (9.42), Duncan (9.23), Deere and others (9.21), Terzaghi (9.108), Jennings (9.47), Goodman (9.29), Coates (9.16), and the Canada Department of Energy, Mines and Resources (9.94).

Structural Discontinuities

The stability of rock slopes depends largely on the presence and nature of defective planes or discontinuities within the rock mass. For the greatest part, the significant physical

and mechanical properties of the rock mass are a function of the attitude, geometry, and spatial distribution of these defective surfaces. The basic principles of rock-slope design are based on

1. The systems of joints and other discontinuities;
2. The relation of these systems to possible failure surfaces;
3. The strength parameters of the joints, which include the properties of both the joint surfaces and any joint infilling materials; and
4. The water pressure in the joints.

Examples of the significance of structural discontinuities in slope design are shown in Figures 9.1 and 9.2.

The stability of rock slopes, therefore, is assessed principally by analyzing structural discontinuities in the rock mass and not by assessing the strength of the intact rock itself. Observations relating to shear failure along discontinuities should be made insofar as the discontinuities affect the cohesion and friction developed with respect to shear along these features. The relations of discontinuities to the direction and inclination of the rock slope and to any factors that might influence any potential surface failure must receive special attention. These statements apply even moreso to faults, regardless of how great or small displacements have been.

Field studies show that rocks are usually jointed in preferential directions. Depending on their modes of origin, however, rocks have joint sets whose characteristics can vary greatly. Wide variations can occur in the average spacing between joints, the nature and degree of joint infilling materials, the physical characteristics of their surfaces, and the degree of their development. One joint set can, therefore, have effects on shear characteristics quite different from those of another set, and the various properties of each of the joint sets must be considered individually in rock-slope design. (Strength evaluations are explained in Chapter 6). The main properties that are associated with structural discontinuities and that require quantitative and qualitative evaluation follow.

Orientation or Position in Space

Orientation or position in space is the most important property. If the orientation of joints favors potential slope failure, the effects of other properties are generally unimportant. Discontinuities dipping out of the slope must be carefully considered. Their potential instability increases proportionally as the strike of the discontinuities approaches that of the slope.

Continuity or Size

Continuity or size is the most difficult property to assess. The strength reduction on a failure surface that contains one or more discontinuities is a function of their size. The average continuity of a particular joint set partly indicates the extent to which the rock material and the discontinuities will separately affect the mechanical properties of the mass (9.23, 9.47); in addition, the average continuity affects the magnitude of possible failures involving these features.

Infilling Materials and Openness

Infilling materials are those materials that occur between the walls of the discontinuities in the mass. The three most important characteristics of infilling materials are thickness, type, and hardness. Jaeger (9.45) notes that, if infilling is sufficiently thick, the walls of the discontinuity will not touch and the strength properties will be those of the infilling material.

Spacing

The spacing of discontinuities partly indicates the extent to which the intact rock and the discontinuities will separately affect the mechanical properties of the rock mass. A rock mass is inherently weaker if spacing is close. Also, greater joint frequency increases the potential for dilatancy.

Asperities

Two orders of asperities are recognized. First-order asperities (waviness) are unlikely to shear off; they affect the shear-movement characteristics along the discontinuity and effectively modify the direction of movement during slope failure (9.21). Undulations or waves on the discontinuity reduce the effective apparent dip of the plane, and this dip angle and not the mean dip angle should be used for calculating the disturbing forces on a potential failure plane (9.68). Second-order asperities (roughness) are much smaller and are likely to shear off during movement; these produce an apparent increase in frictional strength along the discontinuity (9.99).

Previous Shear Movement

Shear displacement on a discontinuity results in breaking through asperities and thus reducing the shear strength from initial peak values to values that approach residual shear strength (9.54).

Rock Type

When a slope consists of several rock types, their combined mechanical behavior may differ considerably from that of the constituent units themselves. Hence, each particular rock type may require individual assessment for its behavior in the slope. Different rock types and the products that result from their alteration have inherently different weaknesses and strengths as a result of their origins and compositions. Hence, the properties of different rock and infilling materials can vary within wide limits (9.107). The characteristics of each rock type significantly influence the friction angle, nature of asperities, and hardness of the walls of the discontinuities.

Rock Hardness

There is a relation between rock hardness and unconfined compressive strength. According to Jennings (9.46), an increase in hardness has a corresponding increase in the shear strength. Infilling material and second-order asperities are sheared through during shear failure along a discontinuity; the shear strength of the discontinuity is a function of the shear strength of the infilling materials and the rock materials that form the asperities.

Origins

The origins of discontinuities will affect their engineering significance in the slope. Faults, as compared to joints, for example, have different origins and accordingly different geometry, spatial distribution, weathering and infilling characteristics, and seepage characteristics.

Groundwater

The presence of water in joints has probably been responsible for more rock slides than all other causes combined. Hence, a thorough knowledge of the character and influence of the hydrogeologic regime is necessary, and a knowledge of the water pressure distribution and the factors that influence it is the most essential. One must consider the controlling influences of texture, stratigraphy, and structure on factors such as flow, permeability, recharge, and storage capacity. Consideration should also be given to environmental factors, such as variations in climatic conditions, that result in periods of either high or low recharge and other variations in groundwater conditions. Further discussions concerning climatic influences are given later.

According to Terzaghi (9.108), Serafim (9.103), and Müller (9.77), water in the slope can affect stability by

1. Physically and chemically affecting the pore water and its pressure in joint infilling materials, thus altering the strength parameters of the materials;

2. Exerting hydrostatic pressure on joint surfaces, thus reducing the shearing resistance along potential failure surfaces by reducing the effective normal stresses acting on them; and

3. Affecting intergranular shearing resistance, thus causing a decrease in compressive strength.

Lithology, Weathering, and Alteration

Before one can completely comprehend the particular problems of stability, one must understand the lithology of the physical properties not only of the rock mass itself but of all the materials in the mass. Usually a slope is made up of a complex of rocks of diverse geologic origins. It may have markedly different sequences of sediments, may be intruded by bodies of igneous rocks, or may be partly metamorphosed. The mass represents an association of several lithologic units whose mechanical behavior is that of an integral whole, which may differ considerably from the individual lithologic units themselves.

A sedimentary rock sequence, for example, is markedly different from an igneous series or a metamorphic complex. Each particular type is characterized by a certain texture, fabric, bonding strength, and macro and micro structures. The most important rock properties are the nature of the mineral assemblage and the strength of the constituent minerals; a rock material cannot be strong if its mineral constituents are weak or if the strength of the bonds between the minerals is weak.

The body of rock or the host rock in which the discontinuities occur directly influences the strength characteristics of the discontinuities, particularly if joint infilling materials are absent. The wall rock of the discontinuities affects the joint strength in two ways (9.5): It affects the frictional properties of the material forming the joint, and it affects the intact strength of the asperities of the joint surfaces. The frictional properties of joints are highly dependent on the proportions of the various minerals that are exposed along their surfaces.

The properties of rock can be altered by weathering, i.e., action by atmospheric elements and conditions. Weathering can adversely affect the deformation properties of rocks and can reduce their ultimate bearing resistance and other strength properties. The effects of weathering are usually estimated qualitatively, but in a thorough slope analysis should be estimated quantitatively.

Moisture can also cause alteration of rocks. An increase of moisture content can cause high swelling pressures in montmorillonite, which occurs in joints either as infilling or as a product of alteration. These high swelling pressures and the low shear strength of montmorillonite can lead to rock falls and, in some instances, rock slides. Changes in the water table can also affect rocks containing soluble minerals, such as rock salt, gypsum, limestone, and dolomite, which are especially susceptible to dissolution and physical alteration. Many slope failures have been attributed to the low strength of moist graphite, talc, chlorite, and other layer-lattice minerals occurring in fault zones.

A fluctuating water table can also contribute markedly to the alteration and periodic changes in the mechanical properties of rocks.

In some rocks, changes in moisture content lead to slaking, a crumbling or disintegration of the rock. Usually shales with higher percentages of clay-size material and mudstones soften on exposure to the atmosphere and revert to a muddy condition when submerged in water.

Climatic Conditions

The effects of climate on the stability of rock slopes in transportation corridors and the various remedial measures that must be taken to accommodate these conditions are important in rock-slope engineering. Daily temperature variations, precipitation, snow, and freeze-thaw conditions, acting either independently or in combination, often cause significant stability problems.

Terzaghi (9.108) noted that the groundwater conditions and hence the effective hydrostatic pressures can vary within wide limits, depending on the climatic conditions and geologic environment. Variations in the position of the groundwater table as a result of seasonal rainfall, sudden heavy storms, and ice on the face of the slope are shown in Figure 9.3 (9.108). Because of the yield of the excavation face, a zone of higher permeability will be close to the face, but the extent of this zone is still largely undefined. If the rock at the toe of the slope is already loaded nearly to failure, the additional water pressure may be sufficient to cause the slope to fail.

From correlations made from several hundred slope failures in Norway, Bjerrum and Jørstad (9.9) show the significance of periods of high rainfall infiltration. They note

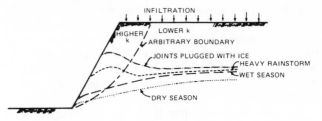

Figure 9.3. Hypothetical possible positions of groundwater table in jointed rock slope (9.108).

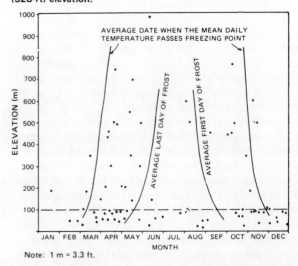

Figure 9.4 Rock falls in eastern Norway in relation to altitude, time of year, and temperature (9.9). Dots indicate rock falls occurring during different seasons below and above a 100-m (328-ft) elevation.

Note: 1 m = 3.3 ft.

Figure 9.5. Correlation of number of rock falls with temperature and precipitation on railway line in Fraser Canyon, British Columbia (9.84).

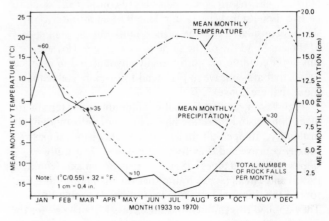

(Figure 9.4) that failures are most prevalent when the water table is high in the spring because of snowmelt and in the fall during heavy rainfall. Some of these failures are also attributed to ice forming on the slope face and causing water pressure to build up in joints.

Figure 9.5 shows comparative analyses by Peckover (9.84) of rock-fall occurrence, rainfall, and temperature in the Fraser Canyon in British Columbia. Maximum rock falls

195

occur in the spring and fall when the mean temperature is about $0°C$ ($32°F$) and frequent freeze-thaw cycles are occurring. When temperatures are above freezing, the frequency of rock falls is a function of degree of rainfall. The number of rock falls originating on the steep rock faces in the Fraser Canyon is probably high because of the absence of snow, vegetation, and soil, the presence of which would insulate the rock from unusually low temperatures and from changes in temperature.

Frost action probably directly or indirectly accounts for more rock falls than all other factors combined. Water undergoes about a 9 percent volume increase when it freezes and exerts tremendous pressure when it freezes in a confined space. According to Reiche (9.97), "Water-filled cracks or joints which terminate downward and which are narrow and perhaps irregular may be converted into essentially closed systems by preliminary freezing of the water in the superficial parts. In such cases the combination of expansion and low compressibility may exert a disruptive force which, if the temperature continued to fall and rock pressure permitted, would approach 30,000 pounds to the square inch at $-22°C$ (i.e., $-7.6°F$)." Freezing temperatures of this order are not uncommon in temperate zones.

Piteau and others (9.93) correlated mean annual movements of an overturning failure with average monthly rainfall, snowfall, and temperature. Although they found that a direct relation existed between movement and rainfall (i.e., groundwater pressure buildup), they found that snowfall and freezing retarded movement. The average monthly rainfall peaked in January, but movement decreased in December, when freezing and snowfall conditions started. The snow blanket prevented rainfall from infiltrating into the slope, and freezing temperatures reduced the availability of free water on the slope.

Slope Geometry in Plan and Section

Most current theories of slope stability consider the slope to be two dimensional (i.e., a unit length of an infinitely long slope is considered to be in plane strain) and the plan radii of the crest and toe of the slope to be infinite. However, this latter condition is not normally encountered in practice; slopes that are concave in plan tend to be more stable than those that are convex.

Highway or railway cut slopes in mountainous terrain are often convex and therefore have a greater tendency toward instability. In open-pit mines, on the other hand, slopes are usually concave and therefore more stable. In a study of slopes of diamond pipes in South Africa, Piteau and Jennings (9.92) and Piteau (9.89) found that the plan radius of curvature of the slopes has a marked effect on slope stability. They applied this finding to the DeBeers Mine to predict the final breakback position, subject to its slope geometry. Lorente de No (9.65) and Förster (9.27) made similar findings.

Horizontal tangential stress concentrations in such a slope can be either compressive or tensile, depending on the slope geometry (9.64). Horizontal stresses tangent to the slope are beneficial in a concave slope, for they create an archlike effect whereby the blocks forming the partitioned rock mass tend to be squeezed together. Compressive stresses substantially improve the shearing strength. All three principal stresses are compressive, and the maximum and minimum principal stresses act in the vertical and radial directions respectively. For the convex slope, the converse is the case, and horizontal tangential stresses are tensile. The maximum principal stress is still vertical, but the minimum principal stress acts in a horizontal direction tangent to the slope, and the slope material is in tension. Cohesion that would normally have occurred is reduced. Since a rock mass is relatively weak in tension, tensile stress concentrations in the slope induce instability, causing unrestrained blocks to slide out.

For an infinite slope in rock (normally specified in rock mechanics when rock strength includes cohesion), a slope can be steeper than that indicated by the angle of friction. Normal stress increases downslope, and the effect of cohesion relative to that of friction decreases. The slope thus flattens to the angle of internal friction. The profile of such a slope, therefore, is theoretically concave from top to bottom. The variation of slope angle with slope height has a small but significant influence on the stress distribution in the slope according to Yu and Coates (9.111) and, therefore, can be expected to have some influence on the stability.

Time Factor and Progressive Failure

Natural rock slopes undergo progressive failure in time by the processes of creep and flow. Hence, it is of considerable importance that one recognize whether the analysis and ultimate design of the slope meet the requirements of short-term or long-term stability. Also, because of progressive failure, one must consider potential maintenance problems and design accordingly. For practical purposes, allowances must be made for some reduction in the strength properties of the rock mass with time (9.76).

Murrell and Misra (9.78) note that time-dependent strains occur when rock material is subjected to relatively high stresses for long periods. Most of the forces involved in such deformations are indeterminate functions of time, being dependent on the effects of the excavation, regional stresses, alteration processes in the mass, physical and chemical action of groundwater, and seasonal variations of temperature and rainfall. These lead to fatigue and opening of cracks with irreversible deformations and progressive weakening of the mass. Therefore, continued movement of a rock slope is cause for concern. Movements are important, for relative displacements along defects in the rock mass tend to reduce the resistance along these defects and may bring about failure.

A slope may be stable when first excavated but, because of gradual deterioration and adjustments toward equilibrium, may become unstable with the passage of time. The time required for deep-seated failure in hard rocks is almost impossible to evaluate. Near-surface failure, such as raveling or detaching of rock segments, however, may develop only a few years after the excavation has been completed. This kind of failure should be considered in the slope design and remedial measures proposed. Soft rocks, such as shale, mudstone, and other types of argillaceous materials, can undergo magnitudes of deformation that may lead to failure within significantly shorter periods, sometimes within several days.

Residual and Induced Stress

The cut slope created by an excavation affects the stresses in a rock mass at the boundary of the excavation. However, predictions of the magnitude of these stress concentrations and their effects on the stability of the slope are complex. To date, results of studies of stress distribution in slopes and the manner in which stresses affect the stability are largely hypothetical. No mathematical or physical models are available to predict the effects of varying the slope excavation geometry and the ultimate variation in stress concentration (9.73). In this regard, probably the most significant advances have been made in using finite element procedures to consider the stress-strain compatibility of the slope.

Contrary to earlier views, bedrock is not a predictable platform in which the only acting forces are vertical due to the weight of the rock. Rocks may also be subject to significant horizontal residual stresses that under certain circumstances could have important influences on the behavior of rocks at excavations. Hast (9.32) notes that this influence has been proved to be quite substantial in deep excavations. Regional stresses, denudation, tectonic uplift, glacioisostatic rebound, and other conditions might also affect the stability of surface excavations in rock, even though less dramatically.

Existing Natural and Excavated Slopes

Slope design should take into account past experience with both stable and unstable slopes. Kley and Lutton (9.51), Lutton (9.66), and Shuk (9.104) show that analyses of both natural and excavated slopes will provide valuable background information for proposed excavation design, particularly those for mountainous terrain.

Usually the angles of both natural and excavated slopes provide conservative estimates of slope angles that can be achieved in slope design. Humans can invariably improve on the slope angles provided in nature by giving careful attention to drainage, artificial stabilization methods, and control of natural slope-forming processes. Whether one is evaluating profiles from surface excavations or natural slopes, the problem is in estimating the degree of the conservatism inherent in the population of slope profiles being considered. This problem is properly resolved by the engineer and the geologist (primarily geomorphologist) complementing each other.

The safety factor of natural slopes is commonly not much greater than unity (9.76). The value of unity, however, applies specifically to that situation in which the most adverse groundwater conditions develop naturally. A hillside will normally have a safety factor greater than one most of the time, but it may become one when the water conditions or the natural disturbing forces, such as recurring earthquakes, are as severe as are likely to occur.

Slopes should not be compared if the general modes of their formation (i.e., types of excavation and slope-forming processes) differ. When slope processes are similar, stable slope case histories can be relied on to predict a lower bound to the design slope angle. The use of slope case histories requires that factors such as slope and failure geometry, geology, and material properties be obtained. Slope monitoring can also be helpful when case histories are used in slope design.

In highway and railway slope problems, the most important factor relating to case history analyses is probably the incidence of failure. This is shown by Piteau (9.90) in a regional slope stability study of the Fraser Canyon. Comparative analyses of incidence of slope failures (rock falls, landslides, debris slides) and of geological factors (geomorphology, structure, lithology, groundwater, river hydraulics) were made to assess those factors controlling slope stability. More than two-thirds of all incidents occurred where the river had been deflected into the bank and had undermined the slope. Severe lateral erosion by the river was the result of either general directional changes in the river or the development of alluvial fans that forced the river into the opposite bank.

Until failure mechanisms of rock slopes are better understood, consistently reliable predictions of rock-slope behavior are not possible. Because natural slopes may provide some clues, as much attention as possible should be given to examining the way in which local slides and deformations develop. Analysis of rock slides in an area can provide excellent information on the mechanics and patterns of their formation.

Coates and Gyenge (9.17) make use of information from the performance and characteristics of existing or previously existing slopes and have developed the principle of incremental design for slopes. This is the "process of extrapolating from the known to the new, or predicting the conditions that result from a change in the present operations." Although this work is basically applied to open pits and the incremental predictions may not be of high accuracy, some of the basic concepts developed can be applied to rock-slope design for transportation routes.

Dynamic Forces

The significance of earthquake vibrations is well documented elsewhere; effects of blasting are considered later in this chapter. The brief discussion here is of the effects of vibrations along transportation corridors due to vehicles.

There is some belief that traffic vibrations, particularly from trains, may be a significant factor with respect to slope stability. However, a comparison of relative masses shows that traffic-induced vibrations are extremely small, even for slopes close to the right-of-way. Therefore, such vibrations can be considered insignificant. According to Peckover (9.84), the incidence of slope failures along railways when the train is passing, for example, is no greater than would be expected from the proportion of total time that the railway is occupied.

FUNDAMENTAL PROCEDURES IN ANALYSIS OF ROCK SLOPES

This section describes the basic approach to analysis of rock slopes as well as the theoretical and analytical process. The process can be summarized as follows: Discontinuities in the rock mass are systematically measured and statistically analyzed to determine their nature and distribution. Estimates are made of the strength properties of the discontinuities. These factors are quantitatively described and theoretically applied to determine the strength along any potential failure plane. Shear strength parameters are assessed, thus allowing the factor of safety of the slope to be calculated and slope design to be considered.

Determination of Structural and Other Relevant Geologic Characteristics

In the analysis of a high rock slope, structural discontinuities of the rock mass are mapped in detail and each feature is quantitatively characterized. The geologic survey is aimed at measuring a sufficient number of joints to allow the data to be analyzed statistically. The statistical analyses and judgment indicate whether the best estimate has been made for the whole population. The entire geologic survey must be conducted so that, from the joint characteristics recorded, the shear strength in the direction of the joints can be numerically assessed by comparison with similar features tested in the laboratory.

Physical access to all discontinuities in a rock mass is not possible. Therefore, maximum information must be extracted from all locations where access is possible. For other locations, information is obtained by a variety of means including exposure mapping, tunnels, trenches, drilling (core logging), terrestrial photogrammetry, aerial photograph interpretation, and various geophysical methods. In exposure mapping, either some variation of detail line mapping or fracture set mapping is generally best to use.

Line mapping methods are discussed by Jennings (*9.46*), Piteau (*9.87*), and Halstead, Call, and Rippere (*9.30*), and fracture set mapping by Call (*9.14*), Mahtab, Bolstad, an l Kendorski (*9.67*), Da Silveira and others (*9.20*), and Herget (*9.36*).

Sources of errors in joint surveys are discussed by Terzaghi (*9.109*), Robertson (*9.99*), and Piteau (*9.88*).

General collection and processing of geologic data are discussed by Knill (*9.52*), the Canada Department of Energy, Mines and Resources (*9.94*), and the International Society of Rock Mechanics (*9.44*).

Geologic conditions vary from project to project, and thus a geologic survey at one site may be entirely different from that at another. Features that should be considered in the survey are coordinates; elevation; rock type and hardness; type of geologic structure; strike (direction of structure surface) and dip (angle of structure surface); dip and strike continuity; thickness, type, and hardness of infilling materials and the proportion of voids and presence of water in the materials; roughness; and waviness (wave shape or interlimb angle). Coordinates, elevation, strike, and dip are used to define position and orientation of the discontinuity in space. These features, plus dip and strike continuity and bearing of the sample line, serve to define intensity. Rock type and hardness, characteristics of infilling material, presence of voids and water, roughness, and waviness are used for assessing frictional and cohesive strength and deformation properties.

The survey data are processed, and the attitude, geometry, and spatial distribution of the jointing are determined by computer analysis to yield structural domains, joint sets and their average properties, and major discontinuities.

After the regional and local geology is assessed, a geologic map is constructed to show the major and minor structural features and the general lithologic distribution. Stereographic projections are easy to use and are of great benefit in evaluations of geometric relations and in showing structural populations. These projections can also be shown on the general geologic map of the area.

Stereographic projection principles and the use of equal-area, equal-angle, and polar projections are documented by Donn and Shimer (*9.22*), Phillips (*9.86*), Terzaghi (*9.109*), and John (*9.48*).

Determination of Structural Domains and Design Sectors

Because both the geologic conditions and the bearing of the proposed cut face can vary from one location to another along a cut, different slope designs and remedial measures may be required at different locations. Therefore, one should analyze individually those parts of the proposed cut slope that are similar in terms of both orientation and physical and mechanical characteristics. Areas of similar geologic characteristics are designated structural domains. Samples of the geologic structural properties within a structural domain will not differ significantly from one part to another. Therefore, the slope stability characteristics and design parameters of the entire structural domain can be determined from sampling only part of it.

Boundaries of straight slope segments that have similar orientations are determined and superimposed on structural domain boundaries to form design sectors. Characteristics of the various design sectors are selected for stability analysis and slope design. Within each design sector, typical joints or design joints, which represent the mean characteristics of the relevant joint sets, are selected for use in the design calculations (*9.94*). Discussions relating to definition of properties of design joints are given by Steffen and Jennings (*9.106*) and Piteau (*9.88*).

Boundaries of structural domains usually coincide with major geologic features, such as faults, shear zones, dikes, sills, geologic contacts, and unconformities (*9.88*). The analyst's ability and experience in assessing the structural geology and the variations in structural characteristics will determine the accuracy and usefulness of the structural domain determinations. Comparative analyses of joint populations within structural domains or between structural domains should be subsequently performed.

Development of a Rock Mass Model Depicting Geologic Structure

After the characteristics of the geologic structural population in each structural domain are defined, a model of the rock mass is developed to depict the three-dimensional relations of the geologic structure. Some workers refer to the rock mass model as a schematic concept or structural picture of the rock mass. Since each structural domain is similar in a statistical sense, a rock mass model is developed for each structural domain. The model can be of a graphical, physical, or mathematical nature or a combination of these.

An essential requirement of the model is that it accurately represent the actual geologic structural population in a statistical sense and that it apply to the entire design sector. A graphical model using the stereographic projection (*9.48*) is usually used, and extensions are often made to mathematical or physical models to determine whether the first boundaries selected are adequate or should be changed. A typical rock mass model using a stereographic net depicting the angular relations between faults that form a potential wedge failure and a proposed cut slope is shown in Figure 9.6. A spatial diagram of these relations is shown in Figure 9.7.

Once the boundaries of the various structural domains are defined, attention is given to delineating joint sets within each domain and determining the characteristics of each joint

Figure 9.6. Typical graphical model or schematic concept of mass on Wulff stereographic net. Angular relations between faults, which form potential wedge failure, and proposed cut slope are shown.

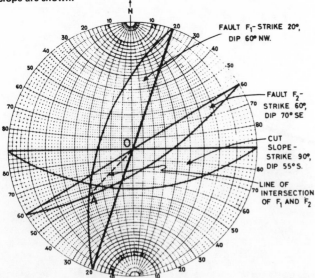

FAULT F₁– STRIKE 20°, DIP 60° NW.

FAULT F₂– STRIKE 60°, DIP 70° SE

CUT SLOPE– STRIKE 90°, DIP 55° S.

LINE OF INTERSECTION OF F₁ AND F₂

Figure 9.7. Three-dimensional diagram of spatial relations of salient features of rock model shown in Figure 9.6.

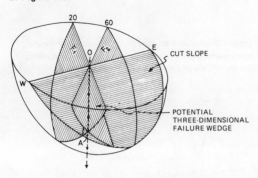

CUT SLOPE

POTENTIAL THREE-DIMENSIONAL FAILURE WEDGE

set. The discontinuities are analyzed by using some form of graphical projection or plot (e.g., stereographic projection or rectangular plot). The structural data and joint sets are defined on the basis of the specified criteria, and the properties of the joint sets lead to the design joint. The percentage of joints in any particular joint set having any one property is determined by relating the number of joints in the set with that property to the total number of joints in the set. Various statistical techniques can be used for this operation (9.94, 9.99).

Determination of Kinematically Possible Failure Modes and Performance of Slope-Stability Analysis

The assumption is made that the surface of failure in the slope consists of a plane or combination of planes, as discussed earlier. The model is investigated to determine what failure modes on planes or combinations of planes (i.e., design joints) are kinematically possible. Four of the basic

failure modes usually investigated are plane, wedge, stepped, and circular. (Discussions of different failure modes are presented in Chapters 2 and 7.) A tension crack for each case can be assumed to exist at the surface. Methods of slope-stability analysis for various failure modes are discussed by Jennings (9.47), Hamel (9.31), Heuze and Goodman (9.37), Hoek and Bray (9.42), Hendron (9.34), and Goodman (9.29).

The analyses consist of the following operations:

1. Estimations of continuity of jointing on potential failure planes;

2. Assessment of the strength of the intact rock;

3. Determination of the effects of the joint characteristics on the strength along joints;

4. Development of the necessary equations of limit equilibrium for the possible modes of failure;

5. Use of various potential failure planes singly or in combination to test for these failure modes; and

6. Determination of the factor of safety of the slope.

An analytical treatment of the relevant data in the stability analysis is described in Chapter 7 and the brief discussion that follows.

Synthesis of Basic Data

The method of analysis is considerably more detailed and vigorous for high rock cuts than for shallow rock cuts. The following procedures may be used.

1. Coefficients of continuity are determined from the lengths of the joints in relation to the probable length of the potential failure surface in the slope. Of all the assessed factors, these figures for continuity are probably the most difficult to determine and accordingly are subject to some doubt.

2. The effects of waviness are assessed, and an angle of waviness is defined from measurements of the wave shape of the joint; this shape is dependent on the length and amplitude of the wave on the joint plane. The angle of waviness is used to modify the apparent dip of the joint with respect to the direction of the slope.

3. The hardness of the intact rock (assessed in the joint survey) is used with an empirical curve to determine a conservative value of the compressive strength of the rock. Cohesive and tensile strength values of the intact rock are estimated on the basis of the compressive strength, and the friction angle of the intact rock is estimated from the rock type.

4. A factor of safety is then determined based on knowledge of the strength parameters applying to failure along the potential surface of failure and the assumption that the Mohr-Coulomb relations apply to shear failure through the intact rock and along joint surfaces. The analysis for failure is based on the apparent dip of the potential failure surface with respect to the strike of the slope.

5. Water pressures are used in the analysis in the same way that they are normally used in soils, but with the further assumption that apparent cohesion and apparent friction parameters are based on effective stresses (9.47).

6. Stability calculations are made to test the different failure modes. For each particular case, the joint sets are ex-

Figure 9.8. Graphical representation of average orientation of main geologic structural features along highway location in Slocan Lake Bluffs area, British Columbia.

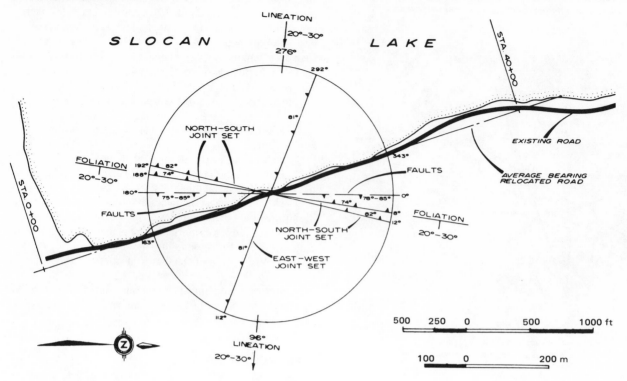

amined in their various combinations until the worst situation is found (i.e., the case that gives the minimum factor of safety for the particular slope angle examined). The procedure is repeated for different slope angles, and the respective factor of safety is determined for each case.

Most rock-slope analysis procedures are not so complicated as these and generally require only an evaluation of the orientation of geologic structure with respect to the bearing and alternative slope angles of the proposed excavation. In such cases the rock mass model can be depicted most suitably by means of stereographic projection techniques shown in Figures 9.6 and 9.7.

Case History of a Typical Rock-Slope Stability Problem

A rock-slope design was required for a proposed highway cut in basically hard gneissic rock along the edge of Slocan Lake, British Columbia. Two dominant steeply dipping joint sets occur in the area: One set is designated the east-west joint set and the other the north-south joint set. Both of these joint sets are shown in Figure 9.1. The orientation of these joint sets with respect to the proposed highway location in plan is shown in Figure 9.8.

The rock mass model in stereographic projection for design is shown in Figure 9.9, including great circles of the average orientation of the east-west joint set and north-south joint set and two alternative cut slopes. One of these slopes is 70°, which is the slope that is to be recommended, and the other slope is 76° (i.e., 1/4:1), which is the slope angle tentatively proposed for preliminary design before the study was initiated. Figure 9.9 shows that the two joint

Figure 9.9. Graphical model in stereographic projection showing angular relations between joint sets, which form potential wedge failure, and proposed cut slope location shown in Figure 9.8.

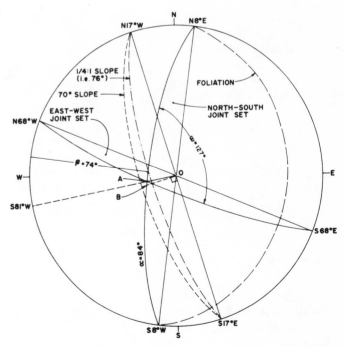

sets form potential wedge failures that have a plane of intersection trending N81°E and plunging 73° (toward the highway). Since this wedge is flatter than the 76° slope, a 76° slope would clearly be unstable.

200

A mechanical stability analysis in this case is somewhat redundant because of the steepness of the potential wedges. For illustrative purposes, however, a simple three-dimensional analysis (9.13) can be carried out to show the minimum angle of residual joint friction (ϕ_r), which would be required to hold up such a wedge if undercut. Consider the following values determined from the rock mass model shown in Figure 9.9:

$\beta = 74°$, average dip of the north-south joint set,

$\alpha = 84°$, angle measured in the plane of the north-south joint set between the strike of this plane and the line of intersection, and

$\omega = 127°$, angle measured in a plane perpendicular to the line of intersection of the two joint sets.

ϕ_r would have to be a minimum of $57.2°$ in order to achieve a factor of safety of unity. Since the value of ϕ_r was estimated to be on the order of $35°$, a $76°$ slope would obviously lead to serious wedge failures. Cable anchors, bolts, or other artificial reinforcement, which would be required in order to prevent failure of a $76°$ slope, would also be uneconomical. Hence, it was recommended that the slopes be cut at an angle of $70°$, which is $3°$ flatter than the line of intersection 0A (Figure 9.9) of the potential wedge failures.

SLOPE DESIGN AND REMEDIAL MEASURES

Remedial measures for rock slopes can consist of stabilization, protection, or warning methods or a combination of these basic methods. These remedial measures and the suggested order in which they should be considered are shown in Figure 9.10 and discussed below.

1. Stabilization methods give a positive solution to the problem in that either the driving forces are reduced or the resisting forces are increased. Because of the complex nature of the rock mass, the effectiveness of these methods is often difficult to assess quantitatively. Stabilization measures reduce the likelihood of rocks moving out of place and generally should be considered first in the remedial treatment of rock slopes. Stabilization also reduces the rate of deterioration, a process that leads ultimately to failure.

2. Protection methods prevent rock materials that have moved out of place on the slope from reaching the roadway and thus offer an additional positive solution to slope-stability problems. The initial cost is usually considerably less than that of the stabilization measures, but the slopes usually require considerably more maintenance (in some cases almost continual maintenance). Protection measures can sometimes be combined effectively with warning measures.

3. Warning methods warn that movements have taken place or that failure has occurred and that a hazard may be imminent in the vicinity of the roadway. These methods have no effect on the source of the hazard and may lead to increased maintenance costs and unnecessary delays when insignificant events accidentally trigger the warning system. Warning methods also tend to provide undue confidence, and sometimes result in relaxation of advisable precautions. Although they may seem economical in some instances, warn-

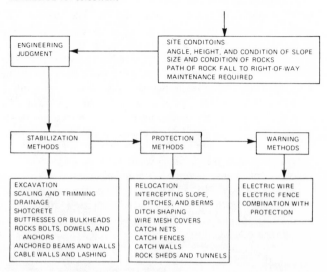

Figure 9.10. Order in which slope treatment methods should be considered for selection.

ing measures are not an effective way to remedy rock-slope problems and generally should be considered only as a last resort. Except for railways in North America, warning methods are seldom used extensively on a permanent, long-term basis by those experienced in rock-slope engineering.

During the planning and construction phases of a project, the considerable advantages in recognizing rock-slope problems that exist or that may develop and in applying the appropriate rock-slope remedial measures at the time cannot be overemphasized. Sound rock-slope engineering at the outset of the project and during the excavation phase will obviate both hazards and much more costly work at a later date. The cost of stabilization and protection measures applied during the excavation phase also can be largely absorbed in the initial construction cost; this is especially true if a contingency item is included in the contract budget.

The following discussions do not include costs of remedial measures, other than in a general way, because costs in one environment or location may be entirely different from those in another. Time, access, and space available for remedial work on highway rock slopes, for example, are often much greater than for similar work on narrow, busy railway lines in mountainous terrain. For this reason, both methods and costs of highway rock-slope engineering work are not always applicable to similar railway problems. Unit costs and other items can also vary widely, depending on local requirements and the skill of the planner.

Remedial measures for rock slopes are discussed by Záruba and Mencl (9.112), Baker and Marshall (9.3), Fookes and Sweeney (9.26), Root (9.100), Mehra and Natarajan (9.72), and Behr and Klengel (9.7). Landslide measures in urban developments are discussed by Legget (9.61) and Leighton (9.62). Crimmins, Samuels, and Monahan (9.19) provide practical information with regard to construction in rock although the reference is mainly to foundations.

Planning and Related Procedures

The stability of a rock slope will change significantly with time because of strength changes, rock deterioration, fluctuating groundwater, and other environmental factors, which generally produce less stable slopes. The most economical overall design, therefore, accounts for this decrease in stability with time and includes consideration of a maintenance and remedial program. Remedial measures for rock slopes along transportation corridors in steep terrain must be integrated with other maintenance tasks. If the route runs for some distance through mountainous terrain, work will probably be sufficient to keep personnel on rock-slope maintenance work full time. Unlike most soil slopes, on which weathering is not an important factor, rock slopes require continual attention if serious problems are to be avoided. A consistent long-term remedial program is required that contains carefully selected priorities. Experienced people are needed for both engineering and construction. Judgment must be based on a detailed knowledge of occurrences at each dangerous location.

Personnel

In setting up a rock-slope remedial and maintenance program for a major transportation route, a project engineer should be assigned full time to the work. The project engineer should have available an experienced engineering geologist and geotechnical engineer and the capability to supervise contract work. If necessary, consultants should be used for highly specialized parts of the work and to provide some personnel on a temporary basis, depending on the magnitude and frequency of the problems. Continuity of personnel should be encouraged, for experience in rock-slope engineering is a most important attribute. An engineering geologist on the team should be able to translate information clearly into engineering terms and integrate information with engineering requirements.

The project engineer must keep in touch with all developments relating to the overall project. The geotechnical engineer and engineering geologist should make detailed inspections, perform slope-stability analysis and design, exercise judgment on specific rock conditions, and help in planning and supervising remedial work. It is particularly important that the staff of the highway, railway, or other transportation agency concerned with rock-slope problems develop sufficient expertise within the organization to critically evaluate work performed by private companies or other agencies. Some of the principal functions of the project engineer and engineering geologist in planning, design, and maintenance are described by McCauley (9.70).

Work Arrangements

To estimate the cost of remedial measures in advance is difficult, for unforeseen conditions are often revealed as the work proceeds. For this reason, decisions on an overall budget required for the improvement of rock-fall hazards are often based on judgment as well as analysis. All concerned should understand that some flexibility is needed in the total budget approved for an annual program so that adjustments can be made in expenditures required at different locations as well as in contract provisions to deal with changed conditions during the work. Because major revisions often are required during the work, rock work should be done on

the basis of proposals or bids invited from contractors known to have up-to-date knowledge of the equipment and techniques required. The proposal or bid should be for a unit price or cost-plus-fixed-fee contract and not for a lump sum. The U.S. National Committee on Tunneling Technology (9.79), although addressing mainly underground work, is a good source of information on sound contracting practices in construction engineering and contract law for rock construction work.

Selection of Priorities

The condition of all rock slopes along transportation routes in steep terrain should be thoroughly inspected once a year or more frequently depending on the severity and implications involved. Inspection can be done by (a) a helicopter survey of inaccessible locations and of the overall conditions of the slopes and (b) a ground survey of hazards at specific locations to determine the need for detailed studies. Based on annual surveys and case histories of the slopes, detailed studies and alternative treatments are planned. Factors to be considered in deciding on locations for remedial work are

1. Maintenance costs, including patrols required;
2. Costs of remedial measures and expected benefits to be gained;
3. Degree of risk to route, determined by considering the amount of traffic, records of past events, measured rock movements, and frequency of clearing rocks from the vicinity;
4. Occurrence of accidents, rock falls, washouts, landslides, and the like;
5. Conditions downhill from the right-of-way, which would indicate how serious an accident or derailment might be; and
6. Views of maintenance and patrol personnel.

The annual program should emphasize stabilizing or protecting the greatest number of locations where rock-fall hazards exist in order to obtain the maximum economical improvement in safety. Large problem areas can often be divided into smaller areas for treatment. The tendency to overprotect the route at locations of recent accidents should be resisted in the interest of obtaining a balanced overall program. Conditions at each particular location may vary significantly, and the application of different remedial measures should be considered accordingly. Details of remedial work must be developed section by section along the route and be based on detailed inspections, existing records, experience, and sound judgment. However, the most careful choosing of hazardous locations and the most thorough analysis in selecting remedial measures will never entirely prevent unexpected rock falls from occurring.

Records

For the efficient planning of rock-slope engineering work, a data storage and retrieval system should be organized and maintained for permanent reference. The following types of records should be maintained:

Figure 9.11. Rock slope before (top) and after (bottom) removal of overhang.

Item	Description
Photographs	Aerial photographs (standard, close-range vertical and oblique stereopairs) and ground photographs
Weather	Daily records of precipitation and temperature from local weather stations and from recorders at selected locations
Traffic delays or highway closure	Time, number of hours, and cause
Rock falls or slides on the right-of-way	Time, location, size of average and largest rocks, volume of material, height and length, source of material, and distance up slope
Removal of fallen material from roadway, ditches, behind walls	Date, location, size of average and largest rocks, volume removed, whether routine or emergency
Stabilization and protection measures	As-constructed plans and continuing inspection records including bolt tensions
Repairs to warning installations	Time, location, maintenance required, effect on installation, size and final location of rocks
Rock movements	From distance-reading and direct-reading instruments
Costs	All remedial and maintenance work done

In addition, a continuing record should be kept of the maintenance performance, the installations, and the measurements conducted at each location requiring attention. For accurate identification, all locations should be recorded to within about 8 m (25 ft).

Photographs

Combined with maintenance records, photographs are essential tools in evaluations of slope conditions and decisions on priorities and types of remedial work required. Photographs should be taken regularly to record the details of rock conditions on slopes both before and after remedial work has been carried out (Figure 9.11). Telephoto lenses can be used for those locations where access is difficult.

Figure 9.12. Oblique photographs of cut bench at Hell's Gate Bluffs, British Columbia, above which excavation was proposed to remedy overturning failure.

Aerial photographs can be useful for studying the condition and configuration of slopes in steep terrain and for giving important clues to the causes and potential sources of slope instability. High-level photographs are generally of minimal use for local slope-design purposes. If possible, low-level photographs at a scale of 1 cm equals 50 to 120 m (1 in equals 400 to 1000 ft) should be obtained. Uses of aerial photographs for engineering purposes are described in Chapter 3 and by Norman (*9.81*) and Mollard (*9.74*).

Aerial oblique or terrestrial oblique single photographs or stereoscopic photographs are frequently used for rock-slope engineering work (Figure 9.11). Stereoscopic photo-

graphs can be used to examine slopes three-dimensionally for purposes of tentative slope-excavation design. Aerial oblique photographs can be taken with cameras mounted on the ends of a 5-m (15-ft) boom carried by a helicopter. Photographs can be taken at any angle from vertical to horizontal and contours can be plotted to a fixed datum. Cracks and features on the order of 5 to 8 cm (2 to 3 in) wide can be detected in this way. Photographs are particularly useful in evaluations of rock slopes that are in steep terrain and not easily accessible from the ground (Figure 9.12). Use of terrestrial photogrammetry in rock-slope engineering is discussed by Ross-Brown and Atkinson (*9.101*).

Monitoring and Inspection

Planning for rock-slope engineering work should include a program of monitoring slope movements and identifying characteristics that indicate changing stability. Detection of general creep or slow translation of highly fractured or soft slope-forming material is important in decisions regarding remedial work. Horizontal and vertical movements of points plotted against time or depth or both provide important information concerning the behavior of a rock slope. Graphs provide a clear indication of the onset of slope failure when plots representing the change in position do not remain linear. When such accelerated movements are evident, the slope must be approaching failure, and measures should be taken to analyze and remedy the situation.

For studies of complex stability problems, the advice of a specialist is needed to determine the most suitable slope-monitoring program. A wide variety of commercial instruments is available for measuring both surface and subsurface movements and for recognizing the characteristics indicative of potential instability. However, the simplest methods and mechanical equipment are not only the most practical but also invariably the most reliable. The various methods of instrumentation are discussed in Chapter 5, in the June 1972 issue of Highway Focus (*9.39*), and by Franklin and Denton (*9.28*), Benson (*9.8*), and Hedley (*9.33*).

The simplest and generally most efficient method of monitoring recently excavated slopes is to measure both horizontal and vertical displacements of protected metal survey pins or hubs set in concrete or grout along a straight line at the crest and toe of the slope. Main control reference points should be located outside the area where movement may occur. If, in the first year, movements appear negligible (other than what might be expected from normal elastic rebound due to unloading), the number of surveys in succeeding years may be decreased until the slope becomes stable. In this work, high precision is essential so that the onset of nonlinear movement can be determined as early as possible and remedial work can be planned accordingly.

Individual cracks, which may be indicative of general instability in an area, can be monitored simply by bridging the crack with a wad of grout or shotcrete or by inserting a wedge in the crack. A regular part of the work program for evaluating and monitoring should involve inspection and maintenance of remedial installations, such as rock bolts, walls, nets, and drains, already in place. Regular inspection is an essential precaution in protecting a substantial capital investment.

Methods of Stabilization

Excavation and Related Design Aspects

For purposes of improving the stability of rock slopes, excavation is used either to reduce the driving forces contributing to failure or to remove unstable or potentially unstable sections of the slope that may lead to failure (i.e., rock falls and slides). Stability, therefore, can be achieved by

1. Removing unstable or potentially unstable material,
2. Flattening the slope,
3. Removing weight from the upper part of the slope,

Figure 9.13. Forty-three-meter (140-ft) presheared through cut involving 130 000 m³ (170 000 yd³) near Britannia, British Columbia, and including cut bench to protect highway from rock falls. Photo is toward south, but sketch is toward north.

4. Incorporating benches in the slope, and
5. Excavating in a manner that minimizes damage to the rock mass.

Excavations made for remedial purposes should provide a permanent solution to the slope-stability problem so that additional excavation work in the future need not be necessary. The principal problems and disadvantages associated with excavation methods lie in their cost. Accessibility after construction may be difficult and, since the slope usually must be excavated from the top downward, mobilization and setup costs can be prohibitive. Also, disposal sites often are limited with the result that waste rock may have to be transported for some distance, unless it can be used for local construction.

Benches

The overall slope angle and different slope angles that may be designed for different parts of the slope are determined by slope-stability analysis methods (Chapter 7). Each cut should be designed to suit the prevailing rock conditions. Slopes containing materials with strength properties that vary significantly in section may require that some parts be flatter than other parts if rock falls and maintenance costs are to be minimized. The use of variable slopes, benches below layers of rapidly weathering rock, and structural support and correct blasting techniques can practically eliminate rock falls in new cuts.

Benches are used in slopes to minimize rock falls onto the roadway (Figures 9.13 and 9.14). For soft rocks (Figure 9.14), such as shale, mudstone, and other argillaceous rocks, benches tend to reduce excessive weathering and erosion and provide rock-fall catchments. Erosion due to groundwater runoff is also controlled since the energy of surface flows is

Figure 9.14. Benches that follow stratigraphy in Niagara Escarpment sedimentary rocks on Highway 403 near Hamilton, Ontario. Design also includes overburden being stripped back 3 m (10 ft) from top of rock and light fence barrier on edge of highway.

Figure 9.14. Benches that follow stratigraphy in Niagara Escarpment sedimentary rocks on Highway 403 near Hamilton, Ontario. Design also includes overburden being stripped back 3 m (10 ft) from top of rock and light fence barrier on edge of highway.

Figure 9.16. Use of multiple benches and controlled blasting techniques at Mica Dam, British Columbia, to maintain rock slope in soundest possible condition and to prevent rock falls.

Figure 9.15. Slope design parameters that are used when benches are incorporated in slopes.

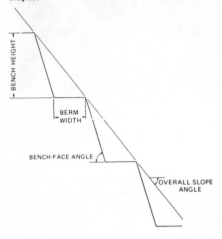

dissipated. Construction safety on benched slopes is usually increased since the hazard of rock falls above workers is reduced. In general, the faces of benches can be considerably steeper than the overall slope angle; hence, any rocks that do fall remain on the benches.

Benches appear to have no effect on the slope with respect to deep-seated failure. Although shear stresses increase with slope height, the direction of the maximum shear stress is supposedly independent of slope height and bench geometry. Benches may require the top of the slope to be moved back and, therefore, considerable additional excavation. However, although they may add to the initial cost of construction, benches may significantly reduce subsequent maintenance costs and thus sometimes more than offset the increased cost of construction. The Colorado Department of Highways, for example, found that the unit cost of cleanup and maintenance work after rock falls is about ten times the unit cost of the original excavation.

Figure 9.15 shows the parameters of a slope incorporating benches, including bench height, berm width, and bench-face angle. These parameters are governed by the physical and mechanical characteristics of the rock mass. Bench height should provide a safe and efficient slope and an optimum overall slope angle. Bench height can be greater

in stronger rock, and the bench face can be terminated at the base of weaker horizons and water-bearing zones. Without affecting the overall slope angle, higher benches generally will allow for wider berms, giving better protection and more reliable and easier access for the regular cleaning of debris. The width of berms should be governed by the size of the equipment working on the bench and by the nature of the slope-forming material, but should generally be no less than 7 m (20 ft).

If the bench faces are inclined, high tensile stresses are less likely to develop near bench crests and thus tension cracks and overhangs are minimized. Avoiding these problems reduces the amount of rock-fall material and increases the safety of the slope. Tension zones in slopes involving benches are discussed by Bukovansky and Piercy (9.11). Design of the bench-face angle should be governed to a large extent by the attitude of unfavorable structures in the slope to prevent excessive rock falls onto the berms. Figure 9.16 shows a high rock cut at Mica Dam in British Columbia, where all three bench design parameters change in different parts of the slope.

If benches are included in the slope design, berms should be equipped with drainage ditches to intercept surface runoff and water from drain holes and other drainage facilities and divert it off the slope and away from problem areas. These ditches should be kept open and free of all debris and ice to ensure adequate performance. Ditch lining (e.g., clay, slush grout, asphalt, polyethylene sheeting) may be required if ditch leakages are anticipated or develop afterward. Berm surfaces should be graded to assist the collection of water in ditches and also to facilitate general drainage in a direction away from potential areas of instability. Care must be taken so that ditches do not create problems by channeling water from one area to another.

Stepped benches may be used on slopes cut in highly weathered rock material to control erosion and to establish vegetation. Figure 9.17 (9.38) shows stepped cut slopes that consist of 0.6 to 1.2-m (2 to 4-ft) high benches with approximately similar berm width and an overall slope angle based on stability analysis. The design objective is that the material weathering from each rise will fill up the step of the bench and finally create a practically uniform overall slope. The steps are constructed horizontally to avoid the longitudinal movement of water, which could cause considerable

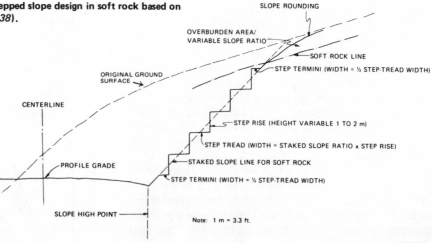

Figure 9.17. Idealized cross section showing stepped slope design in soft rock based on recommendations in several research papers (9.38).

SLOPE ROUNDING

OVERBURDEN AREA/ VARIABLE SLOPE RATIO

SOFT ROCK LINE

STEP TERMINI (WIDTH = ½ STEP-TREAD WIDTH)

ORIGINAL GROUND SURFACE

CENTERLINE

STEP RISE (HEIGHT VARIABLE 1 TO 2 m)

STEP TREAD (WIDTH = STAKED SLOPE RATIO x STEP RISE)

STAKED SLOPE LINE FOR SOFT ROCK

STEP TERMINI (WIDTH = ½ STEP-TREAD WIDTH)

PROFILE GRADE

SLOPE HIGH POINT

Note: 1 m = 3.3 ft.

erosion. Seeding and mulching or other suitable methods of slope stabilization can be readily applied; however, for rapidly raveling slopes, about half of the bench or step width should be filled before seeding is done to prevent smothering of the seed.

Scaling and Trimming

Scaling of loose, overhanging, or protruding blocks is a basic maintenance operation on rock slopes of all sizes along transportation routes in steep terrain. Scaling on the upper reaches of high faces is usually carried out by workers on ropes with hand pry bars, hydraulic splitters or jacks, and explosives (Figure 9.18). Of necessity, this work is slow and intermittent. Mechanical scaling equipment is more efficient and safer, but may have limitations because of severe access problems. Trimming involves drilling, blasting, and scaling to remove small ragged or protruding rock in overhang areas where repetitive scaling would otherwise be required.

Postconstruction scaling and trimming should be carried out on a regular basis. In temperate regions, this is usually started in the spring after the frost leaves the rock. If thorough scaling is performed during excavation, subsequent maintenance and remedial work can be greatly reduced. Before work begins, an engineering geologist and rock foreman should thoroughly inspect each location and make decisions on the rocks to be removed. Scaling and trimming work requires specialized experience, and the performances of different contractors should be compared when additional work is considered. Depending on site conditions, potentially useful tools are bencher drills, gas jackhammers, air-operated scaling tools, suspended powered platforms (spiders), and hydraulic boom cranes (giraffes) for access to low and intermediate slopes. Equipment for scaling and other related activities in lower and higher reaches of slopes is shown in Figures 9.19 and 9.23 respectively.

Blasting Procedures

Rock should be preserved beyond excavation lines and grades in the soundest possible condition. The effects of tension, compression, and shear stresses developed in a rock mass as a result of blasting damage are documented by Bauer (*9.6*), Lang and Favreau (*9.57*), Langefors and Kihlstrom (*9.59*), the U.S. Bureau of Public Roads (*9.12*), Larocque (*9.60*), and Lambooy and Espley-Jones (*9.55*). Uncontrolled blasting results in rough uneven contours, overbreak, overhangs, excessive shattering, and extensive tension cracks in the crest of the slope. Blasting damage, therefore, can lead to significantly higher scaling, excavation, remedial treatment, and maintenance costs. The results of blast shock waves and gases along faults, joints, bed-

Figure 9.18. Typical rock-scaling operation high above highway.

Figure 9.19. Typical hydraulic crane (giraffe) and basket used for scaling lower reaches of slopes.

207

ding, and discontinuities, although not readily apparent on the blasted face, can lead to loosening of the rock. This sometimes occurs well behind the face, allowing easier infiltration of surface water, which may lead to unfavorable groundwater pressures and unnecessary frost action.

Blast-hole patterns and powder loads must be properly balanced so that advantage is taken of the energy released by the explosive and the desired blast effects are obtained with minimum damage to the rock. Control of the degree of fragmentation can also facilitate handling the muck and ensure that the blasted rock is suitable for use as fill. Although guidelines can be specified, the blasting design should be based on practical experience with the rock in question and can best be determined in the field. Special provisions for trial blasts should be made, particularly for large projects, to ensure optimum results for existing operating conditions. The engineer must approve the final blasting design.

On projects involving blasting, claims are commonly submitted by nearby property owners for damages allegedly caused by undesirable detonation by-products such as fly rock, air concussion, and vibration. Evidence is required to assess the validity of claims for blasting damage. A concise outline of the current status for the law concerning damages resulting from blasting (also from slides, runoff, and drainage) is given by Lewis and others (9.63). The project engineer must apply specialized knowledge and expertise on the project to reduce the possibility of valid damage claims and to negate any invalid claims.

In heavily populated areas, fly rock and concussion can be controlled by common sense and good blasting practice. Production blast holes should be adequately stemmed, sufficient collar should be left, and overloading should be avoided. The exposed area to be blasted should be covered with blasting mats or some other suitable blanketing material. Blast vibration measuring equipment is required, however, to assess the potential damage due to ground vibration.

According to McAnuff (9.69):

The particle velocity of earthborne vibration is now generally accepted to be the best measure of damage potential. . . . A particle velocity between 2.8 and 3.2 inches per second is required to reopen or extend old plaster cracks. . . . A peak particle velocity of 2.0 inches per second is safe with regard to plaster cracks. . . . Ground motion particle velocities below 4.5 inches per second are well within the safe range for most engineered structures.

In Figure 9.20, Northwood and Crawford (9.82) show a direct relation between charge and distance and the probability of damage. They indicate that "a simple relationship defining a conservative safe limit is $E^{2/3} = d/10$, where E is the weight of a single charge in pounds and d is the distance in feet."

Before blasting is carried out in populated areas, Ehrlich, Scharon, and Mateker (9.25) recommend as general practice the following procedures:

In addition to the usual examination and description of all structures within a reasonable distance of the blast area in advance of blasting, an effort should be made to inform the public about such things as the characteristics of blasting vibrations, the differences in structural damage caused by blasting from that produced by settling phenomena, the response

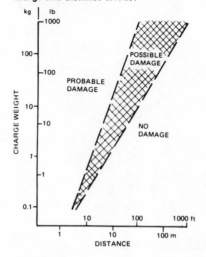

Figure 9.20. Probability of damage versus charge and distance (9.82).

of loose objects, such as wall pictures or mirrors and shelf knick-knacks to vibration, and the human response to vibration. This can be achieved by distributing literature, showing films, conducting lectures at civic meetings, and broadcasting on radio and television.

Ehrlich and others also suggest carrying out a reasonable study of the area to determine the nature and extent of the overburden for purposes of predicting where blast vibrations may be critical and designing the blast accordingly. They also recommend that a "contractor should detonate only at predetermined times and alert the surrounding public by blowing horns or whistles."

Excavation Lifts and Related Procedures

For quality control, each lift of rock excavation generally should not exceed about 10 m (30 ft) in height. Benches this high or lower generally prove to be the best for achieving effective scaling and rock bolting. Also, the accuracy of drilling and, hence, the quality of controlled blasting tend to decrease with increased height of excavation lifts.

Rock bolting, scaling, and similar work, if possible, should be carried out as each successive lift is excavated and completed: This will ensure that the slope above the working area is safe. Careful supervision by qualified personnel is essential to minimize both excavation and future maintenance costs and to maintain the safety of the working area. Unfavorable slope stability, groundwater, and other conditions, though not necessarily of major proportions, may arise during the excavation phase. These should be recognized by the site engineer and appropriate remedial measures provided.

When heavy equipment, such as hydraulic backhoes and tractors with rippers, is used in areas of soft, weathered, or highly broken rock, care should be taken to avoid loosening the final cut face of the slope by equipment operations. Procedures such as line drilling and careful backhoe cutting with special tools are generally advisable.

Surface and Subsurface Drainage

Methods that can be applied to improve either the surface

or subsurface drainage conditions and, hence, increase the stability of the slope should be given high priority in the proposed work. Drainage measures, as compared with other possible measures, frequently result in substantial benefits at significantly lower cost. Often large failures, involving several thousand cubic meters of material, cannot be controlled within practical limits by any means other than some form of drainage. The application of surface and subsurface drainage as part of the general slope design or for stabilization purposes should be considered at the outset of the project because substantial benefits may result at relatively low costs.

Surface Drainage Control

Adequate surface drainage facilities, particularly if the rocks are relatively soft or susceptible to erosion, can substantially improve the stability of a slope where unfavorable groundwater conditions exist. Areas behind the upper portions of unstable slopes should be thoroughly inspected to determine whether surface water is flowing toward unstable areas or into the ground or both. The following methods have been used successfully to control surface drainage (9.40, 9.41):

1. Drain sag ponds, water-filled depressions, and kettles that occur above the working area from which water could seep into unstable zones;
2. Reshape the surface of the area to provide controlled flow and surface runoff;
3. Above the crest of the slope, use concrete, slush grout, asphalt, or polyethylene (Figure 9.21) to temporarily or permanently seal or plug tension cracks and other obviously highly permeable areas that appear to provide avenues for excessive water infiltration (sealing cracks will also prevent frost action in the cracks);
4. Provide lined (e.g., paved, slush-grouted) or unlined surface ditches, culverts, surface drains, flumes, or conduits to divert undesirable surface flows into nonproblem areas; and
5. Minimize removal of vegetative cover and establish vegetative growth.

Drain Holes

Methods for subsurface drainage of slopes include drain holes, pumped wells, drainage galleries, shafts, and trenches. Noteworthy discussions concerning this subject have been presented by Záruba and Mencl (9.112), Baker and Marshall (9.3), Hoek and Bray (9.42), and Cedergren (9.15). Only under special circumstances are subsurface drainage methods other than subhorizontal drain holes used for highway and railway cut slopes; therefore, only a discussion of drain holes is given here.

The purpose of subsurface drainage facilities is to lower the water table and, hence, the water pressure to a level below that of potential failure surfaces. A practical approach that appears to be best suited for most rock-slope problems encountered in highway and railway cuts is to incorporate a system of drain holes to depress the water level below the zones in which failure would theoretically take place in the slope. The drain holes should be designed to extend behind the critical failure zone. Determining the drain-hole design with respect to the geometry of the potential failure mode

Figure 9.21. Temporary cover of polyethylene sandwiched between two layers of mesh to prevent precipitation infiltration into slope at Hell's Gate Bluffs, British Columbia.

provides a reasonable guideline for the preliminary design of the drain-hole system. If for design purposes one assumes a circular failure mode, for example, a reasonably conservative drain-hole design results. In this case, the length of the drain holes is about half the height of the immediate slope needing removal of water. The direction of the drain holes may depend to a large degree on the orientation of the significant discontinuities. The optimum drain-hole design is to intersect the maximum number of significant discontinuities for each meter of hole drilled.

The effectiveness of drains depends on the size, permeability, transmissibility, and orientation of the discontinuities. A drain does not have to produce any noticeable flow of water to be effective; it may have flow only under extreme conditions. Furthermore, the absence of damp spots on the rock face does not necessarily mean that unfavorable groundwater conditions do not exist. Groundwater may evaporate before it becomes readily apparent on the face, particularly in dry climates.

Drain holes usually are inclined upward from the horizontal about 5°. In erodible materials, however, the holes may have to be inclined slightly downward to prevent erosion at the drain-hole outlet due to water flowing out of the drain hole. In this case, a small pipe can be left in the mouth of the drain hole to retard erosion. Spacing of drain holes can range from 7 to 30 m (20 to 100 ft), but 10 to 15-m (30 to 50-ft) spacing generally is used. For high rock cuts, installing drain holes at different levels on the slope may be advantageous to increase the effectiveness of the overall drainage system. For certain conditions on a slope, a series of drain holes may best be installed in a fan pattern so that drill machine setup and moving time is minimized.

Drain holes should be thoroughly cleaned of drill cuttings, mud, clay, and other materials; drain holes not properly cleaned may have their effectiveness reduced by 75 percent. High-pressure air, water, and in some instances a detergent should be used to clean drain holes. In highly fractured ground, care should be taken to ensure that caving does not block drain holes. If caving is significant, perforated linings should be installed so that drain holes remain open. If freezing conditions exist, drain-hole outlets should be protected from ice buildup that could cause blockage. Insulating ma-

Figure 9.22. Shotcrete applied to steep high slope that is prone to minor slides and rock falls.

Figure 9.23. Shotcreting dangerous overhang in extremely steep terrain along tracks of Canadian National Railways in Fraser Canyon, British Columbia: (a) steep highly fractured rock face with serious overhang; (b) shotcrete equipment and buildup of shotcrete at top of picture and under overhang; (c) start of shotcreting operation at tunnel portal; and (d) close-up of shotcrete buildup in overhang shown above.

terials, such as straw, sawdust, gravel, or crushed rock, have been used for this purpose. Electric current has also been used to keep the drain-hole pipe warm enough to prevent ice buildup.

Shotcrete

Shotcrete is a concrete that consists of mortar with aggregate as large as 2 cm (¾ in) in size and that is projected by air jet directly onto the surface to be treated. It is one of the basic methods for treating unstable sections of rock slopes. It is used to prevent weathering and spalling of rock surfaces and to provide surface reinforcement between blocks. The force of the jet compacts the mortar in place. Shotcrete is usually applied in 8 to 10-cm (3 to 4-in) layers, and each layer is allowed to set before successive layers are applied. Shotcrete on rock slopes generally appears to have

Figure 9.24. Example of drainage behind shotcrete (9.1).

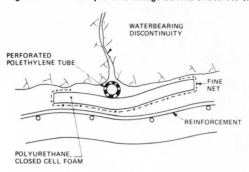

replaced gunite, a similar material that contains smaller aggregate. Rock slopes should be thoroughly scaled to provide the soundest rock condition before shotcrete is applied. An extensive shotcrete application in an area prone to slides and rockfalls is shown in Figure 9.22. Figure 9.23 shows the use of shotcrete for an overhang in extremely steep terrain.

Specifications and discussions of the application of shotcrete are given by the American Concrete Institute (9.2). When shotcrete is applied to an irregular rock surface, the resulting surface configuration is smoother. The shotcrete helps to maintain the adjacent rock blocks in place by means of its bond to the rock and its initial shear and tensile strength acting as a membrane. The result is that a composite rock-shotcrete structure is developed on the surface of the rock. There is no transfer of load from the rock mass to the shotcrete. In that the interlocking quality of the surface blocks is improved (9.10), shotcrete acts as reinforcement and not as support. The more quickly the shotcrete is applied after excavation, the more effective the results are.

Deterioration of shotcrete can result from frost action, groundwater seepage, or rock spalling due to lack of shotcrete bond. Therefore, unfavorable groundwater flows should be drained for long-term stability of the shotcrete cover. Weep holes should be drilled or installed through the hardened shotcrete and into the rock to prevent building of water pressure behind the shotcrete. In Norway and Sweden, semirounded plastic pipes are glued to the rock surface to form surface drainage channels. Small volumes of water are protected against freezing by rock wool, plastic foil, or even heating cables. More universally used are short flexible plastic pipes, which are placed in cracks or holes drilled into water-bearing broken rock; an example of this technique, as described by Alberts (9.1), is shown in Figure 9.24.

Initially dry rock surfaces are preferred in the shotcrete process, although careful control of setting admixtures and nozzle water can give successful applications on wet surfaces. Where alteration products, such as clay or mud, exist on joint or fault planes, care should be taken to clean such surfaces by air or water jet to ensure a good bond between shotcrete and rock. As a rule of thumb, weak material should be removed to a depth at least equal to the width of the weak zone before shotcrete is applied. These areas are where the shotcrete will do the most good and where extra attention will be required in its application.

Shotcrete can be used in combination with steel wire mesh and rock bolts to give structural support and also to form buttresses for small loads. Where shotcrete is applied to mesh, all loose material should be removed from the rock surface and the mesh fabric tightened. Shotcrete can also be used behind anchor beams to provide a uniform contact with the rough or uneven rock surface and, if applied across cracks, may provide a simple means of indicating where movement is occurring. An extensive application of steel fibrous shotcrete to stabilize potential rock falls in basaltic rocks along a railway is described by Kaden (*9.49*).

The most important advantage of shotcrete in treating rock slopes is that it offers a rapid, mechanized, and often uncomplicated solution to rock-fall problems. Various other materials, such as polymers, fiber glass, epoxies, plastic, and rubberized and asphalt compounds, have been either tried or considered to protect rock slopes from the effects of climate and infiltration of groundwater. However, none of these has yet proved entirely successful on a general basis. Polyvinyl chloride, butyl, and neoprene sheeting have been used with limited success in isolated cases, and spray-applied rubberized bitumen coatings have been used with success in retarding slaking of shales (*9.50*).

Support and Reinforcement Systems

Buttresses, bulkheads, and retaining walls are classified as external support systems in that they offer passive resistance to loads imposed by the slope-forming materials that undergo deformation in stages of slope failure and general elastic rebound. Rock bolts, rock anchors, anchored beams, anchored cable nets, and cable lashing are classified as reinforcement systems because they add strength to the rock mass by increasing the general tensile strength and by improving its resistance to shear along discontinuities.

Buttresses and Bulkheads

Buttresses, bulkheads, and other support structures are used for stabilization where failure of overhangs appears to be imminent or where slight cracking or vertical displacement appears to be occurring. Buttresses are designed to take part of the weight of the slope, thus inducing stable conditions and preventing rock falls.

Buttresses, although often costly, are simple, effective, and permanent. They are most effective where overhangs have developed and where excavation to remove the overhangs would be costly because of quantities involved or problems of accessibility. Buttresses are usually constructed at highway or railway level, but they can be just as effective in the upper reaches of the slope. In steep terrain, especially where considerable lateral river erosion has taken place, buttresses or bulkheads are particularly useful. Typical rock buttresses that have been used in such areas are shown in Figures 9.25 and 9.26. Buttresses combined with rock anchors are shown in Figure 9.27.

Retaining Walls

For purposes of rock-slope engineering, retaining walls are used to prevent large blocks in the slope from failing and to control or correct failures by increasing the resistance to slope movement. Retaining walls have the advantage of lessening weathering of the rock slope and of, thus, offering per-

Figure 9.25. Cast-in-place reinforced concrete buttress approximately 15 m (50 ft) high providing stabilization protection of large overhanging slope.

Figure 9.26. Cast-in-place reinforced concrete buttress approximately 15 m (50 ft) high providing stabilization support against possible movement of large block above overhang (note unfavorable discontinuities in the slope).

manent protection. Basic design and use of retaining walls are documented by Peck, Hanson, and Thornburn (*9.83*).

The space along railways and highways is often too narrow for normal gravity or cantilever types of walls, but anchors or rock-bolt tiebacks may be used to overcome this problem. Tied-back walls need only have the strength required for bending and shear resistance between rock bolts. The use

Figure 9.27. Cast-in-place reinforced concrete pillar buttresses supporting overhang on road in Austria. Each buttress has two 9-m (30-ft) long, 356-kN (40-tf) capacity rock bolts.

of rock bolts in combination with a concrete facing wall as a gravity section to stabilize a vertical rock face in sedimentary rock is described by Redlinger and Dodson (*9.96*).

Various types of free-standing and tieback retaining walls include solid cast-in-place concrete walls (*9.35*), prefabricated concrete slabs with anchored waling, vertical concrete ribs to hold precast concrete panels (*9.53*), and steel sheet piling with anchored waling. Figure 9.28 shows an interesting application of a tieback retaining wall formed of galvanized steel members for protection on a high sedimentary rock cut requiring face protection. Anchor bolts were grouted in place at 1.5-m (5-ft) vertical intervals to a minimum depth in the rock of 1.5 m (5 ft). Vertical U-channels of 8-gauge galvanized steel were then mounted on the bolts at 3-m (10-ft) intervals, and horizontal lagging was placed between them. The void behind the wall was backfilled with free-draining material; the water was collected and conducted to a storm sewer.

Steel Reinforcement

Steel reinforcement such as rock bolts and rock anchors reinforce or tie together the rock mass so that the stability of a rock cut or slope is maintained. Rock bolts are commonly used to reinforce the surface or near-surface rock of the excavation or natural slope, and rock anchors are used for supporting large masses of unstable rock. Short reinforcing bars fully grouted into the rock mass are commonly called dowels. Their action, however, is somewhat similar

to fully grouted, untensioned rock bolts.

Steel reinforcement of the rock mass, whether considered as rock bolts or rock anchors, has been carried out with many types of bars, bar bundles, cables, and cable tendons. Both active and passive steel reinforcement systems have been successfully used for stabilizing rock slopes. The active system involves tensioned or prestressed bars or cables that have been anchored at one end within the rock mass by mechanical means or by cement or chemical grouts or by both. The passive system involves untensioned bars that have been fully grouted throughout their length by cement or chemical grouts. The active and passive systems are often compared to prestressed and reinforced concrete, but, since the rock mass is a discontinuous medium, the mechanical properties of the rock mass are different from controlled, man-made concrete.

The main advantage of the active or prestressed anchor system over the passive system is that no movement has to take place before the prestressed anchor develops its full capacity; thus, deformation and possible tension cracking of the slope are minimized. A further advantage of the active system is that a known anchor force is applied, and proof loading can be accomplished during installation of each anchor. Also, the tensioned rock-bolt member can be rechecked, if necessary, at any subsequent time to determine whether the load is being maintained. A greater degree of confidence in the anchor design is thus provided.

General considerations regarding rock-bolt and rock-anchor reinforcement of rock masses are given by Hoek and Londe (*9.43*) and the U.S. Army Corps of Engineers (*9.110*). The reinforcement is usually designed by using the limit equilibrium method of analysis. This method considers only the ultimate failure condition of the potentially unstable rock mass, which is assumed to fail along a probable failure surface or surfaces that have been defined by geologic investigations. For the active or prestressed system, the anchor force required for stability is applied partly as an increased compressive normal stress on the failure plane, thus increasing frictional resistance, and partly as a force resisting the driving force that is causing instability of the slope. The value of the component forces will depend on both the design geometry and orientation of the anchors and the characteristics of the assumed failure surface. For the passive system, the maximum available anchor force is determined from the tensile and shear resistance of the steel cross section at the assumed failure surface crossed by the anchor.

Both systems require adequate anchorage of the steel reinforcing member beyond the assumed failure surface; adequate anchorage involves good bond of the steel to the grout and good bond of the grout to the rock. Long-term corrosion protection of the anchors must also be provided; this sometimes requires special sheathing and grouting for the prestressed anchors. Grout mixes, installation procedures, and testing procedures must be carefully considered by the engineer in design and specifications. In this regard the ability to test load each anchor after installation is a valuable field control that is available when prestressed anchors are used. The various types of slope-failure modes that are judged to be kinematically possible should be considered in the design of the artificial support system. Actual slope-failure conditions in the field may be complex and difficult to analyze; therefore, a considerable degree of engineering

Figure 9.28. Galvanized sheet steel retaining wall that is anchored by shallow rock bolts to prevent failures of high cut slope in sedimentary rock above highway in Hamilton, Ontario.

judgment and experience will continue to be necessary for anchor design.

Analysis of rock-slope deformation is not generally required for transportation routes, except for those involving support of adjacent structures. Theoretical procedures using the finite element method are dependent on the assumed boundary conditions and properties of the rock mass. At the present time, only cases involving small deformations can be analyzed with any reliability. Otherwise, the only procedure is to use as high a factor of safety as possible against ultimate shear failure to minimize undesirable deformation. This problem is discussed by Morgenstern and Eisenstein (9.75).

Rock Bolts

The use of rock bolts is discussed by Lang (9.58) and Lancaster-Jones (9.56). Rock bolts are basically used to reinforce the surface and near-surface rock of the excavated or natural slope. In that it is essentially a tension member, the rock bolt exerts a compressive force, which tends to prevent elastic rebound, frost action phenomena, and general relaxation or exfoliation by keeping the rock in its original position. The rock-bolt system should be designed in such a manner that shear resistance along discontinuities is improved; the rock mass will then acquire significant tensile strength as a result of the operation.

To minimize the decompression or loosening effects associated with recently excavated rock slopes, rock bolts should be installed and tensioned as soon as possible after each lift of an excavation and preferably before the next lift is blasted to inhibit dilation effects of the near-surface section of the excavation. The sooner the rock bolts are installed, the less the subsequent postexcavation movements of the rock mass will be. Figure 9.29 shows a group of large-diameter rock bolts (9.80) concentrated in one area at the base of a slope to maintain the integrity of large fault blocks. Figure 9.30 shows numerous small-diameter rock bolts that have been applied over the entire slope in an attempt to secure smaller individual blocks in the slope.

The method used for transferring load from the head of the rock bolt to the rock depends largely on the condition of the rock. Broken rock may weather away from the bolt head and cause the bolt to lose tension. A choice must be

Figure 9.29 Concentration on bank of 3.5-cm ($1\frac{3}{8}$-in) diameter 14-m (45-ft) long rock bolts at base of section at Hell's Gate Bluffs, British Columbia. Each rod was cold stretched and stress relieved to provide minimum ultimate strength of 1.1 MPa (160 000 lbf/in²). They were tensioned to 712 kN (80 tf) and dropped to 623 kN (70 tf) and then grouted.

Figure 9.30. Broadly distributed 1.6-cm ($\frac{5}{8}$-in) diameter by 1.2 to 3-m (4 to 10-ft) long rock bolts at Hell's Gate Bluffs, British Columbia. These rock bolts were posttensioned to about 53.4 kN (6 tf) to maintain integrity of small individual blocks in slope.

made between protecting the rock face around the head with wire mesh and shotcrete, retaining the rock in place by steel strapping, or distributing the point load by use of concrete pads. Usually the tentative design load does not exceed approximately 60 percent of the ultimate strength of the bolt.

If grouted bolts are used, extreme care should be taken to ensure that the grout does not spread. Excessive grout spread will reduce the permeability of the rock mass and retard drainage by plugging joints through which groundwater would normally have free access. The grout must be pumped into the borehole under low pressure, possibly not exceeding 103 kPa (15 lbf/in²), and other measures must be taken to retard grout spread. If grouted bolts are used in the lower reaches of the slope, free drainage must be maintained.

Rock bolts should be installed by a carefully controlled procedure. The design of the rock-bolt system should be checked and tested during installation for bolt tension and adequate anchorage to ensure reliability and performance of the rock bolts. However, the most important responsibility of the site engineer is to regularly inspect the slope after the rock-bolt installation to guarantee that areas that may become unstable in the future have been sufficiently supported.

Dowels and Perfobolts

Dowels consist essentially of steel reinforcing bars that are cemented into boreholes; they are not subjected to any posttensioning. Unlike rock bolts that provide a form of reinforcement, dowels are in the strict sense a form of rock-slope support. Whereas rock bolts increase both shearing resistance and normal stress, dowels basically increase the shearing resistance across potential failure surfaces. A typical use of dowels for rock slope stabilization is shown in Figure 9.31. Dowels are also used to provide anchor keys and tiebacks for shearing resistance at the toe and flanks of retaining walls, as shown in Figure 9.32. Dowels, together with rock bolts, can fasten small blocky rocks when applied with strapping or bearing pads, stabilize broken rock zones when applied with wire mesh and shotcrete, and anchor buttresses or beams placed below sloping blocks of rock. They can also anchor restraining nets and cables, catch nets, catch fences, cable catch walls, and cantilever rock sheds.

The perfobolt is a form of steel dowel encased within a thin sheet-metal liner filled with low-slump slush grout and perforated. The liner is inserted into a drilled hole in the rock, and a steel reinforcement bar is then inserted into the liner, displacing the grout and forcing it through the perforations. A bond is thus effected with the rock. Perfobolts are particularly applicable in sections of the rock mass in which grouting of standard rock bolts would prove difficult because of open cracks and in which fully grouted artificial support is required. Perfobolts as long as 6 m (20 ft) have been installed; for greater lengths, inserting the reinforcing bar in the liner by hand is difficult.

Tendon and Cable Anchors

Tendon and cable anchors are used to control large failure blocks and are accordingly longer than rock bolts. Anchors usually affect a much larger volume of rock than rock bolts. Although the basic principles that apply to rock bolts are

also applied to rock anchors, rock or cable anchors are generally used more for support than for reinforcement. Anchors can be used to replace retaining walls and other support structures that are normally required for surface excavations. They have also been used effectively to provide intermediate points of support for retaining walls, thus reducing bending moments in the structure and achieving the benefit of requiring a smaller retaining structure.

Compared to their use in mining and hydroelectric developments, large tendon and cable anchors on high rock cuts have had relatively little use on transportation routes. An example of an exception is the considerable success of rock anchors in combination with drain holes in stabilizing progressive failure caused by undercutting the slope on a highway at Windy Point, Australia (9.95). The rock is highly jointed sandstone with silty clay beds that dip out of the slope at about 27°. About 181 440 Mg (200 000 t) of rock were moving, and the anchor design involved 45 anchors, each with an average working load of 166.9 kN (375 000 lbf). The anchor design and drain holes are shown in Figure 9.33 (9.18).

Figure 9.33. Geological section of Windy Point Slide, Australia, showing location of 166.9-kN (375 000-lbf) anchors (average working load) and drain holes (9.18).

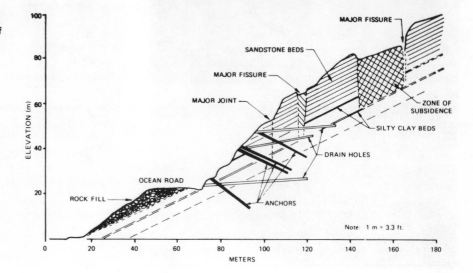

Figure 9.34. Schematic drawing of trial anchor support system including cable anchors, horizontal stringers, and welded mesh between anchor heads (*9.4*).

ANCHOR LOADING PADS
HORIZONTAL STRINGERS
WELDED MESH COVERING SURFACE
GROUTED ANCHORAGE
DEEP ROCK BOLTS OR CABLES

An interesting anchorage system, shown in Figures 9.34 and 9.35, was carried out by Barron, Coates, and Gyenge (*9.4*) at a Canadian mine. The system consisted of deep preloaded cable anchors to support the tentatively unstable ground between the slope face at angle α and some predetermined potential failure plane at angle i. Welded wire mesh was placed over the surface, and horizontal stringers were installed between the cable terminations to prevent local rock falls and to provide bench support.

A practical guide for the use and installation of anchors is given by Coates and Sage (*9.18*), and other useful aspects relating to design and quality control of rock anchors are described by Seegmiller (*9.102*).

Anchored Beams

Anchored beams can be used to distribute the support of rock bolts over a wide area of the slope and, hence, minimize the number of bolts required. The beams can be made of concrete or steel and applied in any direction across a rock face. Anchored beams are particularly useful where

Figure 9.35. Installation of system shown in Figure 9.34: (a) location of two benches where the trial anchor installation is conducted, (b) inserting the cable in the borehole, (c) formwork for horizontal concrete stringer on lower bench and 66-44 welded wire mesh, (d) formwork for anchor pads on top bench and steel bar horizontal stringer, (f) cable anchor tensioned and locked, and (f) completed trial anchor installation on lower bench (*9.4*).

(a)

(b)

(c)

(d)

(e)

(f)

Figure 9.36. Cast-in-place reinforced concrete beams used to distribute point loads from rock bolts seated in the beams (*9.84*). Location is above main line of Austrian Federal Railways.

Figure 9.38. Cable lashing of a large, potentially hazardous, 2-m (6-ft) diameter block about 90 m (300 ft) above highway in Fraser Canyon, British Columbia.

heavy slabs of rock are unstable and where a relatively uniform distribution of point load forces of the anchors or bolts is required. Typical anchored beams installed in Austria are shown in Figure 9.36. Several variations of anchored beams have been used: In Czechoslovakia, for example, anchored concrete frames are used rather than beams. Anchored concrete aprons have been used on heavily jointed rock around bridge piers in Washington State.

Anchored Cable Nets

Anchored cable nets can be used to restrain masses of small loose rocks or individual loose blocks as large as 1.5 to 2.5 m (5 to 8 ft) in diameter that protrude from a rock face. In principle, an anchored cable net performs like a sling or reinforcement net, which extends around the surface of the unstable broken rock to be supported. The cable net strands are gathered on each side by main cables leading to rock anchors, as shown in Figure 9.37.

Beam and Cable Walls

Beam and cable walls can also be used to prevent smaller

Figure 9.37. Application of anchored cables and cable nets.

Figure 9.39. Large, potentially hazardous rock block supporting considerable volume of smaller material secured by two cable strands that were inserted in boreholes through the block and anchored on other side.

blocks of rock from falling out of the slope and may be used on slopes as high as 20 m (60 ft). Beam and cable walls consist of cable nets fastened to vertical ribs formed of steel beams laid against the slope at 3 to 6-m (10 to 20-ft) intervals. The beams are held by anchored cables at the top of the slope and by concrete footings at ditch level.

Cable Lashing

Cable lashing is a simple, economical type of installation for restraining large rocks. As shown in Figure 9.37, this method merely involves tying or wrapping unstable blocks with individual cable strands anchored to the slope. Eckert (*9.24*) describes a case in France in which cables with a capacity of 227 Mg (250 t) and a length of more than 90 m (300 ft) were extended around a mass of broken rock and anchored into sound rock at each side. Lashing using both cables and chains is described by Bjerrum and Jørstad (*9.9*). If rock is too broken to be restrained by individual cables, vertical concrete or steel ribs can be used to spread the points of contact of the cables with the rock. Typical cable lashing of a large block is shown in Figure 9.38. Figure 9.39 shows a unique application of cables in that two cables pass entirely through the block.

217

Methods of Protection

Rock-Fall Characteristics Affecting the Design of Protection Measures

Loose rocks can be kept from reaching the transportation route right-of-way by retaining the rocks on the slope, intercepting falling rocks above the roadway, or directing falling rocks to pass over or under the transportation route without causing harm. Before the best methods can be chosen for a particular location, the characteristics of the rock falls and the general geometry of the slope should be evaluated. The rate at which fallen rock accumulates is also an important factor. Without encroaching on the right-of-way, storage must be provided for rock falls that are likely to arrive during a reasonable maintenance interval. Protection measures must be designed to protect against the various ways in which rock falls reach the right-of-way (9.85).

A rock can arrive at the base of a slope by free falling, bouncing, rolling, or sliding down the slope. For free-falling rocks, the only means of protection are to move the roadway away from the slope and prevent the rocks from bouncing and rolling after landing or to protect the roadway with a rock shed or tunnel. The paths of bouncing rocks are difficult to predict, and interception measures above the right-of-way require high structures, such as a wall or net. Rolling or sliding rocks are easier to intercept since they are in constant contact with the slope; in gullies, their paths are even more predictable.

A rock-fall model developed by Piteau and Clayton (9.91) for a computer program can simulate several hundred rock falls for a slope of specified geometry in a matter of seconds. A coefficient of restitution is used for the rock fall in the bouncing mode, and increased friction is used for the rolling mode to determine the paths of the rock falls for different input velocities and heights. The slope profile is divided into straight segments or slope cells (Figure 9.40), which are numbered consecutively from the top to the bottom of the slope. Rock falls are introduced at different locations above the slope as desired. Probability factors can be assigned to these locations to recognize areas that are either more likely or less likely to be the source of the rock fall. Catch walls at various positions and heights can be assumed in the model to evaluate their effectiveness. In the same manner, the effect of varying the slope geometry by including ditches, benches, and berms can be evaluated.

This rock-fall model was used to relocate a concrete catch wall at the base of an active rock-fall area (Figure 9.51). It was also used to determine whether large blocks located at the crest of a major slide area would reach facilities at the base of the mountainside located some 1500 to 1800 m (5000 to 6000 ft) away from the crest of the slide.

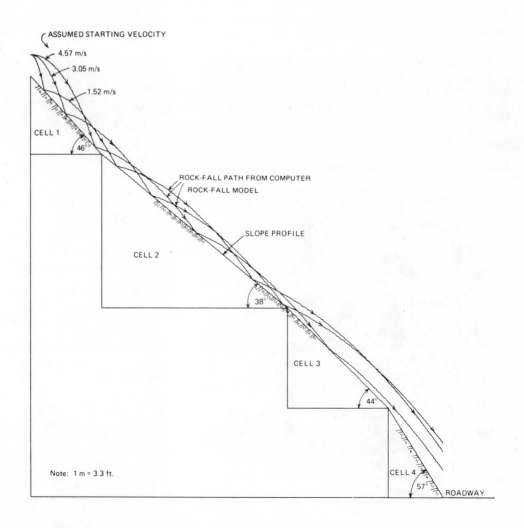

Figure 9.40. Rock-fall paths traced by calcomp plotter based on computer model for typical rock falls at Porteau Bluffs, British Columbia (9.91).

Figure 9.41. Intercepting slope ditch or cross ditch, north of Hope in Fraser Canyon, British Columbia, to catch rolling rocks originating up slope on steep rock faces (note that ditch is shaped partly in talus and makes use of blocks of rock).

Figure 9.42. Shaped berm that is at toe of high rock face and cut in top of high talus slope above Hope-Agassiz highway in British Columbia.

Relocation

Relocation is most effective where rocks are free falling from steep rock faces in proximity to the roadway or where stabilization or some other protection measure is not feasible. Relocation is feasible when space is available and the design criteria are not affected. When combined with proper ditch shaping and possibly with other ditch-level protection, relocation can be the most economical solution and, in some cases, one of the few solutions to the problem. Unless relocation practically eliminates the possibility of accidents, other measures also should be included along with relocation.

Intercepting Slope Ditches and Shaped Berms

Slope ditches are used to intercept rock falls partway up the slope, and shaped berms are used at the top of the slope. These methods ensure that rock falls get caught either before they start to roll or while they are on their way down the slope. In some locations, ditches can be made to intercept rocks and guide them laterally into disposal areas. A typical intercepting slope ditch is shown in Figure 9.41, and a shaped berm at the base of a steep slope is shown in Figure 9.42.

Depending on the nature of the terrain, these methods are generally inexpensive, simple to construct, and easy to maintain. An intercepting slope ditch should only be installed on a slope that can accept the introduction of a ditch without the stability of the entire slope being impaired. The ditch must be designed and built carefully so that the upper slope is not at any time steeper than planned. The ditch should be suitably located to facilitate periodic cleaning by mechanical equipment.

Anchored Wire Mesh

Wire mesh is a versatile and economical material for use in protecting the right-of-way from small rocks. Layers of mesh are often pinned onto the rock surface to prevent small loose rocks from becoming dislodged (Figure 9.43). Figure 9.44 shows that mesh can also be used essentially as

a blanket draped over the rock surface to guide falling rock into the ditch at the base of the slope. The same arrangement can be used on stony overburden slopes to prevent dislodged stones from rolling down the slope. This practice is commonly used on talus slopes in steep mountainous terrain. Mesh can be combined with long rock bolts to provide a generally deeper reinforcement. Mesh in combination with both shotcrete and rock bolts provides general reinforcement and surficial support and retards the deleterious effects of weathering.

Conditions for the use of mesh are particularly suitable if no individual rocks are larger than 0.6 to 1 m (2 to 3 ft) and if the slope is uniform enough for the mesh to be in almost continuous contact with the slope. If large falling blocks in certain areas are likely to dislodge or tear the mesh and present a hazard, rock bolts should be used to reinforce these particular areas. For anchoring mesh at the top of overburden slopes, posts are cast in concrete blocks in hand-dug holes. For bedrock slopes, grouted rock bolts or dowels are used. A cable is slung between the anchorages; the mesh is then fastened to the cable and rolled down the slope, and the vertical seams of the mesh are wired together. For wire mesh blankets, the bottom end of the mesh is usually left a meter or so above ditch level (Figure 9.44) and only a narrow ditch is required. The mesh normally used is 9 or 11-gauge galvanized, standard chain-link or gabion wire mesh. Gabion mesh appears to have an advantage over the standard chain-link materials in that the gabion mesh has a double-twist hexagonal weave that does not unravel when broken.

Protection Methods at Ditch Level

Shaped Ditches

Depth, width, and steepness of the inside slope and storage volume of the ditch are important factors in the design of ditches to contain rock falls. The choice of ditch geometry should take into consideration the angle of the slope that influences the behavior of falling rocks. Ritchie (9.98) evaluated the mechanics of rock falls from cliffs and talus slopes and developed design criteria for ditches. The criteria, which

Figure 9.43. Anchored double-twist, hexagonal wire mesh being fastened on a high rock face to prevent rock falls onto highway.

Figure 9.44. Wire (No. 9) mesh blanket used to control rock falls near Kelso, Washington (9.3). Falling rocks roll down surface of slope under mesh and drop into ditch.

involve the height and angle of slope, depth of ditch, and width of fallout area, are given below (where 1 m = 3.3 ft) and in Figure 9.45, which also shows the nature of rock trajectories for different slope angles. An example of a typical shaped ditch is shown in Figure 9.46.

Figure 9.45. Path of rock trajectory for various slope angles and design criteria for shaped ditches (9.98).

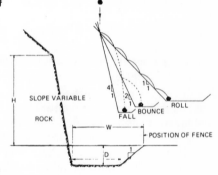

Rock Slope		Fallout Area Width (m)	Ditch Depth (m)
Angle	Height (m)		
Near vertical	5 to 10	3.7	1.0
	10 to 20	4.6	1.2
	>20	6.1	1.2
0.25 or 0.3:1	5 to 10	3.7	1.0
	10 to 20	4.6	1.2
	20 to 30	6.1	1.8[a]
	>30	7.6	1.8[a]
0.5:1	5 to 10	3.7	1.2
	10 to 20	4.6	1.8[a]
	20 to 30	6.1	1.8[a]
	>30	7.6	2.7[a]
0.75:1	0 to 10	3.7	1.0
	10 to 20	4.6	1.2
	>20	4.6	1.8[a]
1:1	0 to 10	3.7	1.0
	10 to 20	3.7	1.5[a]
	>20	4.6	1.8[a]

[a]May be 1.2 m if catch fence is used.

Figure 9.46. Shaped ditch to retain falling rock from nearby vertical slope on highway in Washington State (note steep slope and barrier fence on roadway side).

If lack of space prevents the shaped ditch from being as wide as indicated, limited protection against small rolling rocks can still be provided inexpensively by excavating a

Figure 9.47. Shaped ditches, wire mesh catch fences, and wire mesh catch nets (9.84).

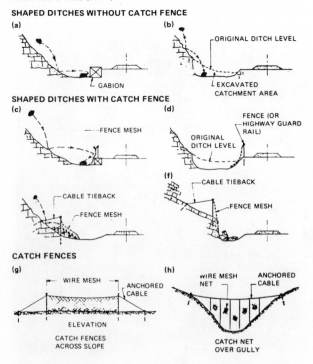

SHAPED DITCHES WITHOUT CATCH FENCE

(a)

(b)
ORIGINAL DITCH LEVEL
EXCAVATED CATCHMENT AREA

GABION

SHAPED DITCHES WITH CATCH FENCE

(c)
FENCE MESH

(d)
FENCE (OR HIGHWAY GUARD RAIL)
ORIGINAL DITCH LEVEL

(f)
CABLE TIEBACK
FENCE MESH

CABLE TIEBACK
FENCE MESH

CATCH FENCES

(g)
WIRE MESH
ANCHORED CABLE
ELEVATION
CATCH FENCES ACROSS SLOPE

(h)
WIRE MESH NET
ANCHORED CABLE
CATCH NET OVER GULLY

Figure 9.48. Special mobile catch fence mounted on flat deck to protect motorists from rock falls during excavation work and scaling in extremely steep terrain.

catchment area, as shown in Figure 9.47b (*9.84*), by forming a windrow of material or by installing a low barrier formed of standard highway guardrail, precast concrete elements, or gabions at the shoulder of the roadway, as shown in Figure 9.47a (*9.84*). Bedrock should not remain exposed in the bottom of ditches, but should be covered with small broken rock or loose earth to keep falling rocks from bouncing or shattering. Mearns (*9.71*) describes how a ditch at grade was filled with a 0.3-m (1-ft) layer of sand, which acted as an energy absorber to prevent rolling rock falls from reaching the road. Several workers have used a similar technique to successfully dissipate energy of rock falls.

Wire-Mesh Catch Nets and Fences

Wire mesh can be effective in intercepting or effectively slowing bouncing rocks as large as 0.6 to 1 m (2 to 3 ft) when the mesh is mounted as a flexible catch net rather than as a standard, fixed wire fence. If suspended on a cable, the mesh will absorb the energy of flying rocks with a minimum of damage to the wire catch net. Catch nets can consist of standard chain-link wire mesh or gabion mesh. This type of barrier has economic potential in some troublesome locations, as shown in Figure 9.47h (*9.84*). It is best to locate catch nets at the lower end of steep gullies where rocks tend to bounce down to the right-of-way. The cable supporting the catch net is anchored to sound rock on either side of the gully with the result that rocks hit the catch net and fall harmlessly at the base of the structure. If there is a steady accumulation of rock, a catch wall on the shoulder of the roadway also may be required.

The principle of the catch fence is similar to that of a catch net. Its purpose is to form a flexible barrier to dissipate the energy of rapidly moving rocks. Various arrangements are shown in Figure 9.47 (*9.84*) for catch fences lo-

cated at or near ditch level. The wire mesh (chain-link or gabion) forming the catch fences is hung on cables supported on posts or strung between posts or trees. Wire-mesh catch fences are usually located on the roadway side of the ditch or at the base of the slope and can be used with or without a shaped ditch. The fence should be suitably situated so that accumulated rocks can be removed easily. Figure 9.31 shows a catch fence, and Figure 9.48 shows a special application of a mobile catch fence for use in scaling operations.

Catch Walls

Catch walls can be used to form a rigid barrier to stop rolling or bouncing rocks as large as 1.5 to 2 m (5 to 6 ft) from reaching the right-of-way. If effective, they usually increase the storage capacity of the ditch so that maintenance intervals can be extended. In many locations, large ditches themselves are not effective for intercepting large rolling rocks and the use of catch walls is advised.

To achieve maximum protection and storage capacity, the catch walls should be located on the side of the ditch closest to the road. In steep terrain, catch walls are commonly used where the right-of-way cuts across postglacial slide areas. Many postglacial slide areas that have been disturbed by excavation are in a constant state of sloughing and readjustment. If broad areas require wall protection, gaps should be left in the catch wall to allow access of maintenance equipment required to remove rock-fall debris.

Concrete Catch Walls

Concrete catch walls are the most widely used type of catch wall in steep mountainous terrain. Concrete walls may be cast in place or precast in short sections and assembled on the site. An efficient precast concrete wall installation is shown in Figure 9.49 (*9.84*). Figures 9.50 and 9.51 show

Figure 9.49. Precast concrete wall that could be used for track protection (9.84).

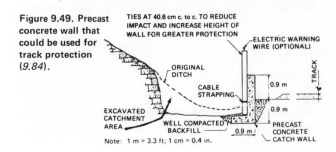

TIES AT 40.6 cm c. to c. TO REDUCE IMPACT AND INCREASE HEIGHT OF WALL FOR GREATER PROTECTION
ELECTRIC WARNING WIRE (OPTIONAL)
TRACK
ORIGINAL DITCH
CABLE STRAPPING
0.9 m
0.9 m
EXCAVATED CATCHMENT AREA
WELL COMPACTED BACKFILL
0.9 m
PRECAST CONCRETE CATCH WALL
Note: 1 m = 3.3 ft; 1 cm = 0.4 in.

221

Figure 9.50. Reinforced cast-in-place concrete wall approximately 2 m (7 ft) high and 1 m (3 ft) wide across a major postglacial slide area along edge of Trans-Canada Highway in Fraser Canyon, British Columbia (note break in retaining wall where large blocks have hit wall).

Figure 9.51. Close-up of concrete catch wall shown in Figure 9.50 (note breaks in wall due to large blocks from postglacial slide debris rolling down slope).

the use of a cast-in-place concrete catch wall in a postglacial slide area, where large rock falls resulted in breaking the wall.

Box Gabion Catch Wall

The box gabion catch wall is a rectangular basket divided by diaphragms into smaller rectangles that are filled with stones. The basket is formed of woven hexagonal steel galvanized wire mesh. Baskets can be placed to be filled individually or wired together in groups and filled accordingly. The wire mesh tends to reinforce the stone in tension. The gabion is a flexible structure that, upon settling or being hit by impact, tends to deflect and deform instead of break. Because gabion walls are highly deformable, differential settlement is not important. Like the gabion mesh, the gabion does not unravel if broken. Gabions make good catch-wall structures, but only recently have they been recognized in North America as a feasible alternative to more rigid concrete walls for protecting the right-of-way from rolling rocks. They prove particularly useful where the right-of-way crosses postglacial slide areas. They can be used efficiently to stop rolling stones as large as 0.6 to 1 m (2 to 3 ft). When adequate filter materials are used as backfill, they provide long-term, free-draining walls. A gabion catch wall is shown in Figure 9.52.

Rail Walls

Rail walls or rail-and-tie walls consist of vertical posts and horizontal members that are extended between the vertical posts. The vertical posts are usually scrap steel rails set in holes, which are hand dug or blasted and backfilled with concrete. The horizontal members are either ties, as shown

in Figure 9.53, or scrap rails cut into 3 to 4-m (10 to 13-ft) lengths, as shown in Figure 9.54. If feasible, cables can be anchored back from the top of the wall to provide resistance to overturning.

Cable Walls

Cable walls consist of steel posts set in concrete, cables strung horizontally between, and a smaller cable or coarse wire woven between the cables to form a crude net or mesh. This type of installation is sometimes used in Europe.

Rock Sheds and Tunnels

Rock sheds and tunnels can be used for protection against rock falls and slides when warranted and when other forms of stabilization and protection are not effective. Although expensive, they can give complete protection and should be considered in areas with serious problems. Maintenance costs are normally negligible. The methods of design and construction of tunnels are dealt with thoroughly in the literature, but the design of rock sheds is not so adequately covered, and experience is required to decide on the most suitable type of structure and the loads to be carried. A rock shed should be able to resist the energy transmitted by the largest rock mass likely to pass over it during its life; therefore, probability analysis should be involved. The energy transmitted will depend on whether rocks are falling free, bouncing, or rolling. High stress concentrations in the structure can be reduced by the provision of a thick cover of loose sand.

If foundation conditions on the outer side of the roadbed are not suitable for footings, heavy rock anchors extending into the upper slope may be used to support a cantilevered shed of the type shown in Figure 9.55. Tunnel portals can also act as sheds to protect the roadway or track, as shown in Figure 9.56. When a rock shed is to be located at the lower end of a gully, wing walls are usually used above the structure to channel material onto it (Figure 9.57). Wing walls should be sufficiently high because the debris will tend to clog and build up when its path is restricted by the structure. For this reason, the slope angle of the roof of the shed should be steeper than the angle of repose of the material to be conveyed over the roadway.

Methods of Warning

Although warning systems do not prevent rock falls, they are necessary on transportation routes where other measures are too expensive or impractical or where a new hazard has developed. In North America, warning methods have been used on railways in mountains to detect rock falls on tracks so that trains can stop before hitting the material.

Patrols

The simplest type of warning method is provided by human patrol. Patrols have the advantages of being reliable and flexible, and their frequency can be adjusted to the demands of traffic and weather conditions. The disadvantages are that they incur continuing costs and require personnel who are willing to work in uncomfortable and often hazardous conditions.

Electric Fences and Wires

There are several variations of electric warning methods. Electric fences are based on the principle that a falling rock large enough to endanger traffic will break or pull out one of the wires and thus actuate a signal to warn approaching traffic. This principle is particularly adaptable to railways on which a signal system to control traffic is already in use. The standard electric warning fence used on railways consists of a row of poles spaced along the uphill ditch line and wires strung between them at a vertical spacing of 25 cm (10 in). Overhead wires, which are supported on members cantilevered out from the top of the poles, as shown in the installation in Figure 9.58, are often required where rock faces are steep and close to the right-of-way.

Figure 9.52. Gabion catch wall along edge of high slope in unconsolidated material on main highway.

Figure 9.54. Rail wall with vertical posts seated in cast-in-place footings (note top of wall anchored by cables to bedrock for support).

Figure 9.53. Rail and tie wall founded on masonry wall to prevent rolling rock from reaching track of Austrian Railway.

Figure 9.55. Cantilevered rock shed supported by deep anchors inserted in upper slope protecting rail line in Switzerland (9.84).

223

Figure 9.56. Tunnel in steep terrain with extended portal shedlike structure to protect Canadian National Railway line from rock falls generated in a slide area (note use of retaining walls of rock blocks below track).

Figure 9.57. Rock sheds to carry rock debris originating in deep gullies over main line of Canadian National Railway in White Canyon in Thompson River Valley, British Columbia.

Figure 9.58. Standard type of railway electric warning fence with wires strung between upright poles and horizontal members that are cantilevered out from top of poles along Canadian National Railway line in Fraser Canyon, British Columbia.

Electric warning fences have some advantages, and the design of each facility can be made to suit the individual sites. One disadvantage is low efficiency; in some instances 80 percent of the alarms have been found to be false. Another disadvantage is that snow-clearing operations, if required, are often impeded and maintenance of the fences proves difficult. These difficulties may be reduced or eliminated by

1. Choosing the spacing of wires according to previous experience at a site,
2. Providing a catchment ditch behind the warning fence,
3. Eliminating the lower wires where the lower slope can be scaled and stabilized, and

4. Supporting overhead wires on a canopy type of frame bolted to the rock face on the uphill side and supported on poles on the downhill side.

A particularly effective type of electric warning system consists of a single wire, anchored at both ends and linked to a warning signal (*9.85*). Such a wire may be fastened around a large unstable rock or across a rock slope above the right-of-way, across a gully where large rocks roll down, or on top of a protective catch wall, as shown in Figure 9.49. The installation is simple, economical, and efficient.

Other Methods

Warning methods that are dependable under all conditions

are undergoing continued study. At this writing, the following are being tried: (a) geophones or vibration meters buried at intervals along the roadway shoulder to pick up vibrations from falling rocks by the Canadian National Railways and Swedish State Railways; (b) television monitoring by the Federal Highway Administration; (c) guided radar by the Canadian Institute of Guided Ground Transport and the Japanese National Railways; and (d) laser beams by the Radio Corporation of America. However, none is known to have been sufficiently developed to be recommended for general use.

REFERENCES

9.1 Alberts, C. Bergförstärkning genom betonsprutning och injektering. Swedish Academy of Engineering Sciences, Stockholm, IVA Publ. 142, 1970, pp. 231-240.

9.2 American Concrete Institute. Use of Shotcrete for Underground Structural Support. ACI, Detroit, Publ. SP-45, 1974, 467 pp.

9.3 Baker, R. F., and Marshall, H. C. Control and Correction. In Landslides and Engineering Practice (Eckel, E. B., ed.), Highway Research Board, Special Rept. 29, 1958, pp. 150-187.

9.4 Barron, K., Coates, D. F., and Gyenge, M. Artificial Support of Rock Slopes. Mines Branch, Canada Department of Energy, Mines and Resources, Ottawa, Research Rept. R228, 1971, 145 pp.

9.5 Barton, N. Review of a New Shear-Strength Criterion for Rock Joints. Engineering Geology, Vol. 7, No. 4, 1973, pp. 287-332.

9.6 Bauer, A. The Status of Rock Mechanics in Blasting. In Status of Practical Rock Mechanics (Grosvenor, N. E., and Paulding, B. W., Jr., eds.), Proc., 9th Symposium on Rock Mechanics, American Institute of Mining, Metallurgical and Petroleum Engineers, New York, 1967, pp. 249-262.

9.7 Behr, H., and Klengel, K. J. Stability of Rock Slopes on the German State Railway. Deutsche Eisenbahntechnik, Germany, 1966, pp. 324-328.

9.8 Benson, R. P. Experience With Instrumentation in Rock Slopes. Acres Consulting Engineers, Niagara Falls, N.Y., 1971, 49 pp.

9.9 Bjerrum, L., and Jørstad, F. A. Stability of Rock Slopes in Norway. Norwegian Geotechnical Institute, Oslo, Publ. 79, 1968, 11 pp.

9.10 Brekke, T. L. Shotcrete in Hard Rock Tunneling. Bulletin, Association of Engineering Geologists, Vol. 9, No. 3, 1972, pp. 241-264.

9.11 Bukovansky, M., and Piercy, N. H. High Road Cuts in a Rock Mass With Horizontal Bedding. In Design Methods in Rock Mechanics (Fairhurst, C., and Crouch, S. L., eds.), Proc., 16th Symposium on Rock Mechanics, Univ. of Minnesota, Minneapolis, American Society of Civil Engineers, New York, 1977, pp. 47-52.

9.12 Bureau of Public Roads. Presplitting: A Controlled Blasting Technique for Rock Cuts. U.S. Department of Commerce, 1966, 36 pp.

9.13 Calder, P. N. Slope Stability in Jointed Rock. Bulletin, Canadian Institute of Mining, May 1970.

9.14 Call, R. D. Analysis of Geologic Structure for Open Pit Slope Design. Univ. of Arizona, Tucson, PhD thesis, 1972, 201 pp.

9.15 Cedergren, H. R. Seepage, Drainage and Flow Nets. Wiley, New York, 1967, 490 pp.

9.16 Coates, D. F. The Stability of Slopes in Open Pits. Proc., 8th Commonwealth Mining and Metallurgical Congress, Australian Institute of Mining and Metallurgy, Melbourne, Vol. 6, 1967, pp. 543-550.

9.17 Coates, D. F., and Gyenge, M. Incremental Design in Rock Mechanics. Mines Branch, Canada Department of Energy, Mines and Resources, Ottawa, Monograph 880, 1973.

9.18 Coates, D. F., and Sage, R. Rock Anchors in Mining: A Guide for Their Utilization and Installation. Mines Branch, Canada Department of Energy, Mines and Resources, Ottawa, Research Rept. R224, 1973.

9.19 Crimmins, R. S., Samuels, R., and Monahan, B. P. Construction Rock Work Guide. Wiley, New York, 1972, 235 pp.

9.20 Da Silveira, A. F., Rodrigues, F. P., Crossman, N. F., and Mendes, F. Quantitative Characterization of the Geometric Parameters of Jointing in Rock Masses. Proc., 1st Congress, International Society of Rock Mechanics, Lisbon, Vol. 1, 1966, pp. 225-233.

9.21 Deere, D. U., Hendron, A. J., Patton, F. D., Jr., and Cording, E. J. Design of Surface and Near-Surface Construction in Rock. In Failure and Breakage of Rock (Fairhurst, C., ed.), Proc., 8th Symposium on Rock Mechanics, American Institute of Mining, Metallurgical and Petroleum Engineers, New York, 1967, pp. 237-302.

9.22 Donn, W. L., and Shimer, J. A. Graphic Methods in Structural Geology. Appleton-Century-Crofts, Englewood Cliffs, N.J., 1958, 180 pp.

9.23 Duncan, N. Geology and Rock Mechanics in Civil Engineering Practice. Water Power, Jan. 1965, pp. 25-32.

9.24 Eckert, O. Consolidation de massifs rocheux par ancrage de cables/Consolidation of Rock Masses by Cable Anchors. Sols-Soils (Paris), Vol. 5, No. 18, 1966, pp. 33-40 (in English).

9.25 Ehrlich, M. E., Scharon, H. L., and Mateker, E. J., Jr. Vibration Analysis and Construction Blasting. In Legal Aspects of Geology in Engineering Practice, Geological Society of America, Engineering Geology Case Histories, No. 7, 1969, pp. 13-20.

9.26 Fookes, P. G., and Sweeney, M. Stabilization and Control of Local Rock Falls and Degrading Rock Slopes. Quarterly Journal of Engineering Geology, Vol. 9, No. 1, 1976, pp. 37-55.

9.27 Förster, W. The Influence of the Curvature of Open Cuts on the Stability of Slopes in Open Work Mining. Proc., 1st Congress, International Society of Rock Mechanics, Lisbon, Vol. 1, 1966, pp. 193-200.

9.28 Franklin, J. A., and Denton, P. E. Monitoring for Rock Slopes. Quarterly Journal of Engineering Geology, Vol. 6, No. 3, 1973, pp. 259-286.

9.29 Goodman, R. E. Methods of Geological Engineering in Discontinuous Rocks. West Publishing, St. Paul, 1976, 472 pp.

9.30 Halstead, P. N., Call, R. D., and Rippere, K. H. Geologic Structural Analysis for Open Pit Slope Design, Kimberley Pit, Ely, Nevada. Paper presented at 97th Annual Meeting, American Institute of Mining, Metallurgical and Petroleum Engineers, New York, 1968, 25 pp.

9.31 Hamel, J. V. Stability of Slopes in Soft, Altered Rocks. Univ. of Pittsburgh, PhD thesis, 1969.

9.32 Hast, N. The State of Stress in the Upper Part of the Earth's Crust. Engineering Geology, Vol. 11, No. 1, 1967, pp. 5-17.

9.33 Hedley, D. G. F. Design Parameters for Borehole Extensometer Measuring Systems. Mines Branch, Canada Department of Energy, Mines and Resources, Ottawa, Internal Rept. MR67/59-LD, 1967.

9.34 Hendron, A. J. Analytical and Graphical Methods for the Analysis of Slopes in Rock Masses. Department of Civil Engineering, Univ. of Illinois at Urbana-Champaign, July 1971, 178 pp.

9.35 Hennequin, M., and Cambefort, H. Consolidation du remblai de Malherbe/Stabilization of Fill at Malherbe. Revue General des Chemins de Fer (Paris), Vol. 85, Jan. 1966, pp. 78-86.

9.36 Herget, G. Surveys of Geological Discontinuities in Connection With the Mechanical Behavior of Rock Masses. Mines Branch, Canada Department of Energy, Mines and Resources, Ottawa, Internal Rept. MR70/38, 1970, 28 pp.

9.37 Heuze, F. E., and Goodman, R. E. Three-Dimensional Approach for Design of Cuts in Jointed Rock. In Stability of Rock Slopes (Cording, E. J., ed.), Proc., 13th Symposium on Rock Mechanics, Univ. of Illinois at Urbana-Champaign,

American Society of Civil Engineers, New York, 1971, pp. 397-441.

9.38 Highway Focus. Vol. 1, No. 3, Aug. 1969.

9.39 Highway Focus. Vol. 4, No. 2, June 1972.

9.40 Highway Research Board. Erosion Control on Highway Construction. HRB, Synthesis of Highway Practice 18, 1973, 52 pp.

9.41 Highway Research Board. Soil Erosion: Causes and Mechanisms, Prevention and Control. HRB, Special Rept. 135, 1973, 141 pp.

9.42 Hoek, E., and Bray, J. W. Rock Slope Engineering. Institution of Mining and Metallurgy, London, 1974, 309 pp.

9.43 Hoek, E., and Londe, P. Surface Workings in Rock. Proc., 3rd Congress, Denver, International Society of Rock Mechanics, Vol. 1, Pt. A, 1974, pp. 613-654.

9.44 International Society of Rock Mechanics. Suggested Methods for the Description of Rock Masses, Joints and Discontinuities. ISRM, 1976.

9.45 Jaeger, J. C. Brittle Fracture of Rocks. In Failure and Breakage in Rock (Fairhurst, C., ed.), Proc., 8th Symposium on Rock Mechanics, American Institute of Mining, Metallurgical and Petroleum Enginners, New York, 1967, pp. 3-57.

9.46 Jennings, J. E. A Preliminary Theory for the Stability of Rock Slopes Based on Wedge Theory and Using Results of Joint Surveys. Univ. of Witwatersrand, South Africa, 1968.

9.47 Jennings, J. E. A Mathematical Theory for the Calculation of the Stability of Slopes in Open Cast Mines. In Planning Open Pit Mines, Johannesburg (van Rensburg, P. W. J., ed.), Proc., Open Pit Mining Symposium, Johannesburg, South African Institute of Mining and Metallurgy, Balkema, Amsterdam, 1971, pp. 87-102.

9.48 John, K. W. Graphical Stability Analysis of Slopes in Jointed Rock. Journal of Soil Mechanics and Foundations Division, American Society of Civil Engineers, New York, Vol. 94, No. SM2, 1968, pp. 497-526.

9.49 Kaden, R. A. Steel Fibrous Shotcrete. Western Construction, Vol. 49, No. 4, April 1974.

9.50 Kilburn, J. Discussion of Papers. In Planning Open Pit Mines (van Rensburg, P. W. J., ed.), Proc., Open Pit Mining Symposium, Johannesburg, South African Institute of Mining and Metallurgy, Balkema, Amsterdam, 1971, pp. 270-271.

9.51 Kley, R. J., and Lutton, R. J. A Study of Selected Rock Excavations as Related to Large Nuclear Craters. U.S. Army Corps of Engineers, Plowshare, PNE-5010, 1967, 157 pp.

9.52 Knill, J. L. Collecting and Processing of Geological Data for Purposes of Rock Engineering. Univ. of Alberta, Edmonton, 1971, 35 pp.

9.53 Kraus, J., and Tyc, P. Sanace Svahu Zeleznicnich Nasypu a Zarezu/Stabilization of Railway Fills and Cuts. Nakladatelstvi Dopravi a spoiu, Prague, 1965, 367 pp.

9.54 Krsmanovic, D. Initial and Residual Shear Strength in Hard Rocks. Geotechnique, Vol. 17, No. 2, 1967, pp. 145-160.

9.55 Lambooy, P., and Espley-Jones, R. C. Practical Considerations of Blasting in Open Pit Cast Mines. In Planning Open Pit Mines (van Rensburg, P. W. J., ed.), Proc., Open Pit Mining Symposium, Johannesburg, South African Institute of Mining and Metallurgy, Balkema, Amsterdam, 1971, pp. 227-234.

9.56 Lancaster-Jones, P. F. F. Methods of Improving the Properties of Rock Masses. In Rock Mechanics in Engineering Practice (Zienkiewicz, O. C., and Stagg, D., eds.), Wiley, New York, 1968, pp. 385-429.

9.57 Lang, L. C., and Favreau, R. F. A Modern Approach to Open-Pit Blast Design and Analysis. Bulletin, Canadian Institute of Mining, June 1972, pp. 37-45.

9.58 Lang, T. A. Rock Reinforcement. Bulletin, Association of Engineering Geologists, Vol. 9, No. 3, 1972, pp. 215-239.

9.59 Langefors, U., and Kihlstrom, B. The Modern Technique of Rock Blasting. Wiley, New York, 1963.

9.60 Larocque, G. A Proposed Method of Establishing Bench Blast Patterns. Mines Branch, Canada Department of Energy, Mines and Resources, Ottawa, Internal Rept. MR71/93-ID, 1971.

9.61 Legget, R. F. Cities and Geology. McGraw-Hill, New York, 1973, 624 pp.

9.62 Leighton, F. B. Landslides and Hillside Development. In Engineering Geology in Southern California, Association of Engineering Geologists, Special Publ., 1966, pp. 149-207.

9.63 Lewis, H., McDaniel, A. H., Peters, R. B., and Jacobs, D. M. Damages Due to Drainage Runoff, Blasting and Slides. National Cooperative Highway Research Program, Rept. 134, 1972, 23 pp.

9.64 Long, A. E., Merrill, R. H., and Wisecarver, D. W. Stability of High Road Bank Slopes in Rock: Some Design Concepts and Tools. Highway Research Board, Highway Research Record 135, pp. 10-26.

9.65 Lorente de No, C. Stability With Curvature in Plane View. Proc., 7th International Conference on Soil Mechanics and Foundation Engineering, Mexico City, Vol. 2, 1969, pp. 635-638.

9.66 Lutton, R. J. Rock Slope Chart From Empirical Slope Data. Trans., Society of Mining Engineers, American Institute of Mining, Metallurgical and Petroleum Engineers, Vol. 247, June 1970, pp. 160-162.

9.67 Mahtab, M. A., Bolstad, D. D., and Kendorski, F. S. Analysis of the Geometry of Fractures in San Manuel Copper Mine, Arizona. Bureau of Mines, U.S. Department of the Interior, Rept. RI 7715, 1973, 24 pp.

9.68 Martin, D. C., and Piteau, D. R. Application of Waviness of Structural Discontinuities to Rock Slope Design. Proc., 29th Canadian Geotechnical Conference on Slope Design, Vancouver, B.C., Canadian Geotechnical Society, Montreal, 1976, pp. VIII.1-VIII.17.

9.69 McAnuff, A. L. Blasting Is an Art as Well as a Science. Canadian Consulting Engineer, Vol. 15, No. 6, 1973, pp. 24-26.

9.70 McCauley, M. L. Engineering Geology Related to Highways and Freeways. In Engineering Geology in Southern California, Association of Engineering Geologists, Special Publ., 1966, pp. 117-121.

9.71 Mearns, R. Solving a Rockfall Problem in Nevada County, California. Highway Research News, No. 49, Autumn 1972, pp. 14-17.

9.72 Mehra, S. R., and Natarajan, T. K. Landslide Analysis and Correction. Central Road Research Institute, New Delhi, India, 1966.

9.73 Merrill, R. H., and Wisecarver, D. W. The Stress in Rock Around Surface Openings. In Failure and Breakage in Rock (Fairhurst, C., ed.), Proc., 8th Symposium on Rock Mechanics, American Institute of Mining, Metallurgical and Petroleum Engineers, New York, 1967, pp. 337-350.

9.74 Mollard, J. D. Photo Analysis in Interpretation in Engineering Geology Investigations: A Review. In Reviews in Engineering Geology, Geological Society of America, Vol. 1, 1962.

9.75 Morgenstern, N. R., and Eisenstein, Z. Methods of Estimating Lateral Loads and Deformations. Proc., Specialty Conference, Cornell Univ., Ithaca, N.Y., Soil Mechanics and Foundations Division, American Society of Civil Engineers, New York, 1970, pp. 51-102.

9.76 Müller, L. Application of Rock Mechanics in the Design of Rock Slopes. In State of Stress in the Earth's Crust (Judd, W., ed.), Elsevier, New York, 1964, pp. 575-598.

9.77 Müller, L. The Stability of Rock Bank Slopes and the Effect of Rock Water on Same. International Journal of Rock Mechanics and Mining Sciences, Vol. 1, No. 4, 1964, pp. 475-504

9.78 Murrell, S. A., and Misra, A. K. Time-Dependent Strain on Creep in Rocks and Similar Non-Metallic Materials. Trans., Institution of Mining and Metallurgy, London, Vol. 71, 1962, pp. 353-378.

9.79 National Academy of Sciences. Better Contracting for Underground Construction. U.S. National Committee on

Tunneling Technology, National Academy of Sciences, Washington, D.C., 1974, 143 pp.

9.80 Nesbitt, M. D. The Stabilization of Rock Slopes at Hell's Gate, British Columbia. Proc., Canadian Good Roads Association, Sept. 1967, pp. 295-309.

9.81 Norman, J. W. The Photogeological Detection of Unstable Ground. Journal of Institution of Highway Engineers (London), Vol. 18, No. 2, 1970, pp. 19-22.

9.82 Northwood, T. D., and Crawford, R. Blasting and Building Damage. Canada National Research Council, Ottawa, CBD 63, 1965, 4 pp.

9.83 Peck, R. B., Hanson, W. E., and Thornburn, T. H. Foundation Engineering. Wiley, New York, 2nd Ed., 1973, 514 pp.

9.84 Peckover, F. L. Treatment of Rock Falls on Railway Lines. American Railway Engineering Association, Bulletin 653, Chicago, 1975, pp. 471-503.

9.85 Peckover, F. L., and Kerr, J. W. G. Treatment and Maintenance of Rock Falls on Transportation Routes. Canadian Geotechnical Journal, Vol. 14, No. 4, 1977, pp. 487-507.

9.86 Phillips, F. C. The Use of Stereographic Projection in Structural Geology. Arnold, London, 1971.

9.87 Piteau, D. R. Geological Factors Significant to the Stability of Slopes Cut in Rock. In Planning Open Pit Mines (van Rensburg, P. W. J., ed.), Proc., Open Pit Mining Symposium, Johannesburg, South African Institute of Mining and Metallurgy, Balkema, Amsterdam, 1971, pp. 33-53.

9.88 Piteau, D. R. Characterizing and Extrapolating Rock Joint Properties in Engineering Practice. Rock Mechanics, Supp. 2, Springer-Verlag, Vienna, 1973, 31 pp.

9.89 Piteau, D. R. Plan Geometry and Other Factors Relating to Natural Rock Slope Stability Applied to Design of DeBeers Mine. Mines Branch, Canada Department of Energy, Mines and Resources, Rept. TB 190, 1974, 120 pp.

9.90 Piteau, D. R. Regional Slope Stability Controls and Related Engineering Geology of the Fraser Canyon, British Columbia. In Landslides (Coates, D. R., ed.), Geological Society of America, 1977.

9.91 Piteau, D. R., and Clayton, R. Discussion of paper, Computerized Design of Rock Slopes Using Interactive Graphics for the Input and Output of Geometrical Data, by P. A. Cundall, M. D. Voegele, and C. Fairhurst. In Design Methods in Rock Mechanics (Fairhurst, C., and Crouch, S. L., eds.), Proc., 16th Symposium on Rock Mechanics, Univ. of Minnesota, Minneapolis, American Society of Civil Engineers, New York, 1977, pp. 62-63.

9.92 Piteau, D. R., and Jennings, J. E. The Effects of Plan Geometry on the Stability of Natural Slopes in Rock in the Kimberley Area. Proc., 2nd Congress, International Society of Rock Mechanics, Belgrade, Vol. 3, 1970, pp. 289-295.

9.93 Piteau, D. R., McLeod, B. C., Parkes, D. R., and Lou, J. K. Overturning Rock Slope Failure at Hell's Gate, British Columbia. In Geology and Mechanics of Landslides and Avalanches (Voight, B., ed.), Elsevier, New York, 1977.

9.94 Pit Slope Manual. Canada Department of Energy, Mines and Resources, Ottawa, 1977.

9.95 Rawlings, G. E. Stabilization of Potential Rock Slides in Folded Quartzite in Northwestern Tasmania. Engineering Geology, Vol. 12, No. 5, 1968, pp. 283-292.

9.96 Redlinger, J. F., and Dodson, E. L. Rock Anchor Design. Proc., 1st Congress, International Society of Rock Mechanics, Lisbon, Vol. 3, 1966, pp. 173-175.

9.97 Reiche, P. A Survey of Weathering Processes and Products. Univ. of New Mexico, Albuquerque, Geology Publ. 3, 1950, 95 pp.

9.98 Ritchie, A. M. Evaluation of Rockfall and Its Control. Highway Research Board, Highway Research Record 17, 1963, pp. 13-28.

9.99 Robertson, A. M. The Interpretation of Geological Factors for Use in Slope Theory. In Planning Open Pit Mines (van Rensburg, P. W. J., ed.), Proc., Open Pit Mining Symposium, Johannesburg, South African Institute of Mining and Metallurgy, Balkema, Amsterdam, 1971, pp. 55-71.

9.100 Root, A. W. Prevention of Landslides. In Landslides and Engineering Practice (Eckel, E. B., ed.), Highway Research Board, Special Rept. 29, 1958, pp. 113-149.

9.101 Ross-Brown, D. M., and Atkinson, K. B. Terrestrial Photogrammetry in Open Pits: Description and Use of the Phototheodolite in Mine Surveying. Trans., Institution of Mining and Metallurgy, London, Vol. 81, pp. 205-213.

9.102 Seegmiller, B. L. Artificial Stabilization in Open Pit Mines: A New Concept to Achieve Slope Stability. Paper presented at Intermountain Minerals Conference, Vail, Colo., American Institute of Mining, Metallurgical and Petroleum Engineers, New York, 1974.

9.103 Serafim, J. L. Influence of Interstitial Water on the Behavior of Rock Masses. In Rock Mechanics in Engineering Practice (Zienkiewicz, O. C., and Stagg, D., eds.), Wiley, New York, 1968, pp. 55-97.

9.104 Shuk, T. E. A Simple Method to Minimize Cost of Rock Slopes. In Estimation of Upper Bounds to Rock Slopes by Analysis of Existing Slope Data, Mines Branch, Canada Department of Energy, Mines and Resources, Ottawa, Rept. 76-14, 1976, pp. 18-34.

9.105 Stacey, T. R. Stability of Rock Slopes in Mining and Civil Engineering Situations. National Mechanical Engineering Research Institute, Council for Scientific and Industrial Research, Pretoria, MEG/4415, 1973, 193 pp.

9.106 Steffen, O. K. H., and Jennings, J. E. Definition of Design Joints for Two-Dimensional Rock Slope Analyses. Proc., 3rd Congress, International Society of Rock Mechanics, Denver, Vol. 2, Part 13, 1974, pp. 827-832.

9.107 Stimpson, B. S. Physical Properties of Rock. Univ. of California, Berkeley, March 1976, 70 pp.

9.108 Terzaghi, K. Stability of Steep Slopes in Hard Unweathered Rock. Geotechnique, Vol. 12, No. 4, 1962, pp. 251-270.

9.109 Terzaghi, R. Sources of Error in Joint Surveys. Geotechnique, Vol. 15, No. 3, 1965, pp. 287-304.

9.110 U.S. Army Corps of Engineers. Rock Reinforcement in Civil Engineering Works. U.S. Department of the Army, EM 1110-1-2907, Feb. 1975.

9.111 Yu, Y., and Coates, D. F. Analysis of Rock Slopes Using the Finite Element Method. Mines Branch, Canada Department of Energy, Mines and Resources, Ottawa, Research Rept. R229, 1970.

9.112 Záruba, Q., and Mencl, V. Landslides and Their Control. Elsevier, New York, and Academia, Prague, 1969, 205 pp.

PHOTOGRAPH CREDITS

Figure 9.32   D. Johnson; courtesy of British Columbia Department
                of Highways (top); D. R. Piteau, Piteau and Associates
                (bottom)
Figure 9.35   K. Barron and others
Figure 9.36   F. L. Peckover
Figure 9.38   D. R. Piteau, Piteau and Associates
Figure 9.39   F. L. Peckover
Figure 9.41   D. R. Piteau, Piteau and Associates
Figure 9.42   D. R. Piteau, Piteau and Associates
Figure 9.43   Courtesy of B. Peirone, Maccaferri Gabions of Canada,
                Ltd.
Figure 9.44   Courtesy of American Hoist and Derrick Company
Figure 9.46   F. L. Peckover

Figure 9.48   D. R. Piteau, Piteau and Associates
Figure 9.50   T. Kirkbride; courtesy of British Columbia Department
                of Highways
Figure 9.51   D. R. Piteau, Piteau and Associates
Figure 9.52   B. Peirone; courtesy of Maccaferri Gabions of Canada,
                Ltd.
Figure 9.53   F. L. Peckover
Figure 9.54   F. L. Peckover
Figure 9.55   F. L. Peckover
Figure 9.56   George Allen Aerial Photos, Ltd.; courtesy of Canadian
                National Railway
Figure 9.57   D. C. Martin, Piteau and Associates
Figure 9.58   D. R. Piteau, Piteau and Associates

Index

The **Transportation Research Board** is a unit of the National Research Council, which serves the National Academy of Sciences and the National Academy of Engineering. The Board's purpose is to stimulate research concerning the nature and performance of transportation systems, to disseminate the information produced by the research, and to encourage the application of appropriate research findings. The Board's program is carried out by more than 300 committees, task forces, and panels composed of more than 3,500 administrators, engineers, social scientists, attorneys, educators, and others concerned with transportation; they serve without compensation. The program is supported by state transportation and highway departments, the modal administrations of the U.S. Department of Transportation, and other organizations and individuals interested in the development of transportation.

The National Academy of Sciences is a private, nonprofit, self-perpetuating society of distinguished scholars engaged in scientific and engineering research, dedicated to the furtherance of science and technology and to their use for the general welfare. Upon the authority of the charter granted to it by the Congress in 1863, the Academy has a mandate that requires it to advise the federal government on scientific and technical mattters. Dr. Frank Press is president of the National Academy of Sciences.

The National Academy of Engineering was established in 1964, under the charter of the National Academy of Sciences, as a parallel organization of outstanding engineers. It is autonomous in its administration and in the selection of its members, sharing with the National Academy of Sciences the responsibility for advising the federal government. The National Academy of Engineering also sponsors engineering programs aimed at meeting national needs, encourages education and research, and recognizes the superior achievements of engineers. Dr. Robert M. White is president of the National Academy of Engineering.

The Institute of Medicine was established in 1970 by the National Academy of Sciences to secure the services of eminent members of appropriate professions in the examination of policy matters pertaining to the health of the public. The Institute acts under the responsibility given to the National Academy of Sciences by its congressional charter to be an adviser to the federal government and, upon its own initiative, to identify issues of medical care, research, and education. Dr. Samuel O. Thier is president of the Institute of Medicine.

The National Research Council was organized by the National Academy of Sciences in 1916 to associate the broad community of science and technology with the Academy's purposes of furthering knowledge and advising the federal government. Functioning in accordance with general policies determined by the Academy, the Council has become the principal operating agency of both the National Academy of Sciences and the National Academy of Engineering in providing services to the government, the public, and the scientific and engineering communities. The Council is administered jointly by both the Academies and the Institute of Medicine. Dr. Frank Press and Dr. Robert M. White are chairman and vice chairman, respectively, of the National Research Council.